Radio Science Techniques for Deep Space Exploration

DEEP SPACE COMMUNICATIONS AND NAVIGATION SERIES

Issued by the Deep Space Communications and Navigation Systems

Center of Excellence, Jet Propulsion Laboratory
California Institute of Technology
Jon Hamkins, Editor-in-Chief

Published Titles in this Series

Radiometric Tracking Techniques for Deep-Space Navigation
Catherine L. Thornton and James S. Border

Formulation for Observed and Computed Values of Deep Space Network Data Types for Navigation
Theodore D. Moyer

Bandwidth-Efficient Digital Modulation with Application to Deep-Space Communications
Marvin K. Simon

Large Antennas of the Deep Space Network
William A. Imbriale

Antenna Arraying Techniques in the Deep Space Network
David H. Rogstad, Alexander Mileant, and Timothy T. Pham

Radio Occultations Using Earth Satellites: A Wave Theory Treatment
William G. Melbourne

Deep Space Optical Communications
Hamid Hemmati, Editor

Spaceborne Antennas for Planetary Exploration
William A. Imbriale, Editor

Autonomous Software-Defined Radio Receivers for Deep Space Applications
Jon Hamkins and Marvin K. Simon, Editors

Low-Noise Systems in the Deep Space Network
Macgregor S. Reid, Editor

Coupled-Oscillator Based Active-Array Antennas
Ronald J. Pogorzelski and Apostolos Georgiadis

Low-Energy Lunar Trajectory Design
Jeffrey S. Parker and Rodney L. Anderson

Deep Space Communications
Jim Taylor, Editor

Radio Science Techniques for Deep Space Exploration
Sami W. Asmar

Radio Science Techniques for Deep Space Exploration

Sami W. Asmar

Jet Propulsion Laboratory
California Institute of Technology
Pasadena, California

WILEY

This edition first published 2022

Registered Office
John Wiley & Sons, Inc., 111 River Street, Hoboken, NJ 07030, USA

Editorial Office
111 River Street, Hoboken, NJ 07030, USA

For details of our global editorial offices, customer services, and more information about Wiley products visit us at www.wiley.com.

Wiley also publishes its books in a variety of electronic formats and by print-on-demand. Some content that appears in standard print versions of this book may not be available in other formats.

Library of Congress Cataloging-in-Publication Data
Names: Asmar, Sami W., author.
Title: Radio science techniques for deep space exploration / Sami W. Asmar, Jet Propulsion Laboratory, California Institute of Technology, Pasadena, California.
Description: Hoboken, NJ, USA : John Wiley & Sons, 2022. | Series: JPL deep-space communications and navigation series | Includes bibliographical references and index.
Identifiers: LCCN 2021053318 (print) | LCCN 2021053319 (ebook) | ISBN 9781119734147 (hardback) | ISBN 9781119734154 (pdf) | ISBN 9781119734161 (epub) | ISBN 9781119734178 (ebook)
Subjects: LCSH: Space vehicles--Tracking. | Space vehicles--Radio equipment. | Radio astronomy. | Solar system.
Classification: LCC TL4030 .A86 2022 (print) | LCC TL4030 (ebook) | DDC 621.3848--dc23/eng/20211103
LC record available at https://lccn.loc.gov/2021053318
LC ebook record available at https://lccn.loc.gov/2021053319

Cover image: Courtesy of Sami Asmar
Cover design by Wiley

Set in 9.5/12.5pt STIXTwoText by Integra Software Services, Pondicherry, India

10 9 8 7 6 5 4 3 2 1

Table of Contents

Foreword

This book is the latest contribution to the Deep SpaceCommunications and Navigation Systems (DESCANSO) series. DESCANSO is a Center of Excellence formed in 1998 by the National Aeronautics and SpaceAdministration (NASA) at the Jet Propulsion Laboratory (JPL), which is managed under contract by the California Institute of Technology. DESCANSO is unrelated to the nearby botanical landscape, Descanso Gardens, but shares its wonderful values of growing and cultivating understanding through inspiration and example. DESCANSO's vision is to help NASA and JPL achieve continuous communications and precise navigation, anytime and anywhere. To do this, DESCANSO supports the development of new technologies, concepts, and systems; sponsors seminars and facilitates interaction and idea exchange; and shares knowledge by publishing books and articles. The DESCANSO book series, authored by scientists and engineers with many years of experience in their fields, lays a foundation for innovation by sharing state-of-the-art knowledge, fundamental principles and practices, and lessons learned in key technologies. A companion article series, the Design and Performance Summary series, captures the design and performance of specific communications and navigation systems used on JPL missions. For science and technology topics more general than communications and navigation, DESCANSO also publishes the JPL Space Science and Technology Book Series.

Jon Hamkins

DESCANSO Leader

Jet Propulsion Laboratory
California Institute of Technology

Preface

Radio signals are used to communicate information between robotic space missions throughout the Solar System and stations on Earth. These signals are altered in their electromagnetic properties between transmission and reception due to propagation effects caused primarily by intervening media as well as forces acting on the spacecraft. When observed for their scientific potential, such alternations can provide very valuable information about the nature and environment of the planetary bodies or Solar System targets under exploration. This also applies to signals transmitted from one spacecraft and received at another, in the case of multi-spacecraft missions.

The media that the radio links propagate through include planetary atmospheres, ionospheres, rings, plasma tori, cometary material, or the solar corona. The Doppler shift to the frequency of the signals caused by the relative motion between the spacecraft and ground stations, or any transmitter-receiver combination, can contain scientific information about the gravitational forces acting on the spacecraft resulting from the bulk mass, density distribution, and global interior structure of the planets or moons, among other effects.

The studies of celestial objects and environments using spacecraft radio signals in the wavelength range of approximately one to a few centimeters constitute the field of *radio science*. Its fundamental scientific observables include the time profiles of the amplitude, frequency, delay, and polarization of the radio signals. From these observables, planetary and solar material as well as aspects of fundamental physics have been investigated in depth for over six decades, leading to numerous published discoveries.

This book introduces the field of spacecraft radio science, showcases its methods, and provides a practical reference document for practitioners and data users. It also captures the historical developments of many relevant technologies and presents a synthesis of information contained in external publications, internal records, and the often-undocumented valuable experiences of specialists in deep space exploration. Many members of the radio science community worldwide contributed to this material or reviewed it.

We address the type of inquiries received from scientists, engineers, technologists, mission designers, spacecraft controllers, ground stations operators, and students. They reflect the diversity of interests and the broad array of disciplines relevant to this field.

Suitable ancillary reading material includes the numerous referenced publications in planetary science, space physics, and remote sensing. The instrumentation chapter illustrates the considerable advances in technologies onboard spacecraft and at the stations of the Deep Space Network and other networks where this field has driven many improvements in clocks and oscillators, phase stability, media calibrations, pointing accuracy, digital receivers, acquisition and operations processes, simulations, predictions, and various other techniques and algorithms.

Research continues to significantly advance in this field, keeping radio science one of the prime scientific components of missions to the Solar System and beyond. Future investigations will build on this knowledge and explore newer technologies in small spacecraft, constellations of spacecraft, optical links, and other innovations we have not yet imagined!

Sami Asmar

Acknowledgments

In addition to journal papers and conference presentations, many historical documents, internal records, and training workshops were used to develop this volume. I thank all who contributed to producing the following material:

- The Radio Science Handbook
- The DSN as an Instrument for Radio Science Research
- Radio Science module in a JPL professional development class on science instruments
- Radio Science Short Course in the 2015 International Planetary Probe Workshop in Germany
- Radio Science Short Course in the 2019 Sardinia Deep Space Antenna Seminar in Italy
- The Cassini Radio Science User's Guide
- Radio Science descriptions in the Planetary Data System archives
- Radio Science white papers in Decadal Surveys
- JPL Radio Science Research and Technology Development Initiatives

The following colleagues from many institutes are acknowledged for their significant contributions in various capacities that advanced the field of radio science over many decades: Sal Abbate, Zubair Abdulla, Chuck Acton, Leon Alkalai, John Anderson, Tom Andert, Hiroki Ando, Chi Ao, Alex Akins, John Armstrong, Aseel Anabtawi, Brad Arnold, Neil Ashby, Dave Atkinson, Georges Balmino, Bruce Banerdt, Jean-Pierre Barriot, Alina Bedrossian, David Bell, Pat Beauchamp, Peter Bender, Jeff Berner, Bruno Bertotti, Al Bhanji, Bruce Bills, Michael Bird, Martin Bloch, Jim Border, Scott Bolton, Jay Breidenthal, Joe Brenkle, Mauro Di Benedetto, Tatiana Bocanegra-Bahamon, Kyle Botteon, Sean Bruinsma, Dustin Buccino, Jennifer Caetta, Kerri Cahoy, Dan Cain, Jim Campbell, Paolo Cappuccio, Nacer Chahat, Raj Kumar Choudhary, Mick Connally, Jim Cutts, Véronique Dehant, Les Deutsch, Ann Devereaux, Dariush Divsalar, Caroline Dumoulin, Daniele Durante, Peter Edenhofer, Chad Edwards, Harvey Elliott, Todd Ely, Anton Ermakov, Paula Eshe, Von Eshleman, Frank Estabrook, Francesca Ferri, Sue Finley, Mike Flasar, Richard French, Bill Folkner, Antonio Genova, Giacomo Giampieri, Gene Goltz, Sander Goossens, Hannes Griebel, Carole Hamilton, Bernd Häulser, George Hajj, North Ham, Ravit Helled, Scott Hensley, Randy Herrera, David Hinson, Mark Hofstadter, Tony Horton, Tay Howard, Luciano Iess, Takeshi Imamura, Bob Jacobson, Jon Jenkins, Doug Johnston, Andre Jongeling, Danny Kahan, Özgür Karatekin, Yohai Kaspi, Mike Klein, Arvydas Kliore, Mike

Kobayashi, Attila Komjathy, Alex Konopliv, Rob Kursinski, Susan Kurtik, Joe Lazio, Sebastien Le Maistre, Jean-Charles Marty, Frank Lemoine, Mike Levesque, Steve Lichten, Rob Lillis, Gunnar Lindal, Ivan Linscott, Jim Lyons, Earl Maize, Rob Manning, Tony Mannucci, Essam Marouf, Steve Matousek, Koji Matsumoto, Lisa Mauger, Erwan Mazarico, Bob Mitchell, Ted Moyer, Greg Neumann, Meegyeong Paik, Marzia Parisi, Ryan Park, Martin Pätzold, Tim Pham, Bob Preston, Trish Priest, Nicole Rappaport, Marc Rayman, Carole Raymond, Tom Rebold, Stefan Remus, Nilton Renno, Nick Renzetti, George Resch, Paul Rosen, Pascal Rosenblatt, Paul Schinder, Davide Serra, Richard Simpson, Bill Sjogren, David E. Smith, Sue Smrekar, Linda Spilker, Tom Spilker, Paul Steffes, Alan Stern, Dave Stevenson, Don Sweetnam, Jim Taylor, Silvia Tellmann, Tommy Thompson, Fraser Thomson, Massimo Tinto, Paolo Tortora, Zaid Towfic, Slava Turyshev, Len Tyler, Tim van Hoolst, Panagiotis Vergados, Ashok Verma, Chris Volk, Hans Volland, Michael Watkins, Greg Weaver, Julie Webster, Mark Wieczorek, Jim Williams, Walton Williamson, Paul Withers, Richard Woo, Gordon Wood, Don Yeomans, Marco Zannoni, Howard Zebker, and Maria Zuber.

I am very grateful to colleagues in the radio science community worldwide for supporting this book and providing valuable insight, feedback, ideas, and inspiration for informative text or illustrative graphics, especially Dave Atkinson, David Doody, Anton Ermakov, Bernd Häulser, James Tuttle Keane, Peter Kinman, Arv Kliore, Virginia Notaro, and Nicole Rappaport. I am especially grateful for the extraordinary collaborators and contributors John Armstrong, Bruce Bills, Alex Konopliv, Essam Marouf, Bob Preston, Stefan Remus, Richard Simpson and Fraser Thomson.

I sincerely thank Joe Yuen, Jon Hamkins, Les Deutsch, Mary Young, Allyson Beatrice, and Joon Park for the review, approval, and production of this volume, along with the team at Wiley & Sons.

My most sincere appreciation and admiration goes to the following colleagues, friends, and role models who have enhanced my career and enriched my life: John Armstrong, Bruno Bertotti, Bruce Bills, Mike Bird, Dustin Buccino, Les Deutsch, Ann Devereaux, Charles Elachi, Antonio Genova, Carole Hamilton, Bernd Häusler, Luciano Iess, Takeshi Imamura, Larry James, Alex Konopliv, Joe Lazio, Ivan Linscott, Essam Marouf, Erwan Mazarico, Meegyeong Paik, Ryan Park, Keyur Patel, Martin Pätzold, Bob Preston, Dick Simpson, Len Tyler, Dave Smith, Mike Watkins, and Maria Zuber.

Author and Contributors

Sami W. Asmar is a physicist and manager at NASA's Jet Propulsion Laboratory, California Institute of Technology, an experimental radio science coinvestigator on several solar system missions, and a technologist leading institutional initiatives to advance radio and laser link utilization for planetary exploration. He has been the project scientist for the GRAIL lunar mission and instrument manager or instrument scientist for many NASA and ESA planetary missions. He also served as the supervisor of the Radio Science Systems Group for many years and is currently a manager at the Interplanetary Network Directorate for strategic partnerships and customer formulation.

He is the current General Secretary, management council chair, and NASA chief delegate to the Consultative Committee on Space Data Systems (CCSDS) composed of 11 member space agencies, 32 observer agencies, and 120 industrial associates.

He invented the Mars Cube One (MarCO) mission and the "Carry Your Own Relay" concept (*NASA Inventions and Contributions Board Award NPO-49544 Mars Relay CubeSat, 2014*) and formulated follow-on concepts, pioneered "communications by Doppler" for mission critical events and other technologies. He has published in leading journals and conferences, taught short courses, and often served as a journal reviewer and conference session chair.

He received three NASA Exceptional Achievement Medals for *Outstanding execution of the Cassini experiment and verification of Einstein's general theory of relativity* (2004); for *Huygens Doppler Wind Experiment* (2006); and for *GRAIL Science Implementation* (2013). He is also the recipient of the Space Ops Exceptional Achievement Medal for *creating a system that has transformed space operations during critical events to acquire information indispensable to the safety and health of costly space exploration missions via Radio Science Systems* (2012, Sweden).

His other awards include a JPL Magellan Award, a European Space Agency Award for Outstanding Contribution to Huygens Probe, a Thomas Paine Memorial Team Award for Advanced Human Exploration of Mars, and numerous NASA group achievement awards.

Contributor **John W. Armstrong** is a radio science investigator and senior research scientist at NASA's Jet Propulsion Laboratory, California Institute of Technology (retired) and a pioneer in the fields of solar studies, experimental tests of the theories of relativity, and experimental noise characterization, among other research fields.

Contributor **Alex Konopliv** is a radio science investigator and senior research scientist at NASA's Jet Propulsion Laboratory, California Institute of Technology, and an expert in the field of planetary gravitational fields and solar system dynamics, among other research fields.

Contributor **Stefan Remus** has carried out radio science research for his doctoral dissertation at the University of the Armed Forces, Munich, Germany, and later as an employee of the European Space agency.

1

Investigations and Techniques

1.0 Introduction

Space agencies send robotic missions to explore the Solar System and deliver the important scientific data obtained by onboard instruments to control centers and science teams via radio communication links. From the early missions, scientists realized the utility of the radio links themselves as fundamental scientific tools for exploration and research. Many phenomena at Solar System targets affect the characteristics of the radio signals in a manner that reveals the targets' scientific properties.

Over the past six decades, scientists have made an extensive set of discoveries about Solar System bodies using spacecraft radio science (RS) investigations and techniques. They have inferred the presence of liquid oceans beneath the icy surfaces of Titan, Enceladus, and Europa, where possible future missions might search for life. They have also profiled the atmospheres and characterized the surface properties of many planetary bodies, key information to also aid potential future human exploration. From the Sun and Mercury to the outer reaches of the Solar System, this field has our increased understanding of the Universe.

Monitoring spacecraft radio signals enables researchers to precisely measure the changes in the link's frequency, amplitude, and polarization, and other electromagnetic wave parameters. With these measurement tools, scientists investigate ionospheres, neutral atmospheres, rings, surfaces, shapes, internal structure, orbital motion, and the dynamics of Solar System bodies. These techniques are also used to investigate the solar corona and fundamental physics, including tests of the general theory of relativity (GTR).

RS will continue to be an essential field for Solar System exploration. Goals have been identified for additional scientific advances in the future enabled by developing innovative instrumentation, advanced calibration techniques, and experimental scenarios that significantly improve the data quality. Future exploration concepts at various planetary bodies include wind profiles from entry probes or balloons, detailed atmospheric structure from radio occultations (ROs), electrical properties and roughness from surface scattering (bistatic radar), as well as interior structure from high-precision planetary gravity fields are among the future exploration concepts at various planetary bodies. The focus will increase on applications of small spacecraft and constellations of spacecraft with crosslinks between them.

The story of RS covers numerous domains in Solar System exploration: science, engineering, technology, signal processing, data analysis, mission operations, and

Radio Science Techniques for Deep Space Exploration, First Edition. Sami W. Asmar.

communication systems. The future outlook for the field is filled with exciting possibilities for additional scientific breakthroughs.

1.1 Historical Background

Wireless communications via radio technology is one of the defining developments of the modern world. Asmar (2010) outlines how, in 1901, Guglielmo Marconi wirelessly received the simple Morse code for the letter S across the Atlantic Ocean and created an alternative to the use of a wire to communicate information. Advancements in this budding technology accelerated over the next three decades, as narrow-band transmitters replaced spark gaps and the device became more popular. United States President Warren Harding installed a radio in the White House and was also the first president to have his voice transmitted by radio. Later, in 1925, President Calvin Coolidge gave the first radio-specific inaugural address broadcast to homes throughout the country.

Within 20 years, video transmission utilizing radio waves brought television to living rooms; then, increases in transmitter power and antenna size allowed nearly global coverage. The invention of the transistor reduced power consumption, device size, and cost. Consequently, the age of small digital devices began, eventually leading to smart phones. Data transfer rates are reaching terabits per second, an enormous leap over Marconi's few dots.

With these technological advances, big antennas were also built and pointed toward the sky. On assignment for Bell Laboratories, Karl Guthe Jansky built a rotating antenna and, in 1933, discovered radio waves emanating from the Milky Way. Inspired by this finding, Grote Reber built the first parabolic radio telescope in his backyard in 1937, which he then used to conduct a radio survey of the sky, confirming the presence of natural radiation. Consequently, the budding field of radio astronomy established a new way of exploring an invisible part of the Universe.

In the mid-1940s, scientists transmitted radio signals toward the Moon and received echoes; this was repeated two decades later at Venus and Mars. This new technique, called planetary radar, revealed variations in the surface scattering properties of the planets and is still in use for the detection of possible hazardous near-Earth objects. Since Marconi's 1901 observation that radio waves propagate beyond the horizon, "radio physicists" investigated the idea of a reflecting layer in Earth's ionosphere. The Nobel Prize for Physics for 1947 was awarded to Sir Edward Appleton who demonstrated the existence of an ionized region capable of reflecting electromagnetic waves and higher region from which the shorter radio waves are reflected and later investigated the reflection of radio waves from meteors. According to the prize announcement, his work was invaluable in allocating suitable frequencies for practical radio communication, broadcasting, and other applications.

In the early 1950s, radio signals were used to study Earth's ionosphere via examining echoes of transmitted pulses and as well as the scattering of radio waves from meteoric ionization (Eshleman and Manning 1954, based on V. R. Eshleman's 1952 Stanford University's dissertation). In that period, the term "radio science" started to appear in some literature to describe the study of Earth's environment using radio waves. The term was later adopted by the planetary science community for investigations based on radio signals transmitted from spacecraft to ground stations. In the

1960s, publication of a research journal titled *Radio Science* began and continues to this day by the American Geophysical Union.

In a parallel historical development, in the mid-1930s, a group of California Institute of Technology (Caltech) students began experimenting with rockets on the campus with Professor von Karman. For safety reasons, they were relocated to an isolated area in an arroyo a few miles away, at a site that would become the Jet Propulsion Laboratory (JPL) in 1944. About 14 years later, the National Aeronautics and Space Administration (NASA) was formed and JPL became NASA's laboratory. By the 1960s, NASA began to launch robotic spacecraft to study the Earth environment and explore the Solar System and beyond.

To enable NASA's deep space missions, large steerable antennas, similar in shape to the early radio telescopes, were developed at JPL to control the spacecraft and receive the valuable scientific data collected during encounters with their distant planetary targets. Unlike communication with satellites in Earth's proximity, deep space communication addresses the challenge of the signal level decrease proportional to the square of the vast distance. A global network of stations for communications, tracking, navigation, and science was named the deep space network (DSN), an engineering marvel.

The Solar System distances are measured in astronomical units (au), the distance from the Sun to Earth. Voyager 1 is the most distant human-made object in space at nearly 150 au, or nearly 23 billion kilometers (km, ~14 billion miles), away. At this distance, its radio signal, produced by a 23-watt (W) transmitter, drops to about 10^{-23} watts per square meter (W/m^2) when it reaches a 70-meter (m) diameter DSN station and still delivers important scientific information. Note that when the DSN transmits a signal to the Voyager spacecraft (the reverse path), it is 20 kW or 1000 times stronger than what Voyager transmits. Traveling at the speed of light, the round-trip radio signal takes about 40 hours when transmitted from the DSN to Voyager 1 and received back at the DSN (Asmar 2010).

1.1.1 The Field of Radio Science

To carry out its increasingly challenging tasks and complement the advancing capabilities of spacecraft radios, the DSN was instrumented with specialized technologies, such as atomic clocks for frequency and timing reference, accurate antenna pointing, and sensitive signal receivers, with some elements cooled to about 5 Kelvin (°K) to minimize the ambient noise (maximize the signal-to-noise ratio [SNR]).

Although other tracking stations and radio telescopes have made significant contributions to the field, the high-resolution RS measurements can only be carried out using instrumented facilities such as those found at the DSN and selected stations of other space agencies.

The modern field of spacecraft RS was made possible by the successive instrumental developments on both ends of the link (onboard spacecraft and at ground stations) that maximized telecommunications performance, increased the system sensitivity to receiving weak signals as well as spacecraft signals with high dynamics in frequency and/or amplitude. Signals transmitted by spacecraft and propagated to ground stations, and in recent cases, from one spacecraft to another, are themselves employed as an additional effective tool to study the planets and their environments.

Many notions came to define this field such as occultations and precision Doppler measurements, illustrated in Figure 1.1, which are representatives of the two broad categories of RS investigations, described in Section 1.2.1. In an occultation experiment, a spacecraft transmits a radio signal to Earth, or to another spacecraft, through the atmosphere of a planet. The signal is perturbed in its electromagnetic properties due to the interaction with the gas material and then examined by scientists to understand the planetary atmospheric structure.

A spacecraft flying near a planet transmits a communications radio signal to a ground station that undergoes an additional Doppler shift. The shift occurs when the planet's gravitational field causes perturbation of the spacecraft's motion. The received signal is examined by scientists to deduce information about the interior planetary structure, as a leading example. The same precision Doppler is a tool for spacecraft navigation.

With these techniques, radio scientists examine changes in the characteristics of electromagnetic waves, such as the frequency, phase, amplitude, line-width, polarization, and round-trip light time, to study planetary atmospheres, ionospheres, rings, and surfaces. They also deduce planetary masses and density distributions from gravitational fields, monitor the solar corona, and test theories of relativity. The signal characteristics are altered due to propagation effects caused by intervening media, scattering surfaces, or due to the additional relative motion caused by the mass and density distribution of the planets or moons, among other effects.

RS experiments have been conducted on almost every Solar System exploration mission through teamwork between academic and government research facilities as

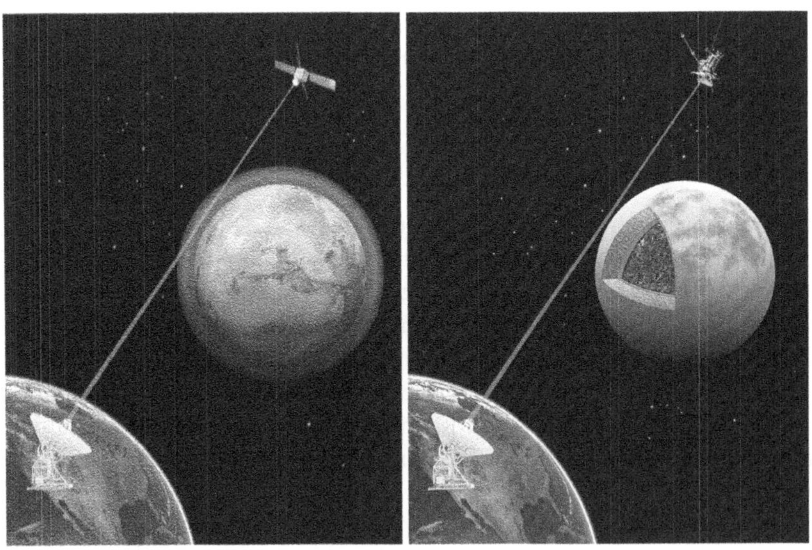

Figure 1.1 Illustration of two representative categories of radio science experiments. On the left, a spacecraft transmits a radio signal to Earth through the atmosphere of a planet. The propagation of the signal occulted by the atmosphere is perturbed by the interaction with the gas material, which reveals the atmospheric structure. On the right, a spacecraft flying near a planet transmits a signal to a ground station that is Doppler-shifted due the perturbation of the spacecraft's motion by the planet's gravitational field, which reveals information on the mass distribution and constrains models of the interior structure.

well as strong international collaboration. These experiments have led to numerous discoveries published in leading scientific journals and conferences.

For six decades, science teams have investigated the interior structure of every planet in the Solar System and several of their satellites, the Moon, and several comets and asteroids; sounded the atmospheres of the planets, Pluto, and several large satellites, as well as the Io Plasma Torus, and Saturn's and Uranus' rings; investigated planetary surface properties via bistatic scattering; investigated the solar corona and the interaction of the solar wind with planetary atmospheres; and conducted tests of the theories of General Relativity.

Selected accomplishments in RS are detailed in Chapters 2–4 and summarized in Appendix 1A along with near-term and long-term future experiments planned by various space agencies to further explore Mercury, Venus, asteroid Psyche, Jupiter's Galilean satellites, and other planetary bodies. These are aligned with community white papers for the Decadal Surveys of the National Academies of Sciences, Engineering, and Medicine, reproduced in Appendices 6A and 6B.

After initially using equipment shared with telecommunications and navigation systems, the RS community introduced science-specific flight and ground instrumentation and supplementary radio links at higher frequencies (e.g., Ka-band on Cassini, Juno, BepiColombo, JUICE and other planned missions). This lowers the noise caused by dispersive media, the electrically charged gas that affects the signal propagating through it as a function of the signal's wavelength. Lower frequencies (e.g., S-band introduced on Cassini for RS only) increase sensitivity to planetary ionospheres under study or ring particle size distribution.

Researchers also developed uplink atmospheric occultation instrumentation with processors onboard spacecraft to enable a higher received SNR, which was utilized during the New Horizons mission's flyby of Pluto (Tyler et al. 2009), and demonstrated as early as 1967 at Venus (Fjeldbo and Eshleman 1969). This was, in effect, a return to the original theoretical concepts of uplink ROs formulated in the 1960s. Onboard RS processors also benefit spacecraft-to-spacecraft crosslinks by achieving global planetary atmospheric occultation coverage significantly faster than downlink spacecraft links to Earth. Furthermore, RS researchers are preparing for the era of laser communications from deep space with plans for optical link science.

1.2 Fundamental Concepts

Solar System exploration missions currently use centimeter–wavelength radio frequency (RF) space links between spacecraft and Earth. Spacecraft transmit to ground station signals carrying the telemetry, the science data acquired by onboard instruments (payload) along with health and status reports. Spacecraft also receives signals from Earth stations carrying updated instructions and sequences called commands (or telecommands).

The basic radio signal used for spacecraft communication is a sinusoidal carrier that is phase modulated with data. Depending on engineering considerations, data are placed on either the carrier itself or on subcarriers (side carriers offset in frequency from the primary carrier). The telemetry sent by the spacecraft to the ground is transmitted in either real-time or is time-sequenced for transmission from data recorders when there is a ground station scheduled in the view period. The commands are also sent via uplink from the ground station to the spacecraft for either immediate execution at reception or at programmed times. For details on communications systems with deep space missions, see Yuen (1983), Doody (2011), and Taylor (2016).

At vast distances, the strength of the radio signals captured on Earth can be extremely weak in some cases, yet still provide useful information that can be critical to mission success. This is accomplished via high-quality flight and ground systems specially developed for deep space communications. Meter-wavelength signals in the ultrahigh frequency (UHF) band are used for proximity links between Mars landers and orbiters, for example.

The types of Solar System exploration missions include the following, each with communication links either directly to Earth, relay via another spacecraft, or both, with implications to possible RS experiments:

- Planetary flybys.
- Planetary orbiters.
- Stationary planetary landers.
- Roving planetary landers.
- Surface-exploring drones or helicopters.
- Sample collectors for return to Earth.
- Planetary atmospheric entry probes.
- Atmospheric balloons or ballutes.
- Formation-flying spacecraft with crosslinks between them.
- Futuristic concepts include sail planes and other aerial vehicles.
- Future concepts of short-lived low-cost small spacecraft exploring targets considered highly risky but have high science return.

For all types of missions, design teams minimize technical risks, conserve the overall mass and power resources, and extensively plan each mission's operational scenarios. They consider factors such as the expected frequency of contact with the ground stations to deliver the acquired data, the proximity to solar conjunctions that can be disruptive to communications, and the availability of relay assets when needed.

In the formulation and developments phases of a space mission, design choices in communications, navigation, and RS systems include

- Science requirements to which design parameters are traceable.
- Link budget parameters such as the transmitter power, antenna size, etc.
- Number of spacecraft antennas and their fields of view.
- Signal carrier wavelength(s) (e.g., S-, X-, Ka-bands, UHF, optical, etc.).
- Regulatory spectrum allocations and channel assignments to prevent radio interference among users.
- Telemetry modulation schemes.
- Data coding schemes.
- Signal polarization.
- Error budget of science parameters.
- Available proven technologies.
- Expected rate and volume of data.
- Consideration of cost.
- Contributions from other space agencies/external partners.

Mission design is the term used to describe the method of delivering spacecraft to their targets and includes maneuvers for interplanetary trajectories and other critical events such as gravity-assist flybys, orbit insertion, and, in some cases, descent and landing. Mission design teams collaborate with navigation (also known as flight dynamics) teams to implement the maneuvers and place the spacecraft on the planned trajectories. A state vector of the spacecraft position and velocity as a function of time is provided to mission users, communications engineers, and radio scientists for planning science operational activities.

Since the design of the spacecraft and the mission affect the quality of RS data, radio scientists participate in the development phases of a project life cycle and provide feedback on the proposed design and implementation possibilities and constraints, as well as performance requirements and specifications that maximize the scientific return of the planned RS experiments.

As previously noted, communication links are also utilized for additional functions. Radiometric measurements of the received radio signals are the primary method of determining and predicting spacecraft trajectories by navigation teams. Precise measurement of the carrier frequency and determination of its Doppler shift establishes the spacecraft relative velocity along the line-of-sight (LOS) between the spacecraft and the tracking station. In addition, the distance between a spacecraft and a station is accurately measured by ranging methods carried out by modulating predefined code onto the carrier signal and measuring the time of its travel at the speed of light from the station to the spacecraft and back (accounting for delays in electronic devices).

Since the distance and velocity measurements, called range and range-rate (RARR), are determined along the LOS, there are cases where a component in the plane of the sky is also important to measure. Radio interferometry across intercontinental baselines at multiple tracking stations are used to determine the position and velocity of the spacecraft in the plane of the sky via a technique called delta differential one-way ranging (delta-DOR), described in Thornton and Border (2003). Delta-DOR measurements and the more general very long baseline interferometry (VLBI) measurements are of high value to RS investigations as they provide astrometry and ephemeris information to aid in optimizing, as well as interpreting the science observations.

Radiometric measurements exemplify the first application of RS techniques. This measurement of perturbations of the expected RARR profiles illustrated how the communications links could be synergistically exploited to probe planetary and interplanetary environments. The intervening media along the propagation path lead to further perturbations in phase and amplitude. While communication engineers treat these perturbations as noise to be corrected for, radio scientists examine them for possible important information embedded in the wave's time of arrival, amplitude, frequency, phase, and polarization, as outlined above.

The interplanetary and ionospheric plasma are dispersive propagation media and the index of refraction depends on the frequency of the signal. Code and timing signals are delayed with respect to the vacuum, affected by the group delay. The phase of the carrier wave itself appears to speed up in the ionosphere for example, which is known as phase advancement. In a descriptive way, the wavelength in a plasma is enlarged with respect to the vacuum case. The absolute value of both effects is the same and depends on the same properties of the electron content.

Compared to telecommunications and navigation tracking applications, RS investigations typically require higher end-to-end link quality, especially higher signal frequency/phase stability. This is achieved via rigorous modeling, calibration, and bounding known sources of error and limiting factors (see Table 1.1) that include

- Uncertainties in the knowledge of the Earth's rotation and polar motion.
- Earth tropospheric path delay.
- Ionospheric and interplanetary plasma dispersion (also a function of proximity to solar conjunction or opposition periods).
- Stability of reference clocks/oscillators.
- Unmodeled nongravitational forces including spacecraft maneuvers interrupting the experimental data arcs, or thruster imbalance that bounds spacecraft dynamical stability.

Table 1.1 Radio science investigations, observables, experiment mode, and limiting factors.

Target [category]	**Science products**	**Observables [radio mode]**	**Limiting factors**
Planetary atmospheres *[Propagation through media]*	• Temperature-pressure profiles • Neutral density • Ionospheric density • Wave properties • Wind velocity • Abundance of key molecules • Shape or equipotential models • Turbulence and scintillation effects • Magnetic fields	• Frequency/Phase • Amplitude • Polarization • *[One-way or crosslinks]*	• Phase stability • Amplitude stability • Signal-to-noise ratio • Fresnel zone size • Absorption • Geometrical coverage • Temporal density • Orbital accuracy • Multipath interference
Planetary surroundings *[Propagation through media]*	• Ring particle sizes • Ring wave structure • Ring density • Torus electron density • Cometary comae • Planetary ejecta	• Amplitude • Phase • *[One-way]*	• Phase stability • Amplitude stability • Signal-to-noise ratio • Fresnel zone size • Absorption • Geometrical coverage • Temporal density • Orbital accuracy
Solar corona and solar wind *[Propagation through media]*	• Electron density • Turbulence and structures • Coronal ejecta • Coronal and solar wind speed • Scintillation effects • Magnetic fields	• Doppler • Range • Amplitude • Polarization • *[One or two-way]*	• Signal-to-noise ratio • Phase stability • Range calibration
Planetary surfaces *[Scattering off surfaces]*	• Electrical properties • Roughness • Material type • Near-surface layering	• Amplitude • Polarization • Doppler • *[One-way or crosslinks]*	• Signal-to-noise ratio • Geographic resolution • System calibrations • Orbital opportunities

(Continued)

Table 1.1 (Continued)

Target [category]	Science products	Observables [radio mode]	Limiting factors
Planetary interiors from gravity [*Remote sensing and rotational dynamics/ geodesy*]	• Planetary bulk mass • Mass/density distribution • Interior structure models • Thickness of layers • Libration • Admittance • Gas giants wind patterns • State of the core • Total and core moments of inertia • Tidal response • Natural oscillations • Size and density of core and mantle • Length of day variations • Mass transport/redistribution • Chandler wobble • Ephemeris	• Doppler • Range • *[Two-way or crosslinks]*	• Frequency stability • Unmodeled nongravitational effects • Trajectory design • Orbital accuracy • View geometry • Length of data arcs • Topography/shape/ equipotential models
Fundamental Physics [*Perturbation in spacecraft motion*]	• Gravitational radiation • Relativistic time-delay • Post-Newtonian parameters • Redshift—Equivalence principle	• Doppler • Range • *[Two-way except redshift]*	• Frequency stability • Calibrations
Atmospheric dynamics [*In situ*]	• Wind velocity • Molecular abundance from absorption	• Doppler • *[Crosslinks or one-way]*	• Frequency stability • View geometry
Atmospheric Density [*In situ*]	• Density from drag effects • Cometary mass flux	• Doppler • *[Two-way]*	• Phase stability • Trajectory

Notes

Target column:

- Planetary refers to all objects in the Solar System: planets, moons, asteroids, comets, etc.
- Planetary surroundings refer to material from planetary rings, plasma tori, comets, ejecta, etc.
- Planetary geodesy includes rotational history of precession, nutation, and polar motion.
- Fundamental physics refers to tests of aspects of the theories of relativity.

Science products column:

- Atmospheric RO determines the total density profile and sometimes specific constituent profiles.
- High temporal density of measurements is important to study atmospheric dynamics.
- High-resolution temporal and spatial densities are important for atmospheric/ionospheric wave studies.
- Wind velocity from RO is derived from the shape of equipotential surfaces and pressure gradients.
- Planetary radiometry and polarimetry also utilize RS instrumentation and techniques for investigations of planetary atmospheres.
- RO technique for in-situ measurements with balloons leads to a derived wind speed and direction indirectly via temperature and pressure by applying criteria such as cyclostrophic conditions.
- Planetary shape models derived from RO can be used to interpret gravity data and interior structure.
- Phase scintillation effects detected in planetary atmospheres and the solar corona.

(*continued*)

Table 1.1 (Continued)

- Planetary ejecta includes volcanoes, plumes, jets, etc.
- Magnetic fields are derived from the Faraday rotation (polarization) in ionospheres and solar corona.
- For the solar corona, space–time localization of the inner heliospheric plasma turbulence can be carried out via multiple radio links.
- Mass transport in planetary atmospheres such as Mars polar material moving seasonally and atmospheric tides and barometric pressure variations have implications to atmospheric occultations as well as gravity field measurements. This is part of the research of geophysical events and atmospheric sensing that also includes seismic activity through surface-atmospheric coupling and topography-induced atmospheric gravity waves.
- Ephemeris improvements are used for cosmological frameworks and studies of cometary jets of material leaving the surface.
- Observations of the redshift contribute to the validation of the Einstein equivalence principle.
- At the gas giants, gravity investigations involve examination of wind patterns and the deep atmospheric structure.

Observables column:

- Observables are the raw data with physical dimensions such as distance, velocity, signal power level, etc. that are input to scientific analysis tools after calibration. There is a lower level of data from the open-loop receiver in the form of the I and Q digital samples of the sine wave propagating in space (with noise).
- Delta DOR is a secondary RS observable that aids in reaching better navigational products and reduces experimental errors.

Limiting factors column:

- Trajectory design refers to proximity to target, speed of flyby, timing of maneuvers, and other events that can change the sensitivity or affect the quality of the RS experiment.
- Orbital accuracy refers to the orbital reconstruction, which is required for RS analysis.

- Unmodeled relativistic effects.
- Suboptimal pointing and attitude control.

Nongravitational forces acting on the spacecraft that can be calibrated via onboard precise accelerometers or other methods include

- Solar radiation pressure.
- Out-gassing of spacecraft components.
- Thermal effects.
- Thruster imbalances.
- Momentum of the high gain antenna (HGA) radiation.
- Factors unique to the target planet such as atmospheric drag, albedo, and infrared radiation.

1.2.1 Categories of RS Investigations

Based on their scientific objectives and primary observing techniques, RS investigations can be grouped into two broad categories; numerous discoveries have been made in each category and their branches as described in the corresponding chapters:

1. Investigations of phenomena affecting the *signal propagation* where the signal travels through, and is occulted by, planetary and/or interplanetary media occulting the path, or scattered by planetary surfaces.

2. Investigations of phenomena affecting the *perturbation of spacecraft motion* due to the gravitational forces, relativistic effects, or in-situ atmospheric densities or winds.

In the first category, scientists investigate the perturbation of the signal's electromagnetic properties along the path between the transmitting asset (spacecraft or ground station) and the receiving asset due to interactions with occulting media or by scattering off planetary surfaces. The effects of scientific interest arise from sounding planetary ionospheres, neutral atmospheres, the solar corona, cometary comae, plasma tori, ring particles, and other Solar System material.

Scattering effects result from the electrical and mechanical properties of planetary surfaces and near surfaces that can reflect a portion of the signal's energy to a receiving asset. This reveals the surface or subsurface reflectivity and roughness. Note that surface scattering, commonly called *bistatic radar*, and RO were historically developed in the same time period by the same early pioneers and share common methods and terminology.

In the second category, scientists investigate the precise perturbations in the relative motion between two assets that produce additional Doppler shifts in the signal frequency. The effects of scientific interest arise from remotely sensing gravitational fields or locally observing *planetary geodesy* via surface landers. Newtonian gravitational forces arising from the bulk mass and density distribution of the planetary targets perturb the motion of the spacecraft and enable radio scientists to determine the gravity fields and deduce the planetary interior structures; relativistic effects are also carefully investigated. This RS category intersects two branches of physics, dynamical forces and electromagnetic propagation, where the motion of the spacecraft is measured from precision detection of the electromagnetic Doppler effect.

Spacecraft throughout the Solar System are always in motion unless they are stationary landers on planetary surfaces such as Mars landers, some of which are equipped with instrumentation that enables the detection of planetary motion, including rotational nutation and precession. Measuring the latter via Doppler tracking of stationary spacecraft aids in constraining the physical characteristics of planetary cores. Other in-situ effects include measuring planetary wind speeds and directions while spacecraft (entry probes) descend through atmospheres and are thus subject to wind forces, and determining the atmospheric density via measuring the drag on spacecraft surfaces, among other examples.

Note that, in the first category, inevitable changes in the signal properties resulting from the relative motion of the assets are quantified or modeled and then removed from the data via calibration techniques described below. In the second category, inevitable changes in the signal properties resulting from propagation through media are quantified or modeled and then removed from the data via calibration techniques described below.

In some cases, there is overlap in scientific interest between the two categories. For example, RO techniques used to generate a shape model of a planetary body can be utilized by gravity researchers to correlate the gravitational field and shape model and study the interior structure. Annother example is investigating the seasonal transport of Mars' polar material via atmospheric occultation and gravity science. Also, ionospheric waves are investigated in some cases for attribution to surface or subsurface effects such as seismic activities.

Representative RS investigations and relevant experimental configurations are outlined in Figure 1.2. Note that in some literature, RO techniques can also be referred to as limb sounding techniques or similar terminology and that gravity investigations are occasionally referred to as celestial mechanics in some literature, including documents of the Galileo mission to Jupiter. Table 1.1 lists most of the known or anticipated RS investigations with associated observables and limits of sensitivity. Table 1.2 provides a traceability matrix for functional and performance requirements of instrumentation 5to be developed to meet future scientific needs discussed in Chapter 6. In the figures and tables, the terms "one-way" and "two-way" are explained in Section 1.4 and graphically illustrated in Chapter 5.

1.2.2 Related Fields

Scientific interpretation of RS data often requires utilization of other data types and collaboration with other science teams. As a result, many RS investigators are often involved in collaborative fields and instruments such as planetary altimeters, radar, imagers, radiometers, accelerometers, gradiometers, polarimeters, and in-situ atmospheric sounders. Furthermore, work in the field of RS is multi-disciplinary and often requires expertise in science, engineering, instrument technology, software tools, operations, and mission support.

There are science investigations that utilize instrumentation and techniques overlapping with RS, sometimes considered a passive branch of this field. Planetary

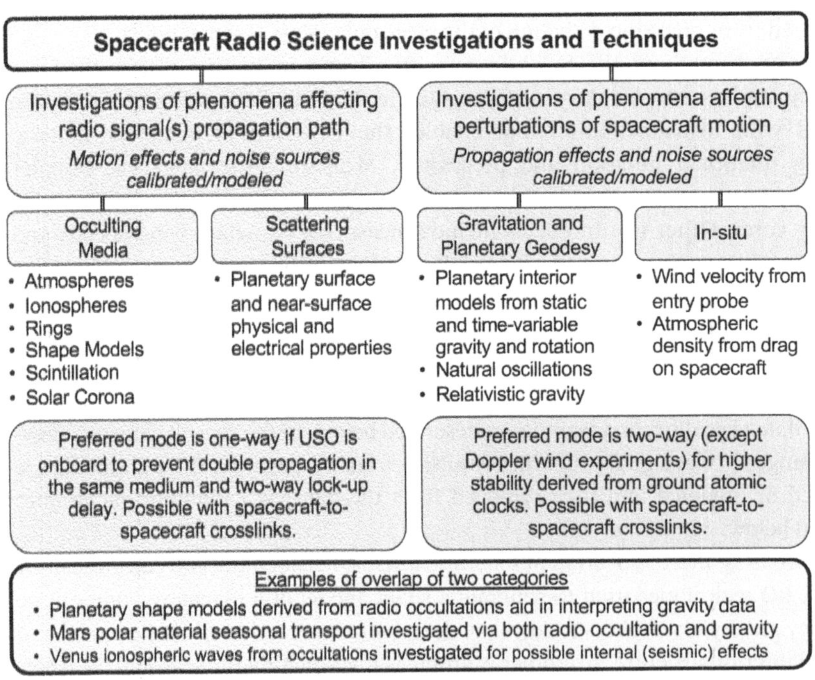

Figure 1.2 Examples of spacecraft radio science experiments grouped into two broad investigation categories and subcategories with representative science products and notes on typical instrumental configurations (USO is ultra-stable oscillator).

Table 1.2 RS functional and performance requirements traceable to science products.

Science targets	Functional requirements	Performance requirements
Planetary atmospheres and rings *via radio occultations*	• Ultrastable links (two links preferred) • One-way or dual one-way if USO; two-way if no USO limited to occultation ingress	• Allan deviation ~1×10^{-13} (~100 s) • Phase noise and spectral purity ~−90 dBc/Hz at 10 Hz offset • Amplitude stability ~0.1 dB over 60 s
Planetary surfaces *via signal scattering*	• One-way or crosslinks; good geometry • Signal power at dual polarizations • Multiple signals for uplink experiments	• Allan deviation ~1×10^{-12} (~100 s) • Amplitude stability ~0.1 dB over 60 s • Precision power measurement calibration
Planetary interiors *via Doppler tracking*	• Precisions two-way Doppler link(s); one-way possible if atomic clock onboard • Shorter wavelength less noisy	• Allan deviation ~1×10^{-14} (~1000 s) • Dynamically stable spacecraft • Calibration of nongravitational forces and propagation media effects
Planetary winds *via Doppler tracking*	• Cross- or one-way Doppler links with stable oscillators at both ends • Signal power if monitoring absorption	• Allan deviation ~1×10^{-11} (~100 s) • Amplitude stability ~ 0.1 dB over 60 s • Multiple links aid in wind vectors recovery
Solar science *via radiometric observables*	• Stable dual Doppler links • Phase at two circular polarizations • Ranging (two-way) for coronal densities	• Allan deviation ~1×10^{-12} (~100 s) • Amplitude stability ~0.1 dB over 60 s • Modelled antenna polarization if monitoring Faraday rotation
Fundamental physics *via radiometric observables*	• Precisions two-way Doppler link(s) • Shorter wavelength less noisy • Precision ranging	• Allan deviation ~1×10^{-14} (~1000 s) • Dynamically stable spacecraft • Calibration of nongravitational forces and propagation media effects

Notes
- Science targets are grouped by both science objectives as well as observation method.
- Functional requirements are the required spacecraft and ground systems capabilities.
- All investigations assume favorable orbits and geometrical configurations.
- USO: Ultrastable oscillator.
- Performance requirements are the engineering specifications to meet known or typical science requirements and can vary depending on the mission and investigation scenarios and constraints.
- Allan deviation (sometimes Allan variance) is a measure of fractional frequency instability for classes of clocks and oscillators.
- Dynamically stability is determined by spacecraft thruster balance or timing of spacecraft maneuver or solar electric propulsion events that disturb the Doppler data quality.
- Precision power calibration is required for comparing small variations in the two polarizations of received signals.

radiometry, for example, can utilize RS-type of open-loop receivers (OLRs) to capture surface brightness temperature in a radiometer mode. For example, Stern et al. (2018) and Bird et al. (2019) described the New Horizons Radio Science Experiment (REX) as an uplink RS instrument with radiometer capabilities (disk-averaged to ±0.1 K) that conducted atmospheric radiometric measurements during the Pluto flyby by analyzing the variation of the RF signals passing through the atmosphere that are received by the spacecraft.

Similarly, the DSN RS OLRs are utilized for related scientific applications such as radiometry and differential interferometry. The latter is a technique used to examine planetary rotational state and tidal deformations by tracking multiple landers or spacecraft from one or multiple stations on Earth in order to suppress many common noise sources and enhances the signals of interest.

Another example is planetary radio polarimetry, a technique used to examine precipitation or water vapor detection in planetary environments, that can also utilize RS instrumentation and data-processing techniques.

Occultations by planetary bodies of natural radio sources, such as pulsars, active galactic nuclei, and masers, can also utilize RS instrumentation and techniques. Withers and Vogt (2017) described occultations of distant astrophysical radio sources as probes of planetary environments, especially magnetic field strength, plasma density, and neutral density. They also examined cases for Jupiter and possible applications to exoplanets occulting their parent stars. They found significant changes in polarization angle due to Faraday rotation occur for radio signals that pass within 10 Jupiter radii, and that significant changes in frequency and power occur from radio signals that pass through the neutral atmosphere.

1.3 Historical Development

The history of RS studies began conceptually with the radiometric measurements for celestial mechanics and radio navigation. The earliest use of the term in the context of planetary exploration goes back to ROs in the early 1960s. Professor Von Eshleman of Stanford University first proposed the method in 1962, building on earlier work with the formulation of mechanisms for radio reflections from meteoric ionization (Villard et al. 1953).

Independently, Dan Cain at NASA's JPL, California Institute of Technology, a navigator studying the effects of refraction in the Earth's atmosphere on the accuracy of the observed Doppler frequency shift, realized the potential for applying sensitive phase measurements to the study of the atmospheres and ionospheres of other planets (Kliore et al. 1964; Cain et al. 1966). Graduate student Gunnar Fjeldbo, who later changed his Norwegian name to Gunnar Lindal, advanced the occultation technique by utilizing a theoretical uplink configuration during his dissertation research with Von Eshleman (Fjeldbo 1964). Fjeldbo/Lindal later joined the Mariner and Voyager science teams during his subsequent employment at JPL, where his colleague Arvydas Kliore was the first RS principal investigator on a planetary mission.

The occultation technique was demonstrated for the first time in 1965 when Mariner IV flew by Mars and determined the salient features of the Martian atmosphere (stories were reported of mission managers needing to be convinced that such an occultation posed no risk to the safety of the mission); the resulting historic occultation profile is reproduced in Figure 1.3. The observations showed that the surface pressure is less than 1% of Earth's; an order of magnitude less than had been previously thought (Kliore et al. 1965; Fjeldbo et al. 1966). Mariner V flew by Venus in 1967 and conducted the second atmospheric occultation experiment (Kliore and Cain 1968; Eshleman et al. 1968).

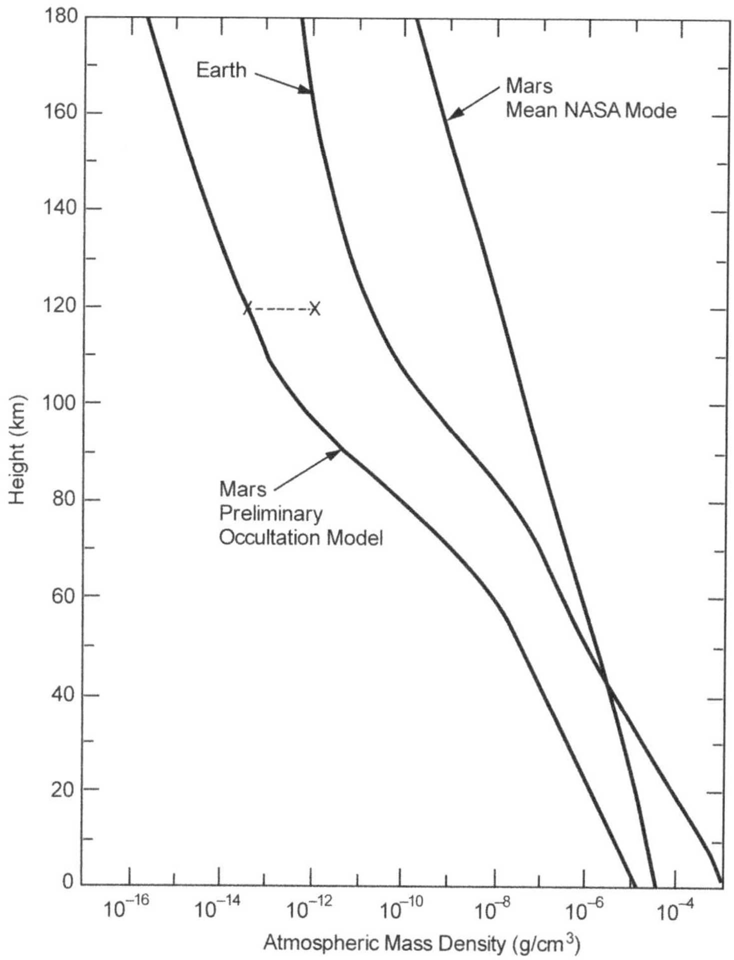

Figure 1.3 A reproduction of the result from the first planetary radio occultation. The technique was demonstrated in 1965 when Mariner IV flew by Mars and determined the salient features of the Martian atmosphere (Kliore et al. 1965; Fjeldbo et al. 1966).

After the successful occultation experiments with the Mariner mission series and as additional scientists and their students trained in this field, planning moved toward other planetary targets and many additional experiments produced notable discoveries. Eshleman et al. (1977) and G. L. Tyler (1987) described the RS experiments with Voyager; Figure 1.4 shows the classical results of the Voyager 2 atmospheric occultations of the four large gaseous planets.

Howard et al. (1992) described the Galileo RS experiments planned at the time (many of which were not carried out due to the Galileo HGA anomaly); Bird et al. (1992a) described the Ulysses Io plasma torus occultation experiment; Tyler et al. (1992) provided an overview of Mars experiments using the Mars Global Surveyor; Pätzold et al. (1993) described RS investigations using the Giotto mission to Halley's Comet; and Pätzold et al. (2016a) described the RS experiments performed with the Rosetta mission to comet 67P/Churyumov–Gerasimenko (67P). Mars Express and Venus Express RS

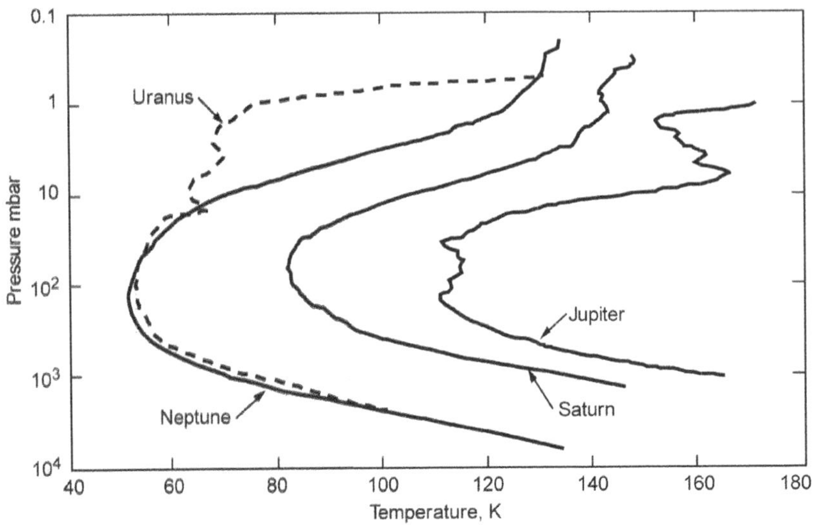

Figure 1.4 Classic results of radio occultation experiments showing the temperature–pressure profiles for the giant planets from radio occultation data acquired with the Voyager 2 spacecraft flybys (Lindal 1992).

experiments are described in Pätzold et al. (2016b) and Häusler et al. (2006), respectively. Kliore et al. (2004) described the Cassini RS experiments for the gravity and atmospheres of Saturn and its large moons, as well as Titan's surface and Saturn's rings.

Additional pioneering work from various space agencies and research centers advanced additional scientific techniques such as:

- Inferring surface morphology from the scattering of the radio signals historically known as bistatic radar (Tyler and Howard 1973; Marouf 1975), first demonstrated with Luna-11 (Yakovlev and Efimov 1966); Yakovlev (2002) describes several experiments on Soviet/Russian missions.
- Techniques for using RO to investigate planetary ring structure (Marouf et al. 1986).
- Radio scintillations reveal small-scale structure in planetary atmospheres (Hinson 1986).
- RO techniques to study planetary ionospheres (Kliore et al. 1967).
- Scintillation studies of solar wind turbulence and structures (Woo 1993).
- Power spectral analysis of the interplanetary plasma (Armstrong and Woo 1981).
- P. Steffes (1985) took the lead on laboratory measurements of the microwave opacity and absorption, especially simulating conditions in the atmosphere of Venus.
- ESA missions utilized RS techniques for extended observations of the solar corona with the Helios mission (Volland et al. 1977; Pätzold et al. 1987; Bird and Edenhofer 1990; Andreev et al. 1997).
- The SELENE mission from the Japan Aerospace Exploration Agency (JAXA) measured the gravitational field of the lunar far side from four-way (see Section 5.2.1.4) Doppler links (Namiki et al. 2009), which utilized a subsatellite with a four-leg link from the ground to the orbiter to the sub-satellite and back.
- Doppler-based vertical wind profiles of the Venus atmosphere from the Venera probe (Kerzhanovich et al. 1983).

- Flow of the Venus winds using measurements of the Vega Venus balloons' position and velocity using Doppler and VLBI observables (Sagdeev et al. 1986a, 1986b; Preston et al. 1986).
- Dual-frequency investigations of planetary ionospheres were performed in the 1970s and 1980s with Luna 19 and 22 at the Moon (Vyshlov 1976), with Mars 2, 4, 5, and 6 at Mars (Savich et al. 1976; Kolosov et al. 1976), with Venera 9, 10, 15, and 16 at Venus (Aleksandrov et al. 1977; Savich 1981; Savich et al. 1986a; Ivanov-Kholodny et al. 1979) and with Vega 1 and 2 at Comet Halley (Savich et al. 1986c; Andreev and Gavrik 1993).
- Venera orbiters' radio sounding of the solar corona (Efimov et al. 1977; Kolosov et al. 1982).

RS experiments have also refined our understanding of post-Newtonian physics. Frank Estabrook pioneered the search of gravitational waves via Doppler tracking (Estabrook and Wahlquist 1975). Later, using the Cassini spacecraft, precision Doppler observations in the cruise phase were used to place upper limits on the strength of low-frequency gravitational radiation (Bertotti et al. 1995; Armstrong 2006). Also using Cassini during its cruise phase, RS tests of aspects of the theories of relativity used Doppler tracking to improve the parameterized post-Newtonian formalism (Bertotti et al. 2003).

Gravitational fields are used in combination with shape models or detailed topography to constrain the interior structure of the planets and other objects in the Solar System. One of the most recent and notable examples of measuring gravity fields is the ultra-high-resolution gravity field of the Moon produced by the GRAIL team (Figure 1.5). This mission has provided an unprecedented view into the interior of the Moon, with a spherical harmonic degree/order-1500 (or higher) field (equivalent to a spatial resolution of a few km), utilizing spacecraft-to-spacecraft links supplemented by links to the DSN (Zuber et al. 2013b; Konopliv et al. 2014b). GRAIL is based on the techniques developed by the GRACE Earth mission (Watkins et al. 1995) and has led to tremendous gain in our knowledge of the Earth's time-varying gravity and the water cycle.

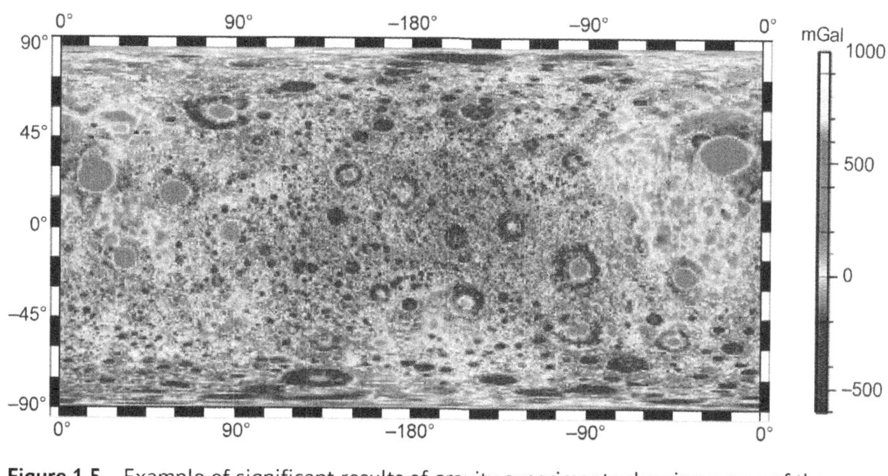

Figure 1.5 Example of significant results of gravity experiments showing a map of the gravitational field of the Moon from the GRAIL mission utilizing crosslinks and links to the DSN. Surface acceleration resolution is better than 1° and on the order of 1500 degree and order in spherical harmonic expansion (based on Konopliv et al. 2014a); see Chapter 3. (Note: mGal is 0.001 Galileo, which is 1 cm^2/s).

In general, the gravity fields of terrestrial planets have been measured to high spatial resolution, while the gravity fields of the gas giants and large moons of the outer Solar System have been measured to low resolution. This is because the resolution of the gravity field is determined by the proximity to the target, duration of measurements, and the overall noise/error budget (see Chapter 3). All the inner planets have been investigated by orbiters at relatively lower altitudes and for extended periods.

Folkner et al. (1997b) pioneered the measurement of planetary precession using the Mars Pathfinder lander. This experiment was also carried out by the InSight mission (Banerdt et al. 2020) and planned for the ExoMars lander mission. Konopliv et al. (2016, 2011) demonstrated that a combination of lander and orbiter data improves the knowledge of the precession; see also Dehant et al. (1993) for an analytical approach. Kuchynka et al. (2014) demonstrated Mars precession measurement obtained from orbiter data only. Derivations of planetary shape models from RO data are discussed in Cain et al. (1972) and Asmar et al. (1999).

RS experiments on atmospheric probes included Doppler wind experiments and other investigations on Venera (Kerzhanovich et al. 1972), Pioneer Venus Probe (Counselman et al. 1980), Venus Balloons (Preston et al. 1986), Galileo Probe (Atkinson et al. 1997), and Huygens (Bird et al. 2005). Rosenblatt et al. (2012) carried out the first in-situ observations of Venus' polar upper atmosphere density using the tracking data of the Venus Express Atmospheric Drag Experiment.

RS investigation of planetary magnetic fields by observing the Faraday rotation was planned to study the Jovian magnetic field with the Galileo mission (Howard et al. 1992) but was not carried out due to an anomaly with the spacecraft HGA. Investigations of the magnetic field in the solar corona via Faraday rotation have been carried out by several teams (Levy et al. 1969; Stelzried et al. 1970; Pätzold et al. 1987; Bird and Edenhofer 1990). Hinson (1984) discuss dual-wavelength radio observations during the occultations of Voyagers 1 and 2 and how they provided data on Saturn's ionospheric irregularities and how magnetic field orientations in Saturn's ionosphere were inferred from the direction of alignment of the anisotropic irregularities.

The use of precision Doppler tracking to model the gravitational fields of planets and moons using techniques described in Chapter 3 was pioneered by investigators that included Bill Sjogren, Dave Smith, Georges Balmino, John Anderson, and many others. One of the earliest historical discoveries was the lunar subsurface mass concentrations, called "mascons" (Müller and Sjogren 1968; Sjogren et al. 1972). Mathematical formulations for gravity research as well as spacecraft navigation applications are described in Moyer (2003).

1.4 Overview of the Radio Science Instrumentation System

Table 1.3 lists the majority of Solar System exploration missions that have carried out RS experiments. A scientific payload for a space mission is selected in a process that starts with the space agency soliciting the community for instruments that meet stated scientific objectives. The instruments are typically self-contained units accommodated onboard the spacecraft (e.g., imager, spectrometer, magnetometer, or altimeter).

Table 1.3 Selected Solar System missions with radio science experiments.

Target	Mission	Experiment
Sun	Helios	SC, FR
	Ulysses	SC, FR, REL
	Other solar missions include Pioneer 5–9, Proba2, SOHO, WIND, ACE, DSCOVR, STEREO, Parker Solar Probe, ESA Solar Orbiter (*planned*), and Aditya (*planned*)	
	Solar conjunctions of Magellan, Galileo, Cassini, and other planetary missions	SC, FR, REL
Mercury	Mariner 10	GRV, SHP
	MESSENGER	GRV, SHP
	Planned: BepiColombo	*GRV, REL, SC*
Venus	Mariner 2	GRV
	Mariner 5	RO
	Mariner 10	RO, GRV
	Venera 4–16	RO (Venera 4)
	Pioneer Venus 1 Orbiter	RO
	Pioneer Venus 2 Probes	DWE, Interferometry
	VEGA 1, 2 Balloons	DWE, Interferometry
	VEGA 1, 2 Landers	DWE
	Magellan	GRV, RO, solar FR
	Venus Express	RO, GRV, BSR
	Completed flybys by many planetary missions	Doppler Tracking
	Planned flybys by BepiColombo, Solar Orbiter	Doppler Tracking
Earth	GRACE and GRACE FO	GRV
	GPS-COSMIC	RO
	GPS-Mir	RO, radio holography
	Flybys by many planetary missions	Doppler Tracking
Moon	*Many historical missions*	
	Pioneer 7	RO (ionosphere)
	Luna series	GRV, RO
	Apollo series	GRV, interferometry
	Clementine	BSR, GRV
	SMART 1	Libration demo, RO
	Lunar prospector	GRV, SHP
	SELENE (Kaguya)	GRV, RO
	GRAIL	GRV
	LRO	GRV, BSR, LR
	Hiten	Doppler Tracking
	LADEE	LASER, radio tracking
	Chang'E 3	Rotational dynamics
	Chandrayaan-1	RO
	Chandrayaan-2 (orbiter)	RO (planned)
	New horizons	RO

(*Continued*)

Table 1.3 (Continued)

Target	Mission	Experiment
	Many planned missions	
Mars	Mariner 4, 6, 7, 9	RO
	Mars 2, 4, 6	RO
	Mars 3	GRV
	Viking 1, 2 Orbiters	
	Mars Global Surveyor	RO, GRV, BSR
	Others include: Pathfinder- Sojourner (*rover*), Phoenix (*lander*), MSL-Curiosity (*rover*), Mars Orbiter Mission, Trace Gas Orbiter	
	Spirit, Opportunity (*entry probe/lander*)	Geodesy
	Mars Express	RO, GRV, BSR
	Mars Odyssey	GRV
	Mars Reconnaissance Orbiter	RO, GRV
	MAVEN	RO
	InSight	Geodesy
	Planned: ExoMars 2020	Geodesy
	Planned: Mars 2020, Hope, MOM-2, MMX	
Jupiter	Pioneer 10, 11	
	Voyager 1, 2	RO, GRV
	Galileo Orbiter	RO, GRV
	Galileo Probe	DWE
	Juno	GRV
	Flybys by: Ulysses, New Horizons, Cassini, etc.	
Saturn	Pioneer 11	
	Voyager 1, 2	RO, GRV
	Cassini	RO, GR, BSR, REL
	Huygens Titan Probe (*lander*)	DWE
Uranus	Voyager 2	RO, GRV
Neptune	Voyager 2	RO, GRV
Pluto	New Horizons	RO, GRV, BSR
Comets	Giotto	Doppler Tracking
	Rosetta	GRV, BSR
Asteroids	Dawn	GRV, BSR, SHP
	Hayabusa 1	Doppler Tracking
	Hayabusa 2	GRV
	NEAR	GRV
	OSIRIS-Rex	Doppler Tracking
	Planned: Psyche	GRV

Acronyms for Table 1.3

Missions	
• ACE: Advanced composition explorer	• MSL: Mars science laboratory
• COSMIC: Constellation observing system for meteorology ionosphere and climate	• Proba: PRoject for Onboard Autonomy
• DSCOVR: Deep space climate observatory	• SMART: Small missions for advanced research in technology
• GRACE FO: Gravity recovery and climate experiment follow-on	• SOHO: Solar and heliospheric observatory
• LRO: Lunar reconnaissance orbiter	• STEREO: Solar TErrestrial RElations Observatory
• MMX: Martian moons exploration	• VEGA: A contraction of Venera and Gallei, the Russian words for Venus and Halley
• MOM: Mars orbiting mission	• WIND: Solar Wind Laboratory

Experiments	
• DWE: Doppler wind experiment	• RO: Radio occultations
• FR: Faraday rotation	• SC: Solar corona
• GRV: Gravity	• SHP: Shape modeling
• REL: Relativity	• BSR: Bistatic radar

Radio scientists typically propose investigations that utilize instrumentation distributed between the spacecraft and ground stations centered on the mission's radio communications and navigation capabilities, as shown in Figures 1.6 and 1.7. Since the performance of the entire distributed instrument affects the quality of the scientific outcome, proper planning, implementation, maintenance, and calibration of each element are required. Calibration can encompass advanced techniques such as multiple radio links, water vapor radiometers, accelerometers, noise-temperature measurements, etc. The limits of experiment sensitivity are based on orbital and geometric considerations, frequency/phase stability of the link, its SNR, amplitude stability, spectral purity, calibration of nongravitational forces, calibration of the intervening media, and accuracy of the trajectory reconstruction.

A key concept to understanding the RS system and properly interpreting the data is the spacecraft transmitter mode. There are three common modes for the spacecraft transmitter operations, one-way, two-way, or three-way as detailed and illustrated in Chapter 5, which also describes other exotic modes and new technologies.

In the one-way mode, the spacecraft transmits a signal generated by an onboard oscillator that is part of its telecommunications subsystem—no uplink from the ground station required. There are two types of spacecraft oscillators; one that is built into every transponder and called the auxiliary oscillator and one that is an optionally added stand-alone instrument with an input line to the transponder called the ultra-stable oscillator (USO).

In the two-way mode, the station transmits a signal to the spacecraft, which receives it and transmits it back to the same station. This step is called coherently "transponding" the signal. The onboard auxiliary oscillator or USO are not involved.

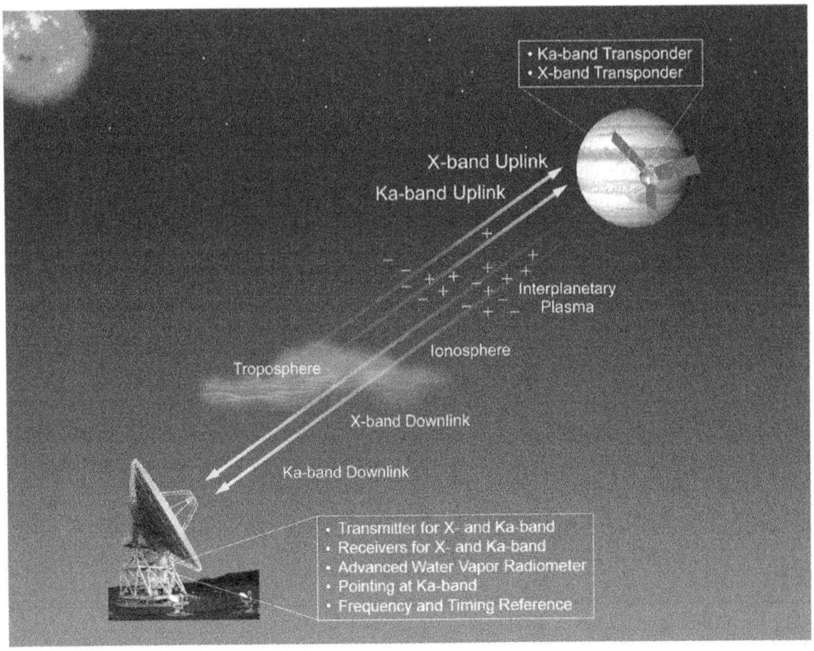

Figure 1.6 Illustration of the Juno spacecraft and a DSN station operating jointly as a gravity science instrument spanning a fraction of the size of the Solar System. The investigation partially calibrates dispersive media via the use of two wavelengths simultaneously and calibrates the Earth's tropospheric effects via an advanced water vapor radiometer. Signals at X- and Ka-bands are simultaneously transmitted from the DSN station to the spacecraft and transponded back to the station. Differencing the two signals isolates the effects of the charged particles. Finally, using a water vapor radiometer, shown adjacent to the station, measures the path delay in the troposphere.

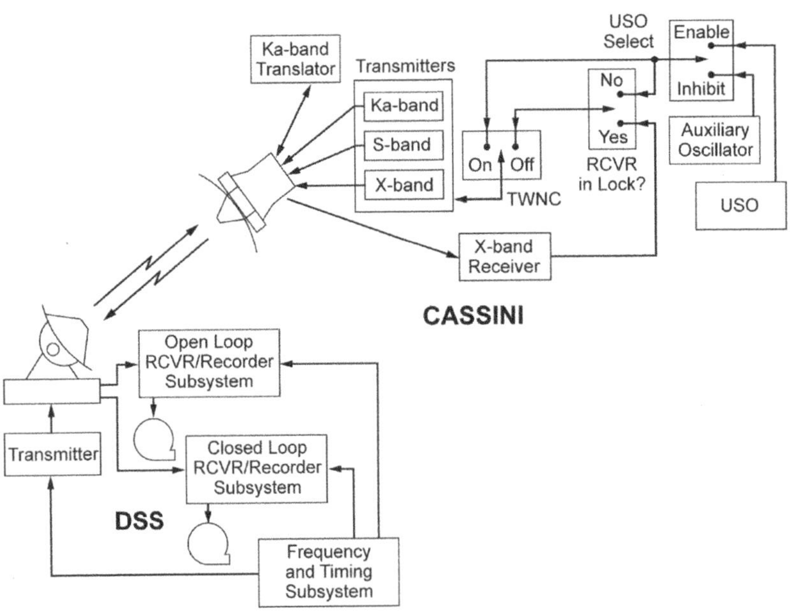

Figure 1.7 Conceptualized primary components of ground and flight communications and RS systems using a deep space station (DSS) and the Cassini mission, which had a radio system with links at three wavelengths (S-, X-, and Ka-bands).

The three-way mode is the same as the two-way mode, except that the station that transmits the uplink to the spacecraft is different from the station that receives the downlink from the spacecraft. This occurs when the round-trip light time travel to a distant spacecraft is longer than the view period of the transmitting station; after the transmission, the Earth has rotated enough such the station is not in view of the spacecraft and another station on a different continent has risen in view. The three-way mode can also occur if two stations at the same location are simultaneously supporting the same spacecraft; only one station can transmit to the spacecraft and this statin receives the returned signal in a two-way mode while the second station receives the same signal at the same time in a three-way mode.

Planning the tracking mode of operations is critically important to RS experiments because this determines the type of clock or oscillator that generates the reference signal. The auxiliary oscillator is the least stable oscillator of the options described above and is not typically useful for RS experiments. The USO can be two or more orders of magnitude more stable and is the preferred reference for occultation experiments. Two-way and three-way signal are referenced to an atomic clock at the ground station, called a hydrogen maser, which is about two orders of magnitude more stable than the USO and is preferred for gravity experiments, as explained more below and detailed in Chapter 5.

1.4.1 Flight System

Elements of the flight system enable RS experiments and affect their ultimate quality. At the heart of the system is the telecommunications subsystem, also known as the Telemetry, Tracking, and Command (TT&C) subsystem, which includes the antennas, transponders, and power amplifiers. There is typically one HGA on a planetary mission, several low-gain antennas (LGAs), and sometimes a medium-gain antenna (MGA). There are also redundant and highly reliable transponders that carry out the reception, transmission, modulation, demodulation, and tracking functions.

Additional transponders or translators (a designation given to transponders without modulation capabilities) can be incorporated in the flight system for RS experiments and handled as science payload elements. For example, as shown in Figure 1.7, the Cassini mission primary link for telecommunications, navigation, and RS is the X-band link supported by two redundant sets of transponders and power amplifiers. RS experiments can also require links at S- and Ka-bands (e.g., Cassini, which carried additional non-redundant radios).

Similarly, for the case of the Juno mission configuration, a separate Ka-band translator augments the X-band transponder in order to enhance the Jovian gravity experiments. Figure 1.6 shows the end-to-end Juno RS instrumentation and how the added Ka-band link enables the calibration of the ionosphere and interplanetary plasma. Tropospheric effects are calibrated via an advanced water vapor radiometer at one DSN station, DSS-25, at Goldstone, California.

The first flight instrument devised and flown specifically for RS enhancement was the USO, called the “single most dramatic improvement in radiometric performance” by G. L. Tyler (1987). A clock and frequency reference, the USO was developed for atmospheric occultation experiments starting with the Voyager mission in order to provide a stable reference to the transponder as it transmits a one-way downlink signal to Earth.

This configuration provides two advantages over carrying out the experiments in the two-way mode. The first advantage is eliminating the double propagation by the uplink and downlink signals in the planetary medium under study, such as atmospheric or ring material, which make the data interpretation more challenging. The second advantage is eliminating the delay due to the time needed by the transponder to lock-up on the uplink signal when the spacecraft is emerging from behind the occulting medium. Although the lock-up time is short (under a minute), for a thin atmosphere like that of Mars, this delay can be comparable to the length of the observation and cause the loss of the egress experiment at the lower atmospheric region.

Based on a thermally controlled quartz crystal, a modern USO model can reach a stability of ~10^{-13} Allan deviation (a measure of clock stability) at integration times between 10 and 100 s. After Voyager, USOs have flown on Galileo and Cassini as well as orbiters at Mars, Venus, and the Moon, and many other planetary and Earth missions listed in Asmar (2012b). In-situ probes, such as the Galileo probe, used a crystal-based USO while the Huygens Titan probe utilized a rubidium-based USO for the faster warm-up time required on short-duration missions.

Other spacecraft elements that can directly impact the quality of RS experiments include the attitude control subsystem, which controls the antenna pointing and overall stability of the spacecraft motion. Non-optimum pointing or attitude jitter can degrade the science experiments, so investigators place strict performance specifications on this subsystem. The attitude control subsystem is also required for special and sometimes complex limb-track maneuvers that optimize RO experiment performance by dynamically pointing the spacecraft HGA toward the limb of the target planet. As discussed in Chapter 5, investigators also account for numerous additional details in dynamical and kinematic nongravitational error sources including the optical properties of the spacecraft surfaces, uncertainties in the location of the center of mass, etc.

1.4.2 Ground System

The DSN includes 70- and 34-m diameter stations located at three continents with capabilities for transmission and reception at various frequency bands allocated by international agreement (see development history in Mudgway 2001). The bands in current usage are S, X, and Ka, corresponding to the wavelengths shown in Table 1.4. The frequencies of the transmitted uplink signals, also known as forward links, are related to the frequencies of the received downlink signals, also known as return links, by multiplicative factors called the turn-around ratios in order to prevent interference between them and keep coherency throughout the complete uplink/downlink chain.

The most common deep space communication band for contemporary missions is X-band, available throughout the DSN and partner networks. Use of Ka-band is increasing. Uplink at Ka-band is currently available at one DSN station (DSS-25 at Goldstone) for RS purposes, with plans for additions at the other complexes, while reception at Ka-band is available throughout the network. S-band capabilities are of lesser usage for planetary missions (deep-space band) but increasing usage for lunar missions (near-Earth band). The DSN is preparing for future optical links as this technology advances and gets demonstrated on future planetary missions.

1.4.2.1 Types of Receivers

The type of receiver at the ground station is a key factor in planning and interpreting RS experiments and data. Reception of the radio signal at the ground stations is

Table 1.4 Band names, wavelengths, and frequencies for links used for deep space, near-Earth, and proximity links relevant to RS and potential RS investigations.

(A) Deep space links (DSN, Telecommunications Link Design Handbook 810-005 Module 201)				
Band	Wavelength (cm)	Uplink (forward) frequency range (MHz)	Downlink (return) frequency range (MHz)	Transponder ratio (downlink/uplink)
S	~13	2110–2120	2290–2300	240/221
X	~3.6	7145–7190	8400–8450	880/749 S–X: 880/221 X–S: 240/749
Ka	~0.9	34,200–34,700	31,800–32,300	14/15 (Cassini) 3360/3599 (Standard) X–Ka: 3360/749
(B) Near-earth links (for spacecraft less than 2 million km from Earth)				
Band		Uplink (forward) frequency range (MHz)	Downlink (return) frequency range (MHz)	
S		2025–2110	2200–2290	
X		7190–7235	8450–8500	
K		22,550-23,150	25,500–27,000	
(C) Relay/Proximity ultra high frequency (UHF) links				
Band	Wavelength (cm)	Forward link (MHz)	Return link (MHz)	
Before Electra payload	~75	437.1	401.58	
Electra (tunable full duplex)	~75	435–450	390–405	

Notes

- Ka-band standard ratio is used for Juno, BepiColombo, and JUICE.
- Electra provides mode for 2-way Doppler coherency.
- Ka-band links for near-Earth missions are called K-band links in DSN documents.

accomplished via two methods. The first is the common closed-loop reception and tracking, in which a phased-locked loop receiver captures the residual carrier signal and precisely measures its phase in a preselected loop bandwidth to track the signal dynamics. Telemetry demodulation can then be performed. Navigation and RS teams use the precise Doppler and range information available in real-time from this type of receiver. However, this receiver has minimum thresholds on signal level and dynamics beyond which it can drop lock.

The second method is open-loop reception in which the receiver has no lock or tracking function. Instead, the OLR, also known as the *Radio Science Receiver* (RSR), down-converts the signal via an oscillator driven by a prediction file based on the expected Doppler profile. It captures the pre-detection signal in a selected bandwidth and sends raw digital complex samples of voltages to the users (such data are ultimately made available to the science community via NASA's Planetary Data System (PDS), as explained in Section 1.6). Users employ their own software to detect the signal from the RSR file and recover the frequency transmitted by the spacecraft for further analysis. For RS experiments, the advantages of the open loop reception method are detailed in Chapter 5 and include capturing the signal during conditions of low SNR, capturing the signal during conditions of high signal dynamics in amplitude or frequency, capturing multipath signal conditions, and preserving the information in the entire spectrum.

1.4.3 Other Ground Stations

The European Space Agency (ESA) currently operates three 35-m diameter stations for deep space tracking known as the European Space Tracking (ESTRACK) network. They are located in New Norcia (Australia), Cebreros (Spain), and Malargüe (Argentina). The ESA stations often support NASA missions and the DSN often supports ESA missions within the framework of a cross-support agreement. ESA plans to expand its development of RS-specific services in the future. Japan, India, S. Korea, Russia, and China also have deep space tracking stations, and additional nations are planning to develop them.

Sensitive radio astronomy observatories are sometimes used for RS or mission-support signal reception when they possess unique advantages over the DSN such as the ability to receive in the UHF band. Planetary missions have benefited from RS collaborations with support of the Green Bank Telescope in West Virginia, Allen Telescope Array in California, Kashima in Japan, Noto in Italy, Parkes in Australia, and Narrabri in Australia. There is potential for collaboration with the Effelsberg Radio Telescope in Germany, the Sardinia Deep Space Antenna (also known as the Sardinia Radio Telescope) in Italy, and the Giant Metrewave Radio Telescope in India.

For RS purposes, the stations of other agencies and radio telescopes have been used to enhance DSN services. For example, an attempt was made to array a DSN station with the Usuda station in Japan for the Voyager Neptune encounter RO (Mizuno et al. 1992).

1.5 Noise, Error Sources, and Calibrations

The success of RS experiments relies on the thorough understanding, modeling, and calibration of error sources RS is the art of the error budget. These have been carefully studied and published for various experiments and are cited in later chapters. Using the Cassini mission as an example, Asmar et al. (2005b) detail the noise budget and achievable accuracy for precision Doppler tracking. The most sensitive achieved fractional frequency fluctuation noise was ~3×10^{-15} at 1000 s integration time, corresponding to better than the outstandingly small 1 μm (micrometer) per second velocity error, in a noise model of fluctuations in the spectral range ~10^{-3}–1 Hz.

In addition to systematic errors, other error sources are due to instrumental or propagation noise. The former are random errors introduced by ground or spacecraft systems and include phase fluctuations associated with finite SNR on the radio links, ground and spacecraft electronics noise, unmodeled bulk motion of the spacecraft or ground station, noise limits of the frequency reference, and antenna mechanical noise. Propagation noise consists of the random frequency/phase fluctuations caused by refractive index fluctuations along the line of sight and is caused by phase scintillation as the radio wave passes through the troposphere, ionosphere, and the interplanetary plasma (Armstrong and Woo 1981, Woo and Armstrong, 1992a, 1992b); see Figures 1.6–1.8.

The free electrons in Earth's ionospheric plasma give rise to both amplitude and phase scintillation. Variations in the Earth's neutral atmosphere, particularly variations in the water vapor content of the troposphere, cause a path delay in addition to amplitude variations. Some of these changes can be calibrated using existing and widely available technologies and models, but the unmodeled components of these distortions can limit the precision of RS measurements. In order to calibrate these effects, two technologies, multi-link systems and water vapor radiometers, have been developed and successfully utilized for the Cassini and Juno precision gravity measurements; variations are utilized for the BepiColombo mission and planned for the JUICE and VERITAS missions.

Calibration of the effect of charged particles is possible through the simultaneous use of two or more radio links at different frequencies. Plasma phase scintillation is dispersive; the refractivity depends on the wavelength and can be isolated with a multi-link system. After correction for the plasma phase scintillation, any remaining phase variation can be attributed to non-dispersive effects. Advanced water vapor radiometers determine the total water vapor content along a LOS by making measurements of the strength of microwave emissions from water molecules in the troposphere. By co-pointing the radiometer with the antenna tracking of a spacecraft, the signal delay is measured in the same path. Investigations that require a precision better than the perturbation of the troposphere on the signal would require this RS instrument, especially at Ka-band links which are less susceptible to charged particles, consequently making the tropospheric effect more pronounced.

For planning future experiments, the quality of RS observations is critically dependent on the relative angular elongation of the Sun (see Figure 3.2 in Chapter 3). Figure 1.8 shows the power spectral density of one-way plasma phase scintillation versus Sun–Earth–spacecraft angle (circles for S-band and crosses for X-band) as well as an approximate level of uncalibrated tropospheric scintillation noise and limits to antenna mechanical noise. Figure 1.9 shows actual data quality as Doppler noise in units of velocity, as a function of the Sun–Earth–Mars angle for three Mars orbiters; it explains the preference for conducting RS experiments at X-band at angles higher than 30°.

The challenge to achieve the required performances is illustrated in Asmar et al. (2013d), which provides quantification of errors for precision Doppler in spacecraft-to-spacecraft links. The integrated scientific measurement system of the GRAIL mission comprises flight, ground, mission, and data system elements. Early modeling and simulation efforts influenced and optimized the design as well as the implementation and testing of each element. Errors are grouped into either dynamic or kinematic sources; the former are forces acting on the spacecraft and perturb their orbit or motion, and the latter are local effects such as thermal variations. Dynamic nongravitational accelerations are due to solar radiation pressure, spacecraft thermal radiation (largest dynamic error), lunar albedo and lunar thermal emission, and acceleration due to unmodeled forces.

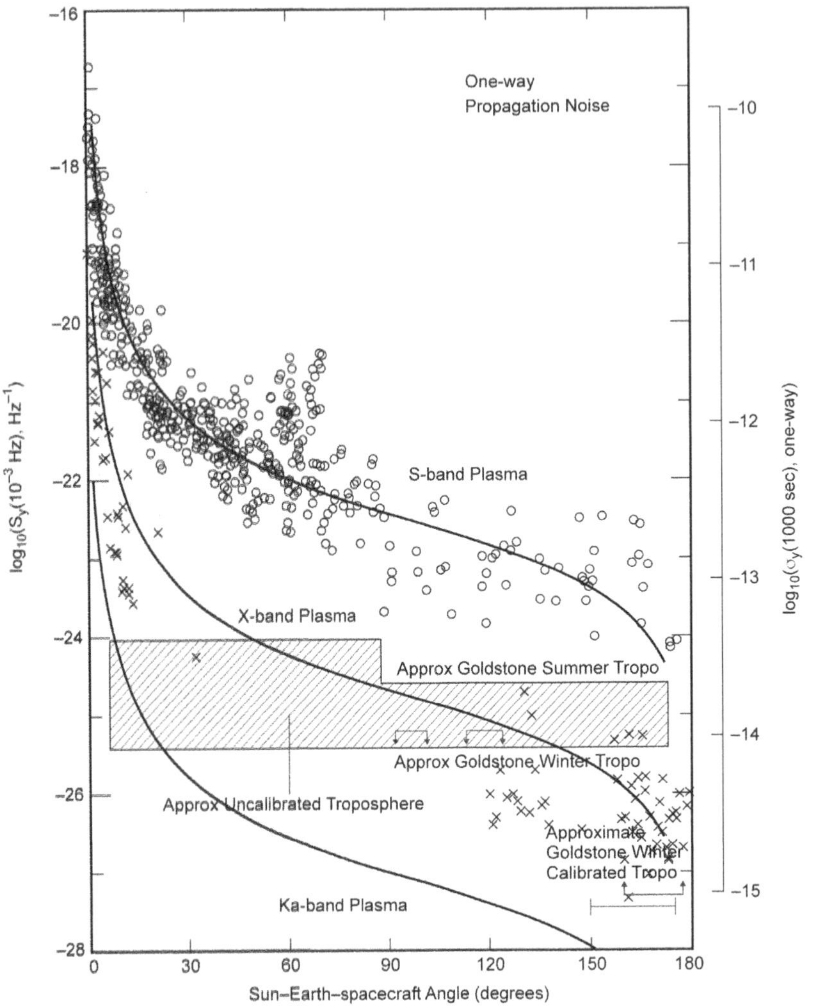

Figure 1.8 Power spectral density of one-way plasma phase scintillation at $f = 0.001$ Hz versus Sun–Earth–spacecraft angle (circles for S band and crosses for X band), approximate level of uncalibrated tropospheric scintillation noise (diagonal-lined area), and limits to antenna mechanical noise (area fenced in by dashed lines). The right-hand scale is the equivalent Allan deviation at 1000 s integration time, assuming a Kolmogorov spectral shape (by J. W. Armstrong in Asmar et al. 2005a).

Kinematic errors directly affect the inter-spacecraft range-rate measurement and spacecraft-to-DSN Doppler observables. Most of the kinematic corrections are estimated as local parameters affecting the data arcs (such as payload bias, drift, and cosine and sine once per revolution estimated amplitudes)—see Chapter 3 for a description of the estimation methodology. The latter result from the spacecraft going between sunlight and shade and the thermal consequences on the spacecraft surfaces. These errors include timing offsets, USO drift in the relevant spectrum, spacecraft attitude jitter, and offset of the phase center from the line connecting the center-of-mass from each spacecraft, which are minimized by actively pointing the spacecraft to align the phase centers. Changes in the temperature of the payload antenna and related hardware are

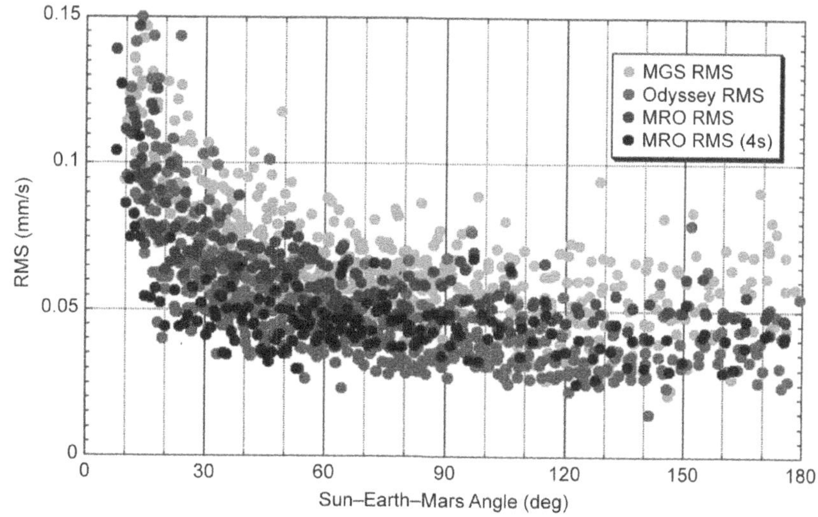

Figure 1.9 Example of the quality of the Doppler observable in terms of the root mean square (RMS) Doppler noise in units of velocity as a function of the Sun–Earth–Mars angle for three Mars orbiters (reproduced from Konopliv et al. 2011b) shows significant degradation at angles of less than 30° at X-band. As shown in Figure 1.8, the noise would be larger at S-band and smaller at Ka-band. MGS is the Mars Global Surveyor and MRO is the Mars Reconnaissance Orbiter.

modeled as a systematic trend in the observables. Amplitudes of the sinusoidal once and twice per orbit signatures depend on the Sun angle. Errors due to sloshing of the fuel in the spacecraft fuel tank are negligible for this particular design.

1.6 Experiment Implementation, Data Archiving, and Critical Mission Support

RS experiments are implemented through a rigorous system engineering process. Science objectives are used to develop the observational design used to allocate resources to instrument components. The need to ensure high confidence in data acquisition is especially crucial when observations represent very rare opportunities such as the Voyager flybys of the Outer Planets or the New Horizon mission to Pluto.

In addition to expertise at academic institutions and other research centers, project-supported teams at JPL function as the focal points of RS experiment implementation. They participate in the planning and design to ensure that spacecraft and ground elements are configured to perform to the required specifications. They perform engineering studies and trade-off analysis, simulations, and coordinate the work of several technical organizations supporting the experiments, as well as participate in mission sequence planning and reviews. The DSN allows trained JPL specialists to remotely configure the RSRs in the network for data acquisition. The acquired science data are validated, calibrated, and delivered along with ancillary information and documentation to science teams as well as archived with NASA's PDS.

Asmar et al. (2014b) describe the PDS data products in detail for the Cassini mission as an example. For each RS experiment, the following types of files are available:

- Science observables files that include tracking and navigation files (TNF).
- Radio science receiver (RSR, also known as OLR) files.
- Calibration and ancillary information that includes Earth orientation parameters (EOP).
- Media calibration troposphere (TRO and ionosphere (ION) data.
- (Advanced) water vapor radiometer (WVR/AWVR) data.
- DSN monitor data.
- Spacecraft engineering telemetry (TLM).
- Spacecraft and planetary kernel (SPK) or orbit ephemeris message (OEM).
- Spacecraft orientation C-kernel (CKF).

The tools, techniques, and expertise developed for RS have proven to be very useful for supporting deep space missions' special applications beyond scientific observations. They are especially suited for searching for or characterizing spacecraft in distress or monitoring the radio signals during critical maneuvers where knowledge of the state of the spacecraft might be uncertain due to high dynamics in signal amplitude or frequency, or due to very low SNR. Under such conditions, the DSN's closed-loop tracking receiver might not achieve or maintain a lock on the signal carrier. The RSR/OLR captures the radio signal's spectrum in the selected recording bandwidth and enables users to apply signal detection algorithms after the event and optimized for the specific signal conditions.

Asmar (2012a) describes the cases of orbit insertion maneuvers and entry, descent, and landing (EDL), a highly critical mission period when regular communication is typically challenging. RS specialists have been able to detect and report in near real-time the successful landing of all Mars landed missions from the Mars Pathfinder, which introduced the concept of communicating by semaphores from deep space (Wood et al. 1997), to Curiosity, as well as the Huygens probe landing on Titan. The information is derived from a Doppler profile, not telemetry, where detection of the carrier signal frequency shift reveals the dynamics such as acceleration or deceleration due to events such as parachute deployment and engine firing. Figures 1.10 and 1.11 show representative events described in Asmar (2012a).

1.7 Radio Science at Home

As noted above, radio signals of known properties were used in the 1950s to study Earth's ionosphere via scattering of radio waves. This led to the creation of the field of RS for studying Earth's environment using radio waves. Since then, there has been a strong link between planetary science and Earth science in this field. GPS RO of Earth's atmosphere has become a very advanced field with crosslink applications to science, as well as weather applications, as an aid in aviation safety, and in supporting the detection and monitoring of volcanic clouds.

For example, Kantha et al. (2013) describe how the Constellation Observing System for Meteorology, Ionosphere, and Climate (COSMIC) project has proven the utility of RO in the extraction of temperature profiles in the global atmosphere for numerical weather forecasting applications, as well as to the extraction of information on the location and intensity of turbulence in the global atmosphere. Such information is of considerable importance to aviation safety. Biondi et al. (2017) reported on studies of the altitude of volcanic clouds and the atmospheric thermal structure after volcanic

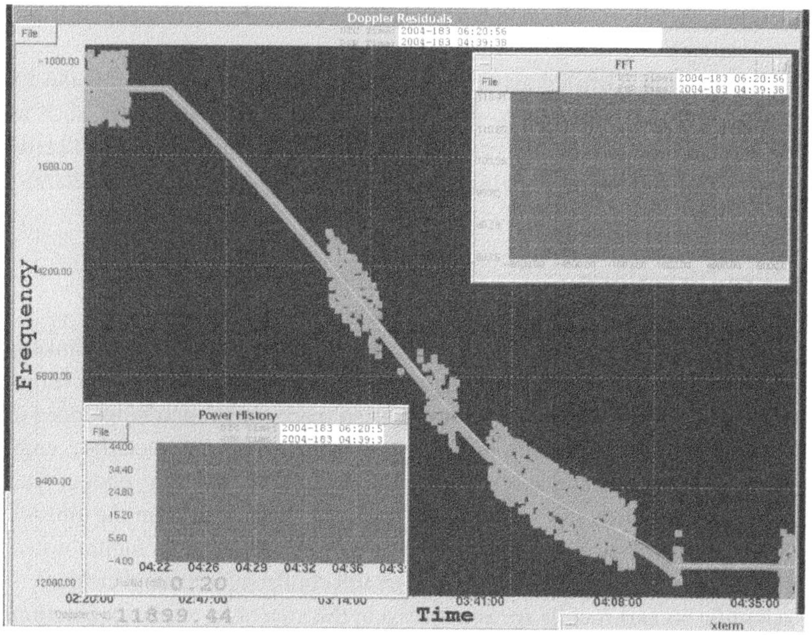

Figure 1.10 Image of the frequency profile as a function of Earth-received time (ERT) of the X-band radio signal received by the DSN from the Cassini spacecraft during the critical Saturn orbit insertion maneuver. During this maneuver, the HGA was pointed in the direction of spacecraft travel to shield the spacecraft from ring particles, and a rear LGA was used for communications, but the LGA link margin was insufficient to receive telemetry. The RSR was utilized to receive the Doppler-shifted carrier signal frequency and display it in real-time in red over a predicted model in yellow. Thick portions indicate occultation by ring material. The lower inset shows the signal's power history, and the upper inset shows a Fourier spectrum at an instant in time (Asmar 2012a).

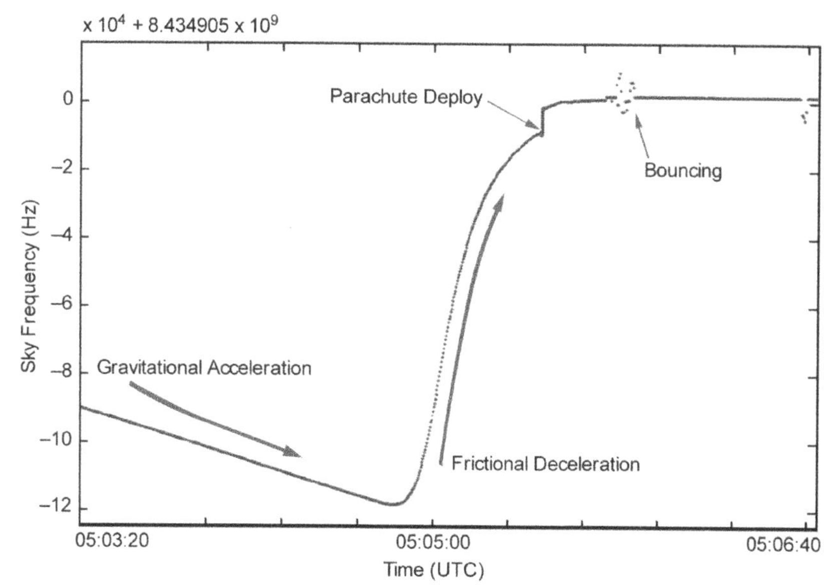

Figure 1.11 Signal observations of the Opportunity Mars rover EDL critical event sequence as captured in the DSN's Radio Science Receiver. The Doppler shift shows dynamical events of the spacecraft (Asmar 2012a).

eruptions using RO profiles colocated with independent radiometer images of ash and sulfur dioxide plumes.

The high resolution of Earth's gravitational field data from the GRACE mission RS crosslink technique has contributed to monitoring important climate issues, such as water storage depletion in Texas (Long et al. 2013), Western and Central Europe (Li et al. 2012), Southwest China (Long et al. 2014), and East Africa (Swenson and Wahr 2009).

1.8 Future Directions

All spacecraft will continue to have telecommunication links, and utilizing these links as tools for high-value science will further enhance the missions. The outlook for future directions in RS investigations and techniques for deep space exploration is described in Chapter 6. The field will continue to thrive as innovations advance in science, engineering, technology, instrumentation, methods, algorithms, mission design, and experiment operations. Furthermore, future observations will benefit from the infusion of new technologies expressly developed to enhance this field of science, or for innovative telecommunications or navigation purposes not yet fully utilized for RS applications.

Future scientific breakthroughs are anticipated in the area of icy world exploration. Europa, Ganymede, and other bodies in the outer Solar System will host dedicated missions which are currently in the development phase. These missions will apply RS techniques proven by the Galileo and Cassini missions to explore the interior structures and detect subsurface liquid layers, as in the cases of Titan and Enceladus. The BepiColombo mission will further explore tests of the theories of general relativity during a long cruise period of numerous solar conjunction periods.

Opportunities for innovations are accelerating as new concepts have been advanced in the area of atmospheric occultations. For example, the traditional one-way downlink from a spacecraft to a ground station can be supplemented or replaced by uplink occultations or spacecraft-to-spacecraft links, for the advantages explained in Chapter 6, as illustrated in Figure 1.12. Similar concepts apply to gravity field measurements and surface scattering research. Special emphasis should also be given to the study of the transportation of momentum in planetary atmospheres, leading to effects such as the atmospheric super-rotation at Venus. Furthermore, Figure 1.13 (also see cover graphic) illustrates an innovation that is likely to significantly alter this field in the future, namely, less costly small spacecraft with capable radio systems that can take additional risks to obtain high-value science. The example in the figure shows the first planetary CubeSat mission, Mars Cube One (MarCO), that accomplished its primary mission in 2018 (Asmar and Matousek 2016). As explained in Chapter 6, this is likely to serve as a baseline of a small spacecraft design for future planetary missions that can carry out RS experiments that advance the field by providing distinct features.

1.9 Summary and Remaining Chapters

This chapter provided a broad overview of the science and technologies that make up the multi-disciplinary field of RS. Scientists investigate radio links between transmitting and receiving radios that can be altered in their characteristics due to

Figure 1.12 An artist's conception of uplink, crosslink, and downlink radio science experimental configurations. On the right side, an uplink signal is occulted by planetary rings and acquired on the spacecraft. The data packets containing the occultation science are then transmitted from the spacecraft to the ground station as telemetry files. The middle link illustrates a traditional one-way downlink occultation experiment where the signal is occulted by the rings prior to arrival at the ground station. On the left side, two spacecraft carry out crosslink occultation and then transmit the data packets to station as telemetry.

propagation effects caused by intervening media or scattering surfaces, or due to the additional relative motion caused by the mass and density distribution of the planets or moons, among other effects. The fundamental scientific observables include the time profiles of the amplitude, frequency, phase, and polarization of the received signals. From these, planetary interiors, atmospheres, rings, and surfaces are investigated along with explorations of the Sun and aspects of fundamental physics.

Readers were familiarized with the concepts of ROs, atmospheric profiles, precision Doppler, gravitational fields, studies of planetary rings, wind and surfaces, signal carrier characterization, effects of the interplanetary plasma, utility of USOs and OLRs, the DSN, and the link configurations known as one- and two-way modes as well as error budgets and calibration techniques.

The remaining chapters detail techniques and accomplishments in the various areas of the field.

Figure 1.13 An artist's conception of the MarCO CubeSat mission to Mars with potential capabilities for RS experiments (Asmar 2015a). The first CubeSats in deep space were intended to provide real-time relay capabilities during the critical landing phase of the InSight mission onto the Martian surface but can serve as platform to plan future planetary RS experiments as shown in Chapter 6 (also see cover graphic).

- Chapter 2 describes the investigation of planetary atmospheres, ionosphere rings, and surfaces using ROs and bistatic radar techniques.
- Chapter 3 describes investigating planetary interiors from measuring their gravitational fields using precision Doppler tracking techniques.
- Chapter 4 describes tests of relativistic theories via precision radiometric observables, as well as and solar studies via occultation of the corona.
- Chapter 5 details the instrumentation and technologies utilized or being developed for this field or related applications that benefit RS investigations.
- Chapter 6 discusses the future outlook for the field as advances continue in various areas.

Appendix 1A

Selected Accomplishments and Planned Observations in Spacecraft Radio Science

1A.1 Selected Accomplishments in Radio Science

- First detection of Martian core and bulk seasonal CO_2 deposition at poles (Mars Pathfinder).
- Vastly improved estimation of the gravity field of Mars, with determination of Love number and correlation with topography [Mars Global Surveyor (MGS), Mars Odyssey, Mars Reconnaissance Orbiter (MRO), and Mars Express (MEX)].
- First detection of planetary gravity field variations other than Earth (MGS).
- First determination of the mass of Phobos (MGS).
- First determination of the nonspherical gravity field of an asteroid [Near Earth Asteroid Rendezvous (NEAR) mission to Eros].
- High-accuracy profiling of Martian atmospheric structure from radio occultations (MGS).
- Determination of height profile and temporal variability of Venus winds (Venera and Vega).
- Detection of ionospheres on the Galilean satellites (Galileo).
- Measurement of Jovian deep atmospheric wind speeds and ammonia concentration (Galileo probe).
- State-of-the-art testing of general relativity and searching for gravitational waves (Cassini).
- Improved estimation of the Mercury gravity field [Mercury Surface, Space Environment, Geochemistry, and Ranging (MESSENGER) mission].
- Measurement of Mars upper atmospheric density from spacecraft drag (several orbiters).
- Determination of surface characteristics of Venus, the Moon, Mars, and Titan from bistatic radar experiments.
- Detection of meteor layers in the Mars and Venus ionospheres [MEX, Venus Express (VEX)].
- Modeling of the mass distribution in the interiors of Jupiter's large moons (Galileo).
- Measurement of height profile of winds on Titan with the Doppler Wind Experiment (Huygens).
- High-resolution measurement of the gravitational field of the Earth via spacecraft-to-spacecraft links [Gravity Recovery and Climate Experiment (GRACE)].
- Measurement of the gravitational field of Saturn and its large satellites and initial evidence favoring subsurface oceans on icy moons in the outer solar system (Cassini).

- Measurement of the atmospheric structure of Saturn and its large satellites (Cassini).
- Profiling of internal dynamics and particle size distribution within Saturn's rings (Cassini).
- First direct measurement of gravity in the lunar farside [Kaguya, also called the Selenological and Engineering Explorer (SELENE) mission].
- Elucidation of the thermal history of the Moon from the high-resolution gravitational field [Gravity Recovery and Interior Laboratory (GRAIL)].
- Detection of buried empty lunar lava tubes with the ultra-high-resolution gravity field of the Moon (GRAIL).
- Determination of the interiors of Titan, Enceladus, Mercury, Phobos, Vesta, Ceres, Lutetia, and cometary nuclei from gravity fields, contributing to understanding their origins and providing key evidence for identifying subsurface oceans on icy moons.
- Sounding of the atmospheres of Titan, Saturn, and Pluto.
- Profiling of the structure of Saturn's rings (Voyager and Cassini).
- Exploration of the surface properties of Pluto (New Horizons), Kuiper Belt Object Arrokoth (New Horizons), and comet 67P/Churyumov-Gerasimenko (Rosetta).
- Investigation of the structure and dynamics of the solar corona and interactions of the solar wind plasma with the planetary atmospheres/ionospheres of Mars and Venus.
- Refined modeling of the atmospheres, surfaces, and interior structures of Mars and Venus.
- Refined modeling of the interior structures of the Earth, the Moon, and Mercury.
- Investigation of the Jovian interior structure via gravity data (Juno).
- Investigation of the Saturnian interior structure via gravity data (Cassini).

1A.2 Planned Observations in the Near-Term

- The Rotation and Interior Structure Experiment (RISE) experiment on board NASA's Interior Exploration using Seismic Investigations, Geodesy and Heat Transport (InSight) mission will continue characterization of the Martian core via radio science (RS) techniques (Folkner et al. 2018).
- The European Space Agency's (ESA) MEX mission will continue investigation of the Mars atmosphere, surfaces, and gravitational field.
- NASA's Juno mission will continue investigation of Jupiter's gravitational field to characterize the giant planet's interior structure (Folkner et al. 2017; Iess et al. 2018).
- NASA's Mars Atmosphere and Volatile Evolution (MAVEN) mission will continue investigation of the Martian atmosphere via radio occultation (RO) techniques (Withers et al. 2018).
- Japan's Akatsuki mission will continue investigation of Venus' atmosphere via RO techniques (Imamura et al. 2011).
- Japan's Hayabusa-2 mission will continue investigation of asteroid 162173 Ryugu (Watanabe et al. 2019).
- NASA's Origins, Spectral Interpretation, Resource Identification, Security, Regolith Explorer (OSIRIS-REx) will continue a gravity experiment at Bennu (McMahon et al. 2018).

- ESA's BepiColombo mission to Mercury will conduct tests of the theories of general relativity at multiple solar conjunction opportunities (Iess et al. 2009b) and use advanced ranging techniques (Paik et al. 2018).
- Juno's extended mission will continue investigation of Jupiter's interior structure, with possible investigation of the tidal gravitational response of Io (degree 2 Love number of Io) as well as the Io plasma torus.
- Additional comparative planetology studies will be conducted based on the results of the characterization of the Saturn and Jupiter interior structures (Kaspi et al. 2020).
- India's Chandrayaan-2 mission will investigate the lunar exosphere as a planned follow-up to occultations studies with Chandrayaan-1 (Choudhary et al. 2016).
- The Belgium-led Lander Radio Science for Obtaining the Rotation and Orientation of Mars (LaRa) experiment, developed as a payload on the 2020 ESA ExoMars mission, will obtain Mars' orientation precession, nutation, and length-of-day variations via coherent range-rate measurements with a European lander (Dehant et al. 2009).
- ESA's ExoMars Atmospheric Mars Entry and Landing Investigations and Analysis (AMELIA) will examine lander trajectory using the radio link to relay orbiters and ground telescope to estimate wind velocity (Ferri et al. 2019).

1A.3 Planned Observations in the Long Term

- NASA's Psyche mission will investigate asteroid 16 Psyche (Elkins-Tanton et al. 2015), including the gravity field of its interior structure.
- NASA's Europa Clipper mission will investigate the gravity field of the Jovian satellite's interior structure (Pappalardo et al. 2017; Verma and Margot 2018).
- BepiColombo will investigate (circa 2025) Mercury's interior structure (Iess and Asmar 2007).
- ESA's Jupiter Icy Moons Explorer mission (JUICE) will investigate the gravity field of Galilean satellites' interior structure (Iess 2013) and Jupiter's atmosphere.
- ESA's Hera mission will investigate Didymos gravity through Juventas crosslink Doppler tracking.
- Experiments at other solar system targets are in various stages of proposal, planning, or development.
- Reanalysis of historical data with improved processing techniques calibration methods, improved trajectory solutions, and/or theoretical models that result in significantly improved scientific outcomes.
- Opportunistic investigations on current and future missions from NASA and partner agencies.
- Expected advances in Earth and cislunar mission capabilities will have significant RS elements with potential scientific applications.
- Continued use of ESA's Trace Gas Orbiter (TGO) nonpolar inclination to improve the resolution of the seasonal variations of the even zonal harmonics and better constrain the CO_2 mass deposits at the Martian polar caps.

2

Planetary Atmospheres, Rings, and Surfaces

2.1 Overview of Radio Occultations

Astronomers have used the occultation method of atmospheric research for over a century since planets naturally pass in front of stars, or occult the stars. First used with visible light, the occultation method has been extended to other wavelengths over time. The development of airborne instruments allowed observation of infrared frequencies blocked by water vapor in the atmosphere. Stellar occultations were brought to the forefront with the 1977 discovery of the rings of Uranus. Since then, the technique has been used to probe the atmospheres of planetary bodies (Elliot and Olkin 1996).

Similarly, since the early robotic exploration of the Solar System, the technique of spacecraft radio occultation (RO) utilized radio signals transmitted from spacecraft, not natural radiation. The resulting measurement investigated the effects of refraction, absorption, and scattering on the amplitude and phase of radio signals by planetary ionospheres, neutral atmospheres, plasma tori, and rings.

As illustrated in Figure 2.1, as a spacecraft travels behind a planet and the line-of-sight to Earth propagates through the planet's atmosphere, the radio link transmitted by the spacecraft and received by an Earth station is perturbed in phase due to atmospheric refraction and sometimes in amplitude due to absorption caused by the constituents of the intervening plasma and neutral gas. RO studies flourished when the quality of both spacecraft and ground radio systems improved to the level that overcame leading error sources. As listed in Table 2.1 almost all planetary missions have investigated target atmospheres and ionospheres using RO either as one of the primary mission objectives or as part of additional scientific opportunities.

The perturbation of the signal measured at the receiving station can be analyzed to produce an atmospheric refractivity profile from which the electron distribution in the ionosphere or temperature-pressure profile in the neutral atmosphere can be derived. In some planetary cases, RO can potentially provide profiles of the atmospheric vertical temperature, pressure as well as and density with geographic and temporal resolution sufficient to model the climate globally and to infer information about atmospheric stability, geostrophic and isothermal winds, geopotential height, and characterization of large-scale atmospheric wave structures such as planetary waves, jet streams, and polar vortices.

Radio Science Techniques for Deep Space Exploration, First Edition. Sami W. Asmar.

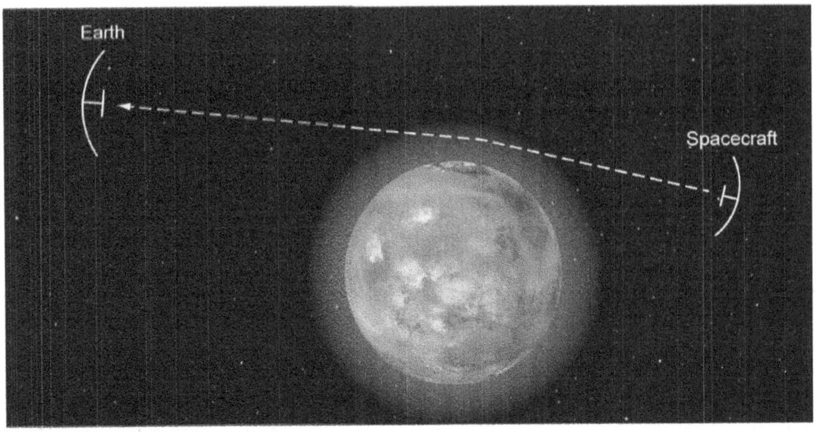

Figure 2.1 Configuration for RS observations at a representative planet. The atmosphere occults a spacecraft moving behind the planet, as seen from Earth, and the signal received at the ground station is perturbed by its interaction with the atmospheric material. The signal phase will vary due to atmospheric refraction and its amplitude can diminish due to absorption by atmospheric constituents.

As discussed in the historical development of Section 1.3, since the 1950s, the effects of refraction by the Earth's atmosphere had been of interest to researchers studying their impacts on radar systems as well as radio reflections from meteoric ionization. Research interest grew into the effects of planetary atmospheric refraction on the accuracy of tracking data collected at deep space tracking stations (Cain 1960, 1964) starting with the 1965 Mariner 4 spacecraft flyby of Mars; navigation teams were concerned about the possible degradation in the quality of Doppler data.

The history of RO studies with deep space mission goes back to the first demonstration in 1965 when Mariner 4 revealed the salient features of the Martian atmosphere. With this demonstration, RO investigations were formalized by NASA. Throughout the 1960s and 1970s, spacecraft were dispatched to study the Inner Planets. An occultation experiment was conducted using Mariner 5 during its flyby of Venus in 1967 (Kliore et al. 1967; Fjeldbo et al. 1971). It acquired the temperature–pressure profile in the neutral atmosphere and the electron density distribution in the ionosphere. Mariner 5 also revealed that the Soviet Venera 4 spacecraft had not reached the surface of Venus (Kliore and Cain 1968; Eshleman et al. 1968). Mariners 6 and 7 flew by Mars (Kliore et al. 1969) and Mariner 9 made repeated orbital measurements of Mars (Kliore et al. 1972). Then, Mariner 10 carried out experiments at Venus and Mercury (Howard et al. 1974a, 1974b; Fjeldbo et al. 1975).

As planetary spacecraft advanced from short flybys to long-term orbits, the two Viking orbiters provided large amounts of important RO data during the 1976–1982 mission to Mars (Fjeldbo et al. 1977; Lindal et al. 1979). From December 1979 to October 1992, the Pioneer Venus spacecraft orbited Venus and provided important data on the structure of the ionosphere (Kliore et al. 1979) and neutral atmosphere (Kliore and Patel 1980; Cimino et al. 1980; Woo et al. 1980), as well as how they changed in response to the solar cycle (Kliore et al. 1991).

Table 2.1 Summary of Solar System radio occultation experiments.

Target	Mission	Year	Downlink
Mercury	*Mariner 10*	1974	S, X
	*MESSENGER**	2008–2009	X
Venus	*Mariner 5*	1967	S, 70 cm*
	Mariner 10	1974	S, X
	Venera 9	1975	32 cm*
	Venera 10	1975	32 cm*
	Pioneer Venus Orbiter	1978–1992	S, X
	Venera 15	1983–1984	5 cm, 32 cm
	Venera 16	1983–1984	5 cm, 32 cm
	Magellan	1991–1994	S, X
	Venus Express	2006–2014	S, X (USO)
	Akatsuki	Since 2015	X (USO)
Mars	*Mariner 4*	1965	S
	Mariner 6	1969	S
	Mariner 7	1969	S
	Mars 2	1971	S
	Mars 4	1974	S
	Mars 6	1974	S
	Mariner 9	1971	S
	Viking 1	1976–1980	S, X
	Viking 2	1976–1978	S, X
	Mars Global Surveyor	1998–2006	X (USO)
	Mars Express	Since 2004	S, X
	Mars Reconnaissance Orbiter	Since 2008	X
	MAVEN	Since 2016	X
Earth	*Mir station to geostationary sat.*	1989–1998	2 and 32 cm
	GPS links to Microlab-1, CHAMP, and COSMIC-2	Since 1995	L1, L2
Moon	*Lunar Prospector**	1998	S
	SELENE	2005	S, X
	SMART-1	2006	S, X, Ka
	Chandrayaan-1	2009	S
	Chandrayaan-2	Planned	S, X
Jupiter	*Pioneer 10*	1974	S
	Pioneer 11	1974	S
	Voyager 1	1979	S, X (USO)
	Voyager 2	1979	S, X (USO)
	Galileo	1990	S (USO)

(Continued)

Table 2.1 (Continued)

Target	Mission	Year	Downlink
Io	*Pioneer 10*	1974	S
	Galileo	1995	S (USO)
Europa	*Galileo*	1995	S (USO)
Ganymede	*Galileo*	1995	S (USO)
Calisto	*Galileo*	1995	S (USO)
Io Plasma Torus	*Ulysses*	1992	S, X
Io Plasma Torus	*Juno*	Planned	X, Ka
Saturn	*Pioneer 11*	1976	S
	Voyager 1	1980	S, X (USO)
	Voyager 2	1980	S, X (USO)
	Cassini	2005–2017	S, X, Ka (USO)
Titan	*Voyager 1*	1980	S, X (USO)
	Cassini	2006–2016	S, X, Ka (USO)
Enceladus	*Cassini*	2006 and 2010	S, X, Ka (USO)
Rings	*Voyager 1*	1980	S, X (USO)
	Cassini	2005–2017	S, X, Ka (USO)
Uranus	*Voyager 2*	1986	S, X (USO)
Rings	*Voyager 1*	1982	S, X (USO)
Neptune	*Voyager 2*	1988	S, X (USO)
Triton	*Voyager 2*	1988	S, X (USO)
Pluto	*New Horizons*	2015	X (USO)
Halley Comet Plasma	*VEGA 1*	1986	C, L*
	VEGA 2	1986	C, L*

Notes

- This table builds on a list that appeared in Yakovlev (2002).
- Also see Table 1.3.
- In addition to the S-band link for communications and RO, Mariner 5 carried an instrument that received two harmonically related frequencies, 49.8 and 423.3 MHz, transmitted from the Stanford radio telescope (Fjeldbo and Eshleman 1969).
- Lunar Prospector and MESSENGER Mercury occultations were for the purpose radius measurement and planetary shape modeling (Asmar et al. 1999; Perry et al. 2011).
- Viking orbiters also measured topography (elevation control points) from occultation data.
- Downlink bands for NASA's deep space missions, S-, X-, and Ka-bands, are listed in Table 1.4.
  - *VEGA C-band: 5.8 GHz; L-band: 0.9 GHz.
  - Mars Reconnaissance Orbiter also carried a demonstration Ka-band radio that was not successful.
- Planetary missions that carried ultrastable oscillators (USO) for the purpose of RO are:
  - Voyager 1
  - Voyager 2
  - Galileo
  - Mars Global Surveyor
  - Cassini
  - Venus Express
  - Akatsuki
  - New Horizons
  - GRAIL
  - LRO
  - JUICE (planned)

The Outer Planets were also explored starting in 1973 when the Pioneer 10 spacecraft passed by Jupiter. This allowed the first occultation measurement of an outer planet where analysis disclosed the existence of an atmosphere of Io (Kliore and Woiceshyn 1976). In 1974, the Pioneer 11 spacecraft probed Jupiter. An occultation experiment was performed during the Pioneer 11 encounter of Saturn, and this experiment yielded the first measurements of the atmosphere and ionosphere of Saturn (Kliore et al. 1980a and 1980b).

The Voyager 1 and 2 spacecraft flew by Jupiter in 1979 and Saturn in 1980 and 1981, respectively. Voyager 2 also flew by Uranus in 1986 and Neptune in 1989. Voyager's RO experiments were described by Eshleman et al. (1977), Tyler (1987), and Tyler et al. (1989) in addition to publications covering results of each flyby. During the Voyager occultations of the atmospheres of the four giant planets, pressure level information was obtained at about 1.0, 1.3, 2.3, and 6.5 bars at Jupiter, Saturn, Uranus, and Neptune, respectively (Tyler et al. 1989; Lindal 1992), as shown in Figure 1.3. The temperature pressure profiles display near adiabatic lapse rates below the tropopause, at about 100 mbar pressure, and considerable structure in the stratosphere. These profiles also give information on the atmospheric constituents, specifically the ratios of hydrogen to helium.

Voyager also studied the ionospheres of Jupiter, Saturn, and Triton. During Voyager 1's near diametric occultation of Titan, its radius was determined to be 2575 ± 0.5 km. It was also shown that nitrogen (N_2) is the major constituent of its atmosphere. The temperature–pressure profile is shown in Lindal et al. (1983); it terminates at a surface temperature of 94 ± 0.7 K. Scintillations were observed at all altitudes below about 100 km, reflecting surprising dynamics in the atmosphere.

Using the occultation data, studies of planetary shapes as well as the magnetic fields of Jupiter and Saturn (Hinson 1984) were also possible. The RO theory was also extended to planetary ring occultation experiments (Marouf et al. 1982) and, during the Voyager 1 encounter of Saturn, a ring occultation experiment led to observations spanning the entire width of the classical ring system, as discussed in Tyler (1987).

During the Ulysses flyby of Jupiter in February 1992, which provided a gravity assist on its path to a polar orbit of the Sun, an occultation experiment was conducted at the Io plasma torus. The spacecraft transmitted signals at 2.3 and 8.4 GHz (or S- and X-bands). The signals were received, recorded, differenced, and integrated in order to derive the columnar electron density of the Io plasma torus (Bird et al. 1992a). The measurements agreed with models of the torus based on Voyager data implying that the amount of gas injected from Io is similar to that observed during the Voyager era. Conversely, the torus seemed to be less extended out in the centrifugal equator, implying a smaller plasma temperature than predicted.

In 1991 and 1992, the Magellan spacecraft also conducted occultation experiments to study the atmosphere of Venus, obtaining highly accurate profiles of atmospheric refractivity and absorptivity (Steffes et al. 1994; Jenkins 1993; Jenkins et al. 1994). These measurements, in turn, provided profiles of ionospheric density, atmospheric temperature, pressure, and sulfuric-acid-vapor abundance. A spacecraft limb-tracking maneuver, plus the high effective isotropic radiated power, made it possible to probe the atmosphere down to an altitude of 35 km above a nominal surface at the 3.6-cm wavelength and down to 33 km at the 13-cm wavelength.

Tyler et al. (1992) provided an overview of the Mars Global Surveyor (MGS) experiments. Perturbations of the 3.6-cm wavelength (X-band) radio path by the atmosphere during periods of atmospheric occultation provided the vertical temperature–pressure structure $T[p(r)]$ to accuracies at the surface of $\Delta T \approx 0.4$ K and $\Delta p \approx 2$ Pa, and ~10 K and ~0.6 Pa at altitudes of 40–50 km systematic sampling of the atmosphere in combination with the accuracy of the MGS radio system supported studies of the large-scale dynamics of the atmosphere, including seasonal variations of the atmospheric fields and embedded waves such as Kelvin and Rossby waves. Terminator region ionospheric electron density profiles were also obtained much of the time. Figure 2.2 shows example results from MGS occultations from Hinson et al. (2001).

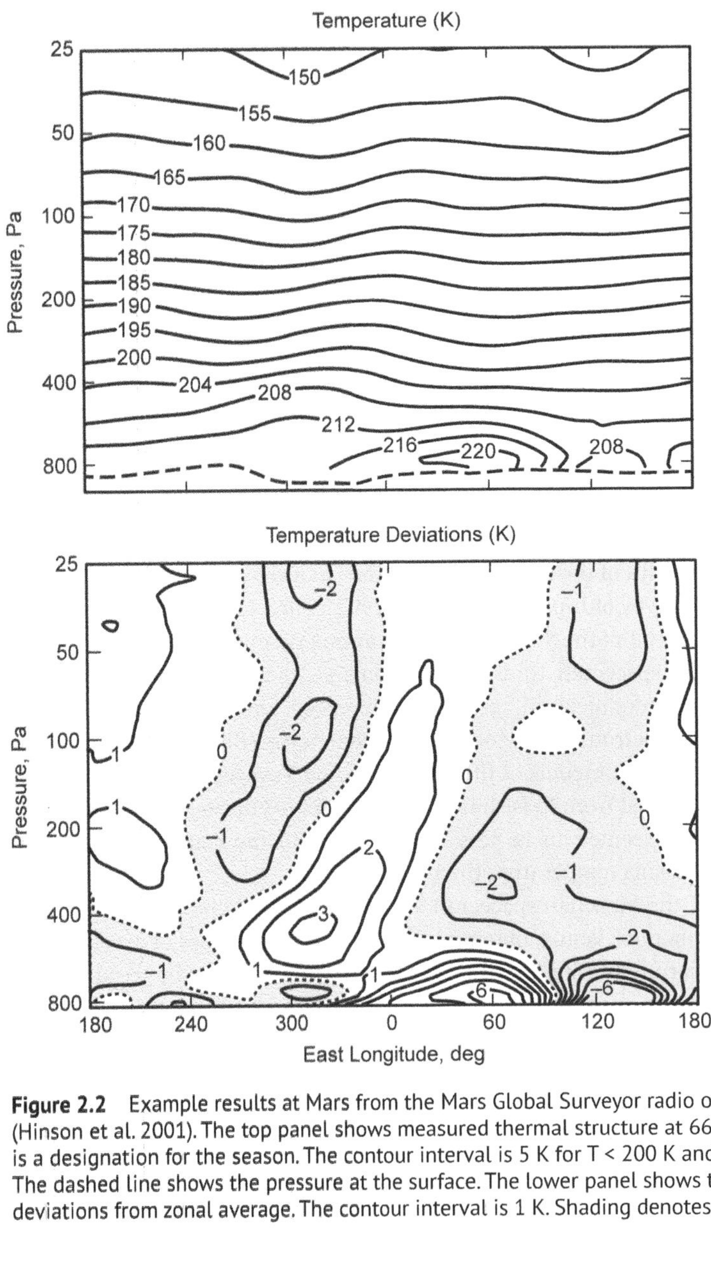

Figure 2.2 Example results at Mars from the Mars Global Surveyor radio occultations (Hinson et al. 2001). The top panel shows measured thermal structure at 66°N, Ls = 75°, which is a designation for the season. The contour interval is 5 K for T < 200 K and 4 K for T > 200 K. The dashed line shows the pressure at the surface. The lower panel shows the temperature deviations from zonal average. The contour interval is 1 K. Shading denotes negative values.

Hinson et al. (2014) presented results from RO measurements with the Mars Reconnaissance Orbiter. One set of measurements sounded the tropics in northern summer at a local time ~1 h before sunrise. Some of these profiles contain an unexpected layer of neutral stability with a depth of ~4 km and a pressure at its upper boundary of ~160 Pa. The mixed layer is bounded above by a temperature inversion and below by another strong inversion adjacent to the surface. This type of structure is observed near Gale Crater, in the Tharsis region, and at a few other locations, whereas profiles in Amazonis Planitia and Elysium Planitia show no sign of a detached mixed layer with an overlying inversion.

Hinson et al.'s (2014) numerical simulations demonstrated that water ice clouds can generate this distinctive type of temperature structure through their influence on radiative transfer at infrared wavelengths. A second set of measurements were carried out at mid-to-high northern latitudes in spring, when carefully coordinated observations by the MRO Mars Climate Sounder are also available. Within 21 comparisons, the average temperature difference is less than 1 K where the aerosol opacities are smaller than 10^{-3} km^{-1}, at pressures of 10–50 Pa, whereas it increases to ~2 K where the aerosol opacities exceed this threshold, at pressures of 50–300 Pa. The standard deviation of the temperature difference is ~2 K, independent of pressure. This set also provided unique information about the stability of the annual CO_2 cycle and the dynamics near the edge of the seasonal CO_2 ice cap.

Galileo experiments were described by Howard et al. (1992) with significant results from Hinson et al. (1997) on the Jovian ionosphere, Kliore et al. (2002) on the detection of an ionosphere of Callisto, Kliore et al. (1997) on the detection of an ionosphere of Europa, and Hinson et al. (1998) on the Galileo RO measurements of Io's ionosphere and plasma wake.

Cassini experiments are described in some detail in the following sections as part of illustrative examples of atmospheric RO techniques. An overview of the Cassini RS experiments is found in Kliore et al. (2004).

2.2 Neutral Atmospheres

Following discussions in Asmar et al. (2014b), Figure 2.3 shows typical RO observational configuration. The spacecraft transmits a signal that is refracted by the atmosphere under study prior to reception at a ground station. As the spacecraft moves, the ray probes deeper into the atmosphere until it is absorbed or occulted by the planet's surface. Since this is a one-way downlink observation, the stability of the signal depends on the quality of the reference oscillator onboard the spacecraft, which are sometimes equipped with ultrastable oscillators (USO) to improve the quality of the link (see Chapter 5).

Between the spacecraft and ground station, the ray is refracted in the planet's atmosphere. α is the bending angle, ρ is the ray-asymptotic distance, and p is the center of refraction; unit vectors are also shown in Figure 2.3. The position of the spacecraft when the photon was emitted is $\boldsymbol{r}_S$, and the position of the antenna on Earth when the photon was received is $\boldsymbol{r}_E$.

The experimental geometry is computed iteratively from the ephemerides of the spacecraft, the center of the planet, and the center of the Earth relative to the Sun,

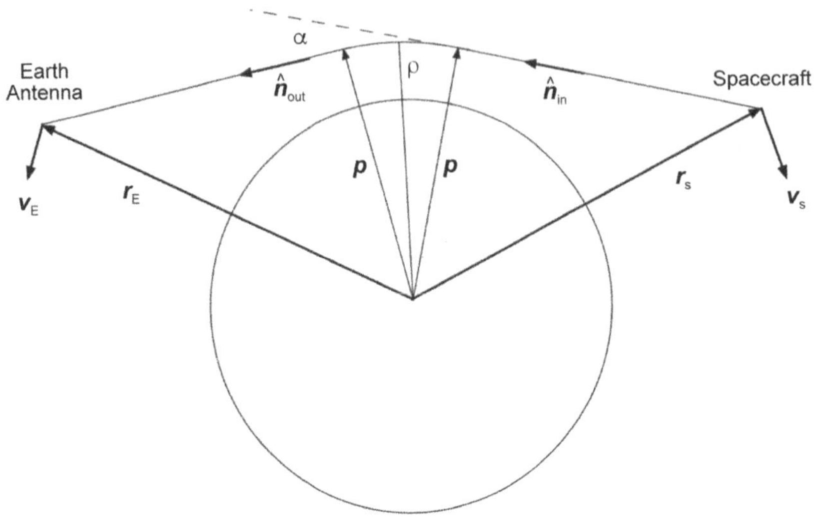

Figure 2.3 Geometry of a radio occultation observation. Between the spacecraft and ground station, a transmitted ray is refracted in the planet's atmosphere.

taking into account the exact location of the receiving station. Analysts combine knowledge of the spacecraft position and velocity (state vector), the assumption of spherical symmetry, and the measured Doppler shift to determine the bending angle as a function of ray-asymptotic distance. The bending angle is used to compute the refractivity profile of the atmosphere and then a temperature–pressure profile.

According to P. Schinder (contributor to Asmar et al. 2014b), this method relies on successful inversion of the observed changes in the signal's frequency and amplitude and the inversion technique has been firmly established for the case of spherical symmetry where the atmospheric structure depends only on the radial distance from the center of the planet. A spherical assumption provides a good first approximation for the case of oblate planets, where the center of refraction is determined by the local radius of curvature at the occultation location. Principles of derivations for this technique are described in Tyler (1987). Lindal (1992, Appendix) lays out the technique for an oblate body assuming solid-body rotation for the atmosphere.

Two different numerical techniques can be used for the occultation data inversion. The more general technique is ray tracing, which uses geometrical optics to trace rays from the spacecraft to the receiving antenna, simultaneously solving for the structure of the atmosphere as each successive ray is traced. When the target atmosphere is spherically symmetric, a much faster and simpler Abel transform directly connects the bending angle between the ray leaving the spacecraft and the ray arriving at the ground station with the index of refraction of the gas. For example, the Abel technique is used for Titan occultations while ray tracing is used for Saturn occultations because Saturn is significantly oblate and its atmosphere supports planetary scale zonal winds, which further complicate the atmospheric structure.

After the Appendix on Ray Tracing Inversion in Lindal (1992), the frequency of the photon received at the DSN, fR, may be calculated from:

$$f_R = f_T \frac{\sqrt{1-(v_1/c)^2}}{1-(v_1/c)\cos\phi_1}\left(1+\frac{\Delta U_N}{c^2}\right)\frac{1-(v_2/c)\cos\phi_2}{\sqrt{1-(v_2/c)^2}}, \tag{2.1}$$

where c is the velocity of light in a vacuum, v_1 is the velocity of the spacecraft at the time the photon leaves toward Earth, ϕ_1 is the angle between the spacecraft and the photon's velocity vectors, v_2 is the velocity of the station at the time the photon is received, and ϕ_2 is the angle between the velocity vectors of the arriving photon and the station; these are in the inertial reference frame at rest with respect of the center of mass of the planet under study at the time the photon passes through the atmosphere. f_T is the frequency transmitted from the spacecraft antenna as measured by an observer at rest on the spacecraft; this frequency is determined from the tracking data before and after the atmospheric measurements. $\frac{\Delta U_N}{c^2}$ accounts for the gravitational frequency shift expected from the general theory of relativity where the numerator denotes the difference between the Newtonian gravity potentials at the spacecraft and at the tracking station and is positive, blue shift, due to the dominating effect of the Sun's gravity field for missions at the Outer Planets.
Lindal (1992) describes the application of this formulation to the Voyager 2 occultation of the Neptune atmosphere (result shown in Figure 1.3). Another example is the Cassini Titan ingress occultation of June 22, 2009, described in Schinder et al. (2012), where Lindal's equation is simplified to:

$$f_s = f_r\left(\frac{\left(1-2\Phi_S - v_S\cdot\hat{n}_{in}/c\right)}{\left(1-2\Phi_E - v_E\cdot\hat{n}_{out}/c\right)}\frac{\sqrt{1-2\Phi_E-\left(|v_E|/c\right)^2}}{\sqrt{1-2\Phi_S-\left(|v_S|/c\right)^2}}\right), \tag{2.2}$$

where f_s is the spacecraft's transmitted frequency and f_r is the received frequency for the same photon one light time later; v_s is the spacecraft velocity at the time of transmission, v_E is the velocity of the receiving antenna at the time of reception, Φ_E is the gravitational potential at the receiving antenna including the potentials of the Earth and the Sun, and Φ_s is the gravitational potential at the spacecraft, including Saturn, Titan, and the Sun. $\hat{n}_{in}$ is a unit vector tangent to the ray path as the photon leaves the spacecraft, and $\hat{n}_{out}$ is a unit vector parallel to the ray path when the same photon arrives at the receiving antenna one light-time later; in vacuum $\hat{n}_{in} = \hat{n}_{out}$, a condition which will not be true once the ray path enters the atmosphere. The dimensionless potential $\Phi = GM/Rc^2$, where M is the appropriate mass and R the appropriate distance. In the case of Titan, Φ_s contains the sum of the potentials of Titan, Saturn, and the Sun at the position of the spacecraft.

The spacecraft-transmitted frequency was applied to each successive frequency in the equation, but instead of solving for f_s, analysts used the initial f_s to solve for the frequency predicted to be received f_r. This prediction was subtracted from each measured sky frequency creating Doppler residuals. Since the spacecraft emitted a constant frequency in its rest frame, f_s should be the same throughout the occultation, and the Doppler residuals should ideally be zero until the atmosphere is reached. Small deviations from the ideal include small errors in the ephemeris, the small drift of the USO, and the fact that the propagation path to the DSN antenna includes Earth's atmosphere. Figure 2.4 shows the signal frequency obtained for this Titan occultation experiment and Figure 2.5 shows the Doppler residuals obtained after accounting for these three effects.

2.2.1 Abel Inversion

The assumption of spherical symmetry imposed upon the atmosphere of the planet simplifies the analysis and makes the inversion analogous to a central force problem. The force is the gradient of the index of refraction, which tugs on the ray and changes its direction as it traverses the atmosphere. Because the force is central, each successive ray is confined to a plane containing the spacecraft, the center of Titan, and the receiving station on Earth (P. Schinder, private communications). Successive ray paths will be in different planes because the spacecraft and DSN antenna are continually moving; however, planar geometry can be used to obtain the bending angle within each plane.

With the assumption of spherical symmetry, it can be shown (Phinney and Anderson 1968) that there is a unique Abel transform relating the bending angle to the refractivity of the atmosphere. There is an analog of total angular momentum conservation from which the minimum radius of the ray periapsis can be obtained without ever determining the actual ray trajectory. The path of the ray is neither required nor determined using this method.

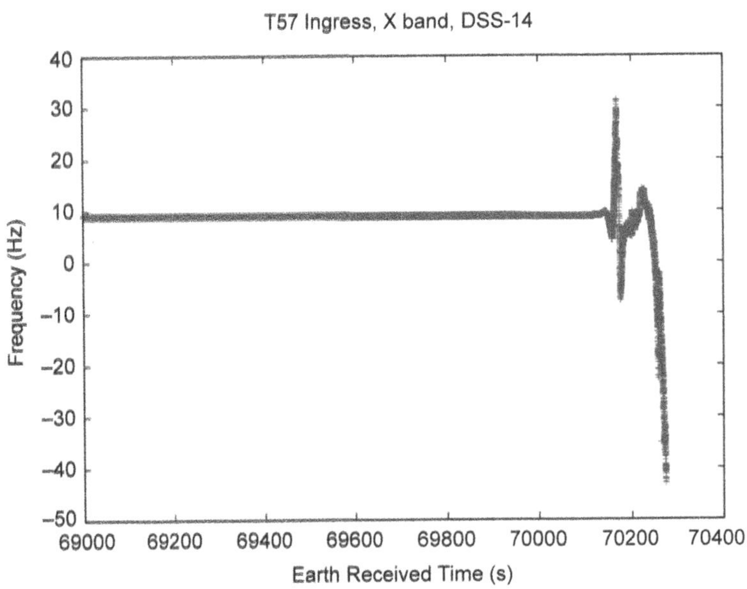

Figure 2.4 Frequency of the Cassini carrier at the output of the X-band Radio Science Receiver (RSR) during the occultation ingress observation designated T57I. RSR complex samples (I/Q pairs) were recorded at 16 kilohertz (kHz) (file S51TIOI2009173_1845NNNX14RD.1A2). Data were processed using 512-point fast Fourier transforms (FFT), with running averages of more than 8 transforms slipped by one transform per calculation (0.032 s per output point, 0.256 s per independent point). Horizontal axis gives time since Universal Time Coordinate (UTC) midnight in seconds at the 70-m Deep Space Station at Goldstone, California (DSS-14) receiving antenna.

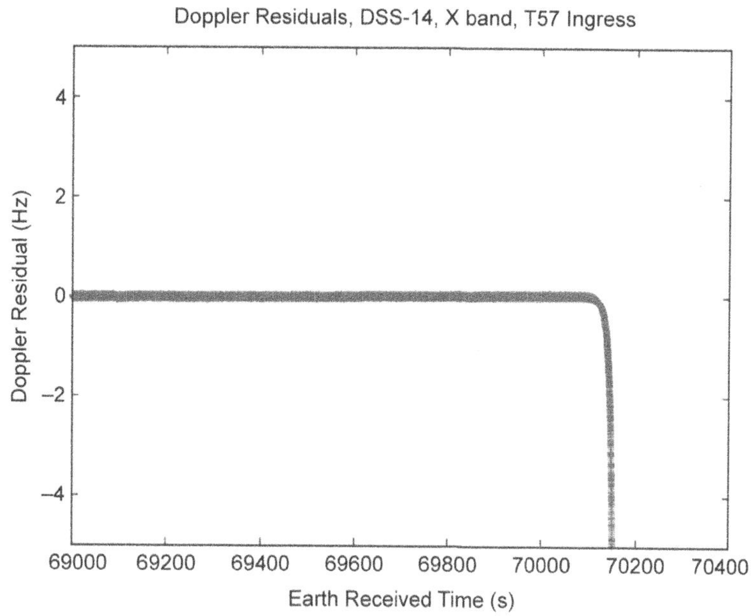

Figure 2.5 Doppler residuals for T57I, DSS-14, X-band.

The Abel integral relates the refractivity and bending angle to the impact parameter, which is the product of the radius and the index of refraction at the ray's closest approach. The bending angle is computed from the geometry and the Earth-received frequency by invoking several conditions. The primary condition that can be imposed due to the spherical symmetry is the ray trajectory that remains in a single plane; that plane is defined by the spacecraft, the center of the target, and the receiving DSN antenna. The angle between the spacecraft velocity vector and the initial ray direction is determined from the frequency recorded on the Earth. The ray must eventually intersect the ground station to be recorded.

The unit vector perpendicular to the solution plane is

$$\hat{z} = \frac{r_s \times r_E}{\left| r_s \times r_E \right|}, \tag{2.3}$$

where r_S is the vector position of the spacecraft and r_E is the position of the receiving antenna both with respect to the target. Both can be derived from the ephemeris by solving the light-time equation. Some conditions can be imposed on the unit vector in the initial direction of the ray. First, it can be constrained to lie upon a known cone because both the Earth received frequency and the velocity of the spacecraft are known:

$$v_s \cdot \hat{n}_{in} = \left| v_s \right| \cos \beta, \tag{2.4}$$

where $\cos \beta$ is determined from the frequency record by

$$\left| v_s \right| \cos \beta = -\left(\frac{f_s}{f_r} \left(1 - v_E \cdot \hat{n}_{\text{out}} \right) - 1 \right). \tag{2.5}$$

Here, f_s is the spacecraft oscillator frequency in its rest frame, f_r is the frequency recorded at the antenna, v_E is the velocity of the antenna, and $\hat{n}_{out}$ is the unit vector in the final direction of the ray after it has passed through the atmosphere (second-order relativistic effects are omitted in this description for clarity). The other condition that can be immediately imposed is that $\hat{n}_{out}$ is perpendicular to p.

Additional constraints are provided by considering the vector that is perpendicular to the outgoing ray asymptote. The vector r_E from the center of the target to the antenna on the Earth and the vector p from the center of the target to the point on the outgoing ray asymptote such that p is perpendicular to the asymptote, satisfies the following conditions:

$$\begin{aligned} r_E \cdot p &= p^2 \\ p \cdot p &= p^2, \\ \hat{z} \cdot p &= 0 \end{aligned} \tag{2.6}$$

where p is the impact parameter, which can be determined geometrically. Since the ray path must be symmetric about the origin, the impact parameter for the incoming ray must be the same as that for the outgoing ray. The spacecraft's velocity vector is projected onto the ray plane:

$$v_s' - v_s - \left(v_s \cdot \hat{z}\right)\hat{z}. \tag{2.7}$$

The angle between the spacecraft's velocity vector and its radius vector is

$$\gamma = \cos^{-1}\left(\frac{v_s' \cdot r_s}{\left|v_s'\right|\left|r_s\right|}\right). \tag{2.8}$$

Then,

$$p = \left|r_s\right| \sin\left(\left|\alpha - \gamma\right|\right). \tag{2.9}$$

The subsequent analysis described in Schinder et al. (2012) can be used to obtain an expression for the bending angle derived from the solution of six nonlinear equations,

$$\alpha = \cos^{-1}\left(\hat{n}_{in} \cdot \hat{n}_{out}\right). \tag{2.10}$$

The following form of the integral from Lipa and Tyler (1979) was used to compute the index of refraction as a function of the impact parameter:

$$n(r) = \exp\left[\frac{1}{\pi}\int_r^{r_1} \ln\left(\frac{p}{p_1} + \sqrt{\left(\frac{p}{p_1}\right)^2 - 1}\right) d\alpha(r)\right]. \tag{2.11}$$

Here, $n(r)$ is the index of refraction, and p_1 is the impact parameter of the "surface" of the atmosphere or ionosphere, a point outside of which the index of refraction is assumed to be zero; and r, the radius of the ray periapsis, the closest point on the ray to the surface of the planet, found by an analog to angular momentum conservation $r = p/n$.

The refractivity N is related to the index of refraction by $N = (n - 1) \times 10^6$. Once the atmosphere's radial refractivity profile has been determined, the temperature and pressure as a function of altitude are easily computed. Titan's atmosphere is assumed to be primarily N_2 with a small admixture of methane (CH_4) that is constant throughout the atmosphere and is 0.014 by number (1.4%). The density of the atmosphere (in kilograms per cubic meter, kg/m^3) is

$$\rho = \frac{f_{CH_4} M_{CH_4} + f_{N_2} M_{N_2}}{f_{CH_4} r_{CH_4} + f_{N_2} r_{N_2}} m_A N, \tag{2.12}$$

where $f_{CH_4} = 0.014$ is the fraction of CH_4 by number, $f_{N_2} = 1 - f_{CH_4}$ is the fraction of N_2 by number, $M_{CH_4} = 16.0426$ is the mass of CH_4 in atomic mass units, $M_{N_2} = 28.006$ is the mass of N_2 in atomic mass units, $m\text{A} = 1.66053 \times 10^{-24}$ g is the atomic mass unit, $r_{CH_4} = 1.63 \times 10^{-17}$ is the refractivity per molecule of CH_4, and $r_{N_2} = 1.1 \times 10^{-17}$ is the refractivity per molecule of N_2. Once the density is found, hydrostatic equilibrium

$$\frac{dP}{dz} = -g(z)\rho(z) \tag{2.13}$$

and an equation of state can be used to find the temperature profile. The Bender equation of state for Titan's atmosphere (Bender 1973) is adopted, which is more realistic than the ideal gas law for N_2 at such low temperatures. Finally, the temperature–altitude profile shown in Figure 2.6 is obtained. Note that the absolute temperatures obtained from using the ideal gas law would be several degrees different from these results.

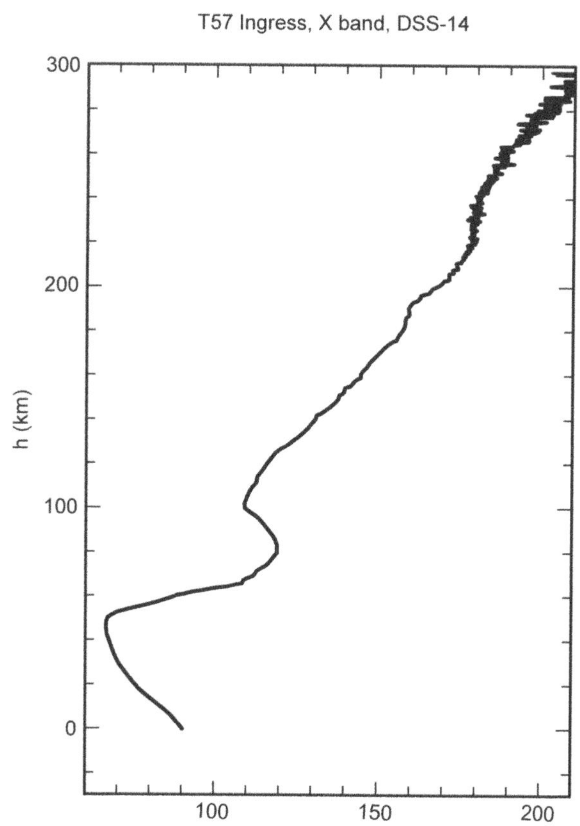

Figure 2.6 Temperature–altitude profile for T57I, using the Bender equation of state (Bender 1973).

2.3 Ionospheres

Nagy and Cravens (1988) suggested that the presence of an ionosphere implies the existence of an atmosphere, or in other words there must be a significant gaseous envelope surrounding any Solar System body in order for an ionosphere to exist.

Venus has daytime and night-time ionospheres at the positions probed by RO. The main layers are thin by terrestrial standards, with the night-time peak concentration of electrons being about two orders of magnitude below that of the daytime peak. Above the night-time peak were several scale-height regimes extending to a radius of at least 7500 km, and probably to 9700 km, from the center of Venus. Helium and hydrogen at plasma temperatures of 600–1100 K seem to be indicated in the regimes from 6300 to 7500 km, with cooler molecular ions in lower regions. Above the daytime peak a sharp plasmapause was discovered, marking a sudden transition from appreciable ionization concentrations near Venus to the tenuous conditions of the solar wind. This may be indicative of a kind of interaction of the magnetized solar wind with a planetary body that differs from the two different kinds of interaction characterized by Earth and by Moon.

For Venus and for Mars, the magnetic field of the solar wind may pile up in front of the conducting ionosphere, form an induced magnetosphere that ends at the plasmapause, above which any ionosphere that tends to form is swept away by the shocked solar wind that flows between the stand-off bow shock and the magnetopause. The neutral atmosphere was also probed, and a surface reflection may have been detected, but the data have not yet been studied in detail. Results are consistent with a super-refractive atmosphere, as expected from Soviet measurements near the surface (Florenski et al. 1977). Thus, two unusual features of Venus can be described in terms of a light trap in the lower atmosphere and a magnetic trap in the conducting ionosphere.

The RO technique described in the preceding section also applies to investigations of ionospheric layers. Charged particles interact with the propagating signal in a dispersive manner, that is, as a function of the signal's wavelength. Thus, occultation experiments using two or more transmitted frequencies are ideal for ionospheric occultations since the dispersion can be characterized by differencing the signals. All nondispersive effects in the Doppler profile, such as trajectory, gravitational, and nongravitational forces and errors, drop out in the differencing.

For ionospheric occultations, Doppler data are typically available through the DSN receiver tracking system described in Chapter 5. However, better data quality and flexibility in processing favor the use of measurements from the DSN's open-loop receiver (OLR), also referred to as the Radio Science Receiver (RSR), described in Chapter 5. The processing steps to convert from raw RSR data samples to ionospheric density profile are summarized below.

The data user needs to provide either a Fast-Fourier Transform (FFT) algorithm or a digital phase-locked loop (PLL) algorithm that produces, from the raw data, the recorded signal's frequency and power as a function of time within the RSR output sample stream; see, for example, Lipa and Tyler (1979) and Paik and Asmar (2010). The steering function of the RSR described previously is included in the recording, and these values are added to the RSR frequencies to obtain the sky frequency, the actual frequency sensed by the receiving antenna. The steering compensates for the RSR tuning and follows predictions generated in advance based on the predicted spacecraft trajectory.

Residuals are the differences between measured sky frequencies and the calculated frequencies based on values of all known contributing factors after the fact. The calculation parallels the calculation used in deriving the steering values, but it uses the best a posteriori spacecraft trajectory, Earth rotation, Solar System ephemrides, and USO frequency. This can be done by using spacecraft trajectory information contained in the trajectory file called the SPK kernels (in the SPICE NAIF system) provided by the navigation team and USO test results acquired separately from science observations. It is best to create a model of the sky frequency profile from the SPK file and then difference it from the observed profile to generate the residuals containing the science information.

Once the signal frequency, frequency residuals, power, and spacecraft trajectory data are available for a given occultation, standard techniques can be applied to determine the ionospheric profile of the atmosphere. In its simplest form, ionospheric analysis does not require an inversion process and is much more straightforward than neutral atmospheric analysis.

For example, Kliore et al. (2008) show the ionosphere of Titan; see Figure 2.7. The first four sets of ROs of Titan's ionosphere were obtained by the Cassini spacecraft between March 2006 and May 2007. These occultations occurred at middle and high latitudes, at solar zenith angles from about 86–96 deg. The main ionospheric peak was seen, as expected from modeling and previous observations, near 1200 km, with a particle density of ~1–3 × 10^3 cm^{-3}. A consistent feature near 1000 km was also seen, and one of the polar observations found a significant (~3 × 10 cm^{-3}) layer in the region of 500–600 km. This layer also is seen in other observations with a density varying from about 0.7 to 1.7 × 10^3 cm^{-3}, suggesting a variable production source (or sources) for this peak.

2.4 Rings

During a ring occultation observation, Cassini transmitted three coherent sinusoidal signals simultaneously through the rings of saturn at S-, X-, and Ka-bands. They were generated from one onboard USO, allowing measurement of their amplitude, frequency, and phase, as well as their relative phases. After Asmar et al. (2014b), this section follows Marouf and Wong (2012) and Marouf et al. (1982, 1983) that describe the ring occultation data processing. References in one-sided occultations mean either ingress or egress, but not both, and references to Prime Missions mean Cassini's originally planned three-year science acquisition period, which was divided into many Revolutions (or Revs), as opposed to the three extended periods.

During a typical Cassini ring occultation experiment, a 70-m DSN station is usually used to receive the X- and S-band signals and a 34-m station is used to receive the Ka-band signal, and X-band is used as a backup. For any given band, the received signal can be modeled as a narrow-band signal of the form where ω_c is assumed to be known. The signals are down-converted in frequency to the center of the recording bandwidth (BW). Typical sampling rates used for ring occultations are 1, 16, and 50 kHz. The RSR encodes and preserves full information about the values of local oscillator frequencies used for the down-conversion, so that sky frequency may be recomputed after the fact.

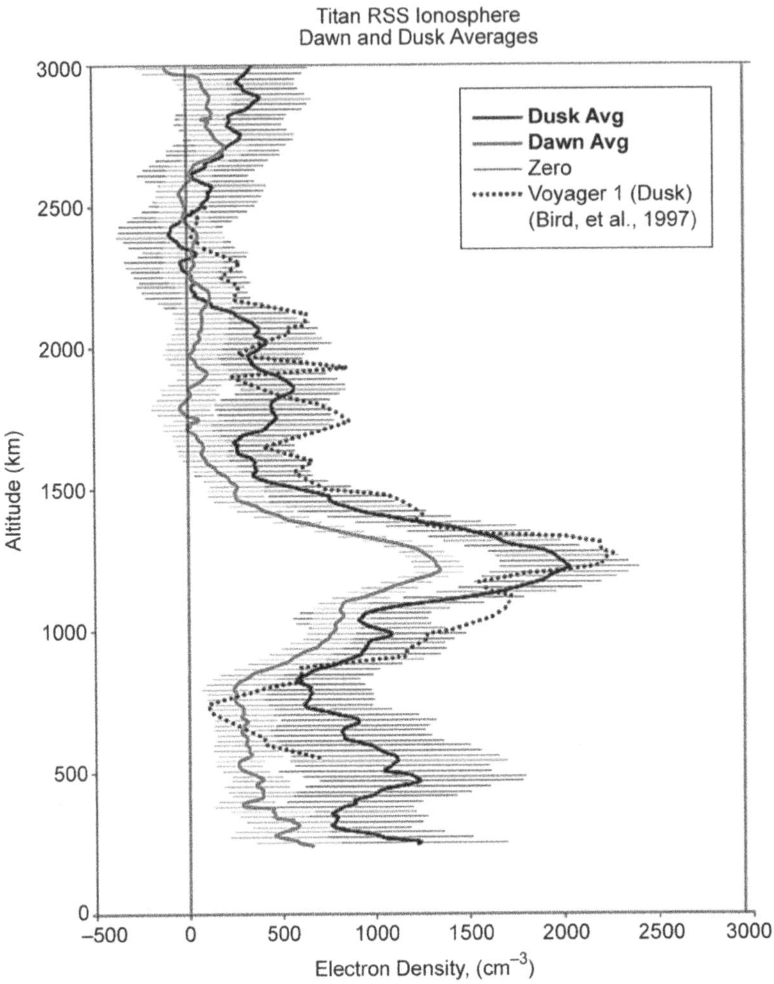

Figure 2.7 From Kliore et al. (2008), observed dawn, dusk, and dawn-averaged electron density profiles. For Titan observations are flybys 12 and 14 (T12 and T14) at the southern mid latitudes; the dusk observations are also in sunlight, and the dawn observations are in the dark. The Voyager 1 profile is also shown for comparison.

$$s(t) = I_m(t)\cos(\omega_c t) - Q_m(t)\sin(\omega_c t) = \mathrm{Re}\left\{\left[I_m(t) + iQ_m(t)\right]\exp(i\omega_c t)\right\}, \qquad (2.14)$$

The recorded (measured) samples encode the $I_m(t)$ and $Q_m(t)$ values using a uniform time grid. The pair (I_m, Q_m) can be modeled by the complex number

$$I_m(t) + iQ_m(t)$$

and the samples are usually referred to as the measured complex signal. Earth-received time (ERT) UTC is used for time tagging the samples. The time history of the complex signal during an occultation provides valuable information regarding ring structure and particle physical properties.

2.4.1 Ring Occultation Observables

As a transmitted signal passes through the rings, the ring material perturbs it and two signal components can be identified in the measured spectrum of the perturbed signal as shown in Figure 2.8. The first is the direct signal, a narrow spectral line that is the remnant of the coherent incident sinusoid after being attenuated and phase shifted by ring material. The time histories of estimated average power and phase change of the direct signal provide a measured extinction and phase shift profile. The extinction is characterized in terms of an optical depth and, with the help of a reconstructed spacecraft trajectory, is used to compute an optical depth versus radius profile of the ring system as a function of wavelength. Since the initial profile is diffraction-limited, accurate measurement of the signal phase allows reconstruction of the profile to remove diffraction effects (Marouf et al. 1986).

The second component is the near-forward scattered signal, a frequency-broadened signal that originates from incoherent superposition of the signals scattered by ring particles located within the intersection of the main lobe of the Cassini high-gain antenna and the rings plane (antenna footprint). The Doppler shift caused by the relative motion of the spacecraft, ring particles, and ground station broadens the spectrum of this signal component. The time history of the scattered signal spectra (Figure 2.9) can be used to determine the size distribution of large (>1 m) particles as well as other physical properties (Marouf et al. 1983; Zebker et al. 1985).

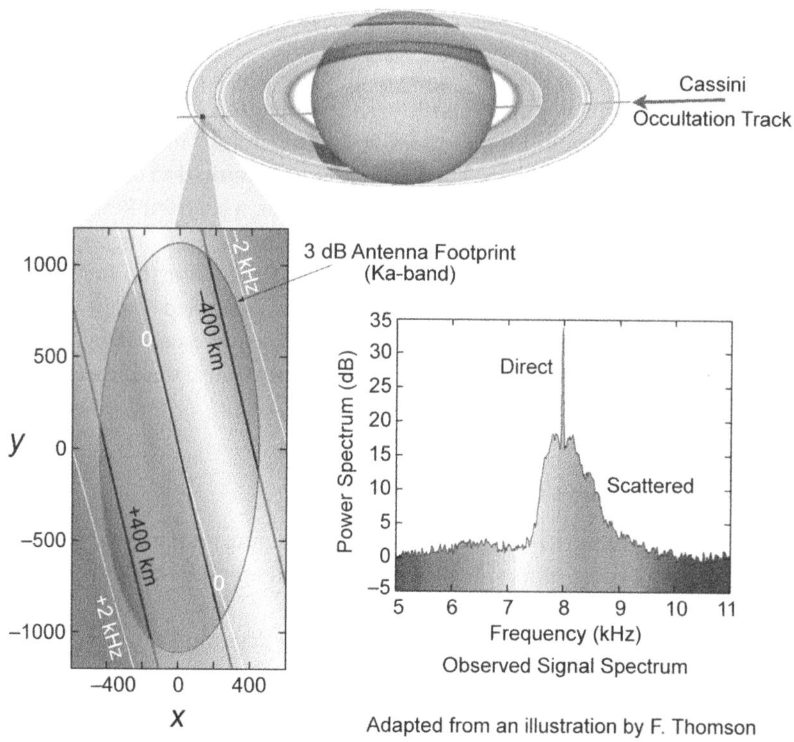

Figure 2.8 Illustration of ring radio occultation direct and scattered signal, and of Doppler mapping over the Ka-band footprint of the Cassini high-gain antenna.

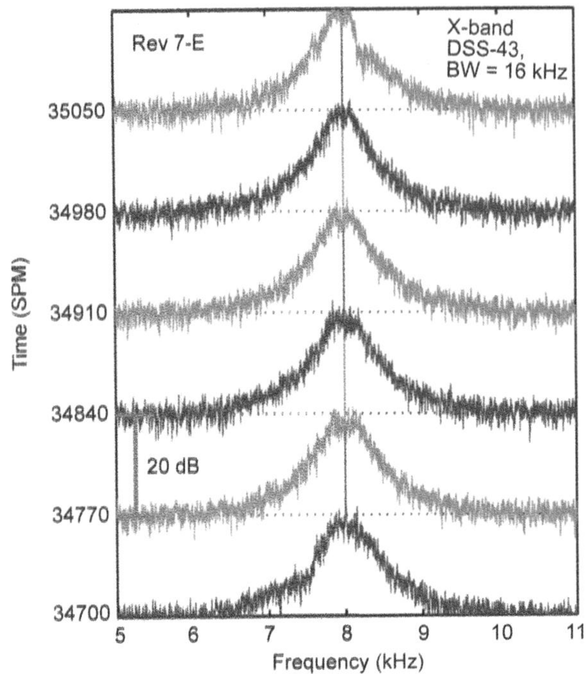

Figure 2.9 Example time sequence of Ring A spectra used to estimate the frequency offset of the direct signal from the center of the recording bandwidth. SPM is seconds past midnight on May 03, 2005.

Twenty-four one-sided ROs of Saturn's rings were observed during the Cassini Prime Mission. For such occultations, the spacecraft trajectory is chosen so that the path of the radio signals transmitted from Cassini to an observing ground receiving station would be interrupted by the ring system. The Prime Mission occultation observations sampled a broad range of ring opening angles, B. In particular, a set from Revolutions (Revs) 7–14 of mostly diametric occultations captured relatively large angles, 19.5–23.5 deg, allowing detailed profiling of the optically thick Ring B for the first time. Another set from Revs 53–67 of mostly chord occultations sampled small angles, 6–10 deg. The limited intermediate range of angles, 14–15 deg, was sampled by occultations from Revolutions 28, 44, and 46. These 24 occultation cuts also sampled a broad range of ring longitudes, both inertial and Earth-relative.

2.4.2 Ring Occultation Analysis

Reliably extracting ring structure from the recorded data requires reliable estimation of the variations in phase of the direct sinusoidal signal and amplitudes of the direct and scattered signals as functions of time.

Accurate estimation of the signal phase uses the measured deterministic offset of the frequency of the direct signal from its expected position at the center of the recording bandwidth to adjust the phase of the acquired complex samples. This is accomplished by compensating the phase of the measured in phase and quadrature (I/Q) samples as follows:

$$I_c(t) + iQ_c(t) = \left[I_m(t) + iQ_m(t)\right]\exp\left[-i\psi(t)\right], \tag{2.15}$$

where the phase function $\psi(t)$ is computed from

$$\psi(t) = \int^{t} \hat{f}_{\text{offset}}(\tau)d\tau + \psi(t_0). \tag{2.16}$$

$\hat{f}_{\text{offset}}(t)$ is an estimate of the frequency offset, and $\psi(t_0)$ is an arbitrary constant. Usually, a large fraction of $\hat{f}_{\text{offset}}(t)$ comes from inaccuracies in the predicted spacecraft trajectory used to compute the predicted Doppler shift $f_{dp}(t)$ needed to tune the RSR. A spacecraft trajectory reconstructed after the experiment can be used to compute a more accurate Doppler shift, $f_{dr}(t)$. The difference is used as part of the estimate above. $\hat{f}_{\text{resid}}(t)$ denotes any remaining residuals:

$$\hat{f}_{\text{offset}}(t) = \left[f_{dr}(t) - f_{dp}(t)\right] + \hat{f}_{\text{resid}}(t). \tag{2.17}$$

The Cassini egress ring occultation Rev7-E is used to illustrate the analysis. A sequence of power spectra is derived from the measured $I_m(t) + iQ_m(t)$ samples; a representative subset of those spectra is shown in Figure 2.9, where both the direct to Earth and scattered signal components are clearly visible. Each spectrum is searched for the frequency of its peak value near the center of the recording bandwidth; this peak is presumed to be the direct signal, but it can include scattered signal or noise where the ring has large optical depth.

Least squares fitting of model spectra of sinusoidal signals to the measured spectral values around the peak can be used to refine the position of the peak (Lipa and Tyler 1979). The desired $f_{\text{offset}}(t)$ is then computed as the offset of the estimated peak frequency from the center of the recording bandwidth (BW = 1 kHz) in this case. The noisy green curve in the upper panel of Figure 2.10 depicts $f_{\text{offset}}(t)$ computed for Rev7-E over a time span that covers Rings C, B, then A, followed by a free-space baseline. The estimated profile is very noisy and almost random in optically thick regions of Ring B; it is less noisy in Ring A, and it is much less noisy in Ring C, the Cassini Division, and outside Ring A.

Also plotted in the upper panel of Figure 2.10 is the difference between the reconstructed and predicted Doppler shifts, $f_{dr}(t) - f_{dp}(t)$, which accounts for the trend and a large fraction, but not all, of $f_{\text{offset}}(t)$. The difference, $f_{\text{resid}}(t) = f_{\text{offset}} - (f_{dr} - f_{dp})$, is the undulating curve plotted in the lower panel of Figure 2.10, here computed using coarser resolution to mitigate the effect of the noise on the frequency estimates. The most optically thick parts of Ring B still yield unreliable estimates (the green part of the plot). Least squares fit of the other more reliable parts of the curve (cyan color) yields the heuristic estimate $\hat{f}_{\text{resid}}(t)$. Potential causes for $\hat{f}_{\text{resid}}(t)$ could be systematic errors in the reconstructed spacecraft trajectory and/or other unmodeled long-term effects. The upper panel of Figure 2.10 shows a comparison between the computed smooth estimate (solid red curve) $\hat{f}_{\text{offset}}(t) = \left[f_{dr}(t) - f_{dp}(t)\right] + \hat{f}_{\text{resid}}(t)$ with the measured (noisy green curve). The agreement is good.

The process described above is critical for ensuring that the phase of the calibrated signal $I_c(t) + iQ_c(t)$ is primarily due to the ring material, instabilities in the USO reference phase, and any other unknown short-time effects that are not possible to correct for.

The second requirement for reliable computation of ring properties is accurate determination of the free-space signal level $P_0(t)$ used to normalize the signal level through the occultation period.

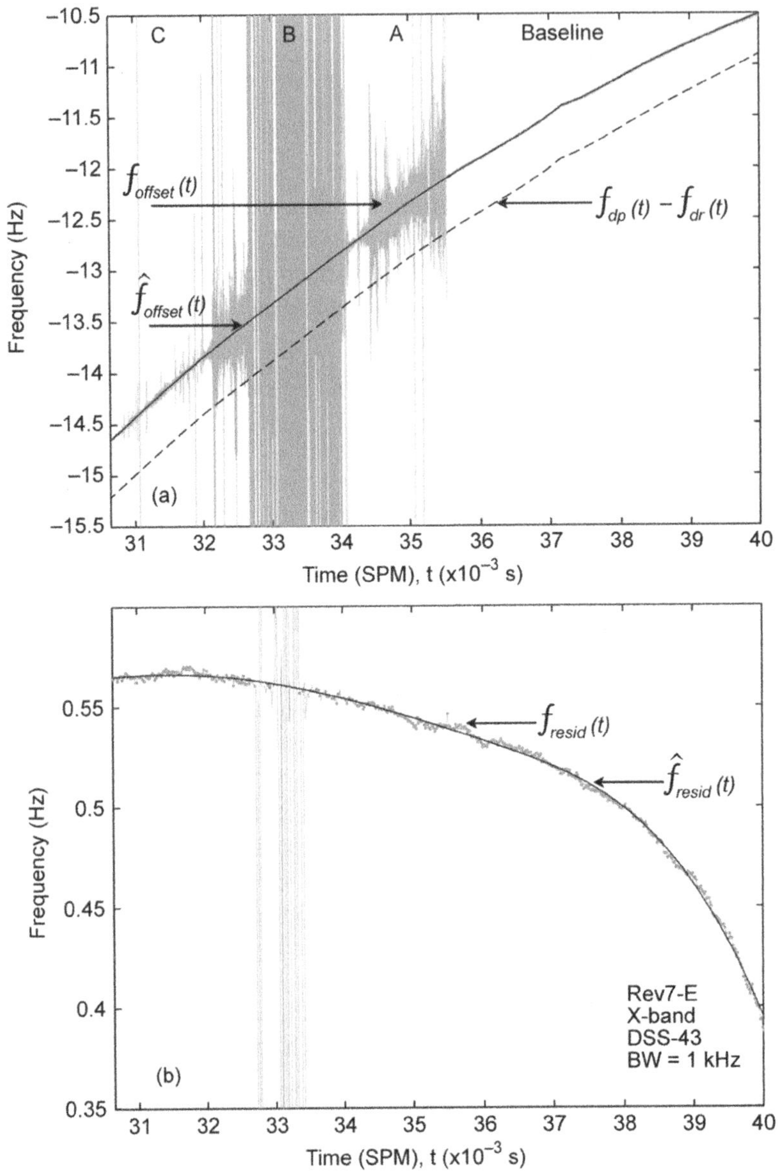

Figure 2.10 Example estimations of direct signal frequency offset (upper) and residual signal frequency (lower). SPM is seconds past midnight May 03, 2005.

Direct measurement of $P_0(t)$ is possible only during time periods when the line-of-sight from the spacecraft to the Earth receiving station is clear of the rings. All ring occultation observations were designed to allow measurement of $P_0(t)$ over a sufficiently long period of time exterior to Ring A. The geometry of some radial occultations enabled measurements of $P_0(t)$ in the gap between Ring C and Saturn's ionosphere. When possible, measurements of $P_0(t)$ over known gaps within the rings were also used.

Least-squares spline fitting is used to obtain $P_0(t)$ estimates for the ring-free periods. The fitting yielded values $\hat{P}_0(t)$ everywhere throughout the observation period, as the example in Figure 2.11 illustrates. The calibrated diffraction-limited complex ring transmittance is defined as

$$\hat{T}_R(t)+i\hat{T}_I(t)=\left[I_c(t)+iQ_c(t)\right]/\sqrt{\hat{P}_0(t)} \tag{2.18}$$

The frequency-calibrated and normalized samples, $\hat{T}_R\ (t)+i\hat{T}_I\ (t)$ is the *calibrated data* and provides the starting point for all subsequent data analysis steps including computation of optical depth profiles and scattered signal power spectra. The spectra shown in Figure 2.10 (BW = 16 kHz) were computed using calibrated data and standard power spectral estimation techniques (~2 s averaged spectra, each centered on the time listed in Figure 2.10).

As Figure 2.11 clearly shows, there are both short- and long-term variations in $P_0(t)$. The long-term variations are primarily caused by changes in the elevation angle of the DSN receiving antenna during the few to several hours of an experiment and systematic (spacecraft and/or ground) antenna pointing errors. The short-term fluctuations are typically caused by one or more terrestrial factors including turbulence in the atmosphere, clouds, wind gusts (causing mechanical vibration of the ground receiving antenna), or fluctuations in the receiver gain or thermal noise.

Complex samples of the calibrated data are uniformly spaced in time. To translate the measurements to ring profiles, the reconstructed spacecraft trajectory is used to calculate the ring intercept radius as a function of time. The calculations are completed using a quasi-inertial Sun-centered reference frame with the light time included.

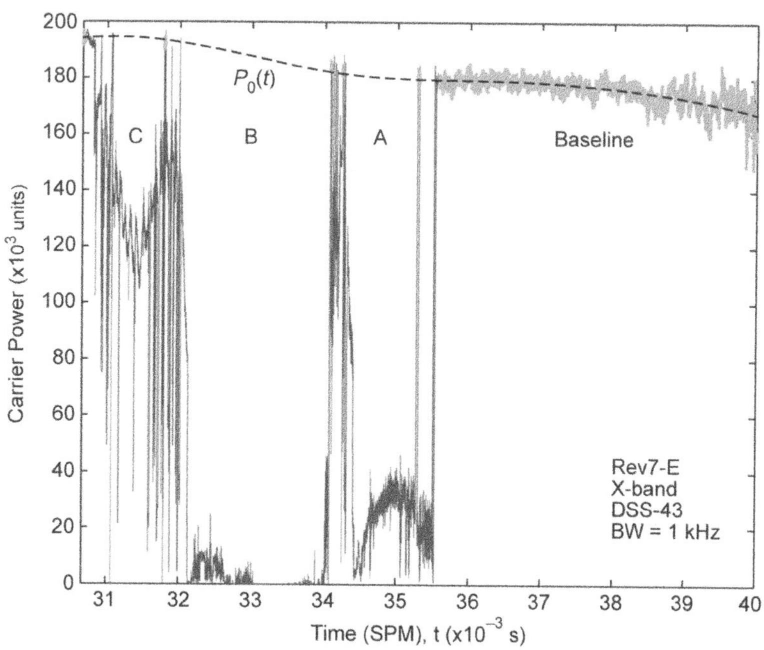

Figure 2.11 Example estimation of free-space signal power $P_0(t)$. SPM is seconds past midnight May 03, 2005.

The procedures were developed for analysis of the Voyager ring occultation experiment and are documented in Rosen (1989). Ring radius accuracy better than several hundred meters is achievable in principle; the primary source of error is the orientation of Saturn's pole. The calculations here are based on the pole orientation in the planetary constants (PCK) files produced by the Cassini Navigation Team. In turn, the collective set of occultations may be used to improve future estimates of pole orientation and precession rate.

The radial velocity of the ring-plane intercept point is not constant, and data that are sampled uniformly in time translate into samples that are nonuniformly spaced in ring radius. Any of several standard numerical interpolation algorithms may be used to interpolate the calibrated data onto a uniform radius grid. Care must be exercised, however, to ensure that interpolation errors are minimized. Below the notation

$$\hat{T}(\rho_0)=\hat{T}_R(\rho_0)+i\hat{T}_I(\rho_0) \tag{2.19}$$

is used to denote calibrated complex ring transmittance samples at radius ρ_0 where the sampling is uniform in radius. As described in detail in Marouf et al. (1986), $\hat{T}(\rho_0)$ provides the starting point for removing the diffraction effects and recover high-resolution profiles of ring optical depth. A summary of the steps and an example are provided below.

2.4.3 Ring Diffraction Correction

RO observations of planetary rings are diffraction limited. The measured ring transmittance $\hat{T}(\rho_0)$, observed at time t and corresponding to a ring radius ρ_0, differs from the true value $\hat{T}(\rho_0)$, which would be observed just behind the ring plane, because of diffraction effects. The relationship is in the form of the Fresnel transform,

$$\hat{T}(\rho_0)=\frac{1-i}{2F}\int_{-\infty}^{\infty}T(\rho)\exp\left[i\frac{\pi}{2}\left(\frac{\rho_0-\rho}{F}\right)^2\right]d\rho, \tag{2.20}$$

where F is the Fresnel scale of diffraction, which depends on the observation wavelength and geometry (Marouf et al. 1986, Eq. 6). In principle, $T(\rho)$ may be recovered from $\hat{T}(\rho_0)$ by using the inverse Fresnel transform

$$T(\rho)=\frac{1+i}{2F}\int_{-\infty}^{\infty}\hat{T}(\rho_0)\exp\left[-i\frac{\pi}{2}\left(\frac{\rho-\rho_0}{F}\right)^2\right]d\rho_0. \tag{2.21}$$

In practice, quadratic phase approximations of the transform kernel (Marouf et al. 1986, Eqs. 15 and 16), sample resolution of $\hat{T}(\rho_0)$, and signal-to-noise in the measured values set limits on the $T(\rho)$ recovery.

The normal optical depth profile and the phase-shift profile are computed using the definitions

$$\begin{aligned}\tau(\rho)&=-2\mu_0\ln\left(\left|X(\rho)\right|\right)\\ \phi(\rho)&=\tan^{-1}\left[X_I(\rho)/X_R(\rho)\right],\end{aligned} \tag{2.22}$$

where $\widehat{X}(\rho_0)=\widehat{T}(\rho_0)+\hat{n}(\rho_0)$ and $X(\rho)=T(\rho)+n(\rho)=X_R(\rho)+iX_I(\rho)$ have been used to denote the noisy measured and recovered values, respectively, B is the ring opening angle, and $\mu_0=\sin(|B|)$.

2.4.4 Data Decimation and Profile Resolution

Practical aspects of implementation of the Fresnel inversion procedure must be considered. Clearly, the infinite range of ρ_0 in (2.23) must be replaced by a finite data "window" of width W, which sets a lower limit (smallest detectable structure size) $\Delta R_W = 2F^2 / W$ on the reconstructed resolution (Marouf et al. 1986). To preserve information about structure of spatial scale equal to or larger than some desired ΔR_W, the diffraction-limited data $\widehat{T}(\rho_0)$ must be sampled at a rate equal to or larger than twice the highest spatial frequency (in accordance with the sampling theorem). Thus, to resample $\widehat{T}(\rho_0)$ at a reduced sampling rate, the data must first be low-pass filtered using a bandwidth commensurate with the lower sampling rate and then decimated so that the sample spacing is $\Delta R_W / 2$ (or smaller) to prevent artifacts due to aliasing from corrupting the data.

Another potential source of corruption of the reconstructed ring profile is the artificial truncation of the data to finite width W. The sudden jump in the value of data samples near the window edge, from finite to zero value, introduces artifacts in the reconstructed $T(\rho)$, especially in optically thin ring regions. The artifacts can be mitigated by multiplying the data by a tapered window function that gradually decreases to zero close to its edges (Marouf et al. 1986, Eq. 18). Parameters of the window function are chosen to minimize artifacts over the reconstruction dynamic range (optical depth range). Unfortunately, this is achieved at the cost of degradation in spatial resolution by a factor close to 2. The degraded resolution is usually referred to as the effective resolution $\Delta R_{\text{eff}} =\sim 2\Delta R_W$; its initial value ΔR_W is referred to as the processing resolution.

Additional factors that limit achievable resolution include uncertainty in knowledge of F, finite USO phase noise, and computational complexity of the inversion process (Marouf et al. 1986).

An example of a diffraction reconstruction is shown in Figure 2.12. The main ring feature shown is the Huygens Ringlet located within the Huygens Gap. Also shown is the so-called Charming Ringlet near the outer edge of the gap. The blue curves depict the diffraction-limited X-band profiles $\hat{\tau}(\rho_0)$ and $\hat{\phi}(\rho_0)$ at 250 m resolution. Edges of both ringlets as well as the outer gap edge are smoothed by diffraction; diffraction oscillations are visible in the vicinity of the Huygens Ringlet edges. The red curves depict the profiles $\tau(\rho)$ and $\phi(\rho)$ reconstructed using a processing resolution $\Delta R_W = 500$ m, yielding profiles of effective resolution ΔR_{eff} close to 1 km.

2.4.5 Signal-to-noise Ratio-resolution Tradeoff

Note that the best achievable radial resolution for a given ring feature strongly depends on the optical depth of the feature, that is, on the measurement signal-to-noise ratio (SNR). Finer spatial resolution requires data sampled at a higher rate or, equivalently, data captured in a larger bandwidth, which are thereby noisier. Since the thermal noise is additive, the SNR can be improved using coherent averaging over a number of data samples, but at the expense of resolution.

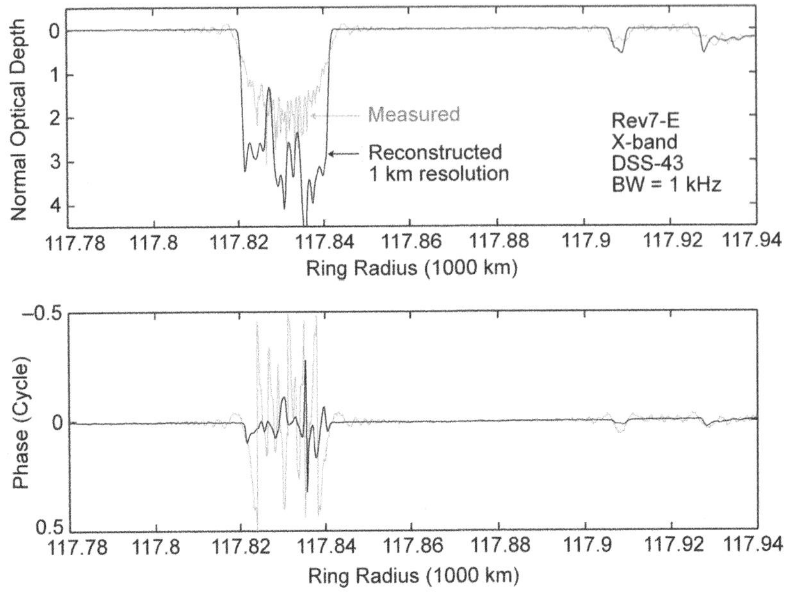

Figure 2.12 Optical depth and phase profile reconstructions from diffraction-limited Rev7-E ring occultation observations of May 03, 2005. The Huygens and Charming Ringlets are at radii 117,830 ± 11 km and 117,908 ± 3 km, respectively. The outer edge of the Huygens Gap is at 117,927 km.

There is always a trade-off between finer resolution and uncertainty in the reconstructed profile that is analytically characterized by a parameter known as the threshold optical depth τ_{TH}, roughly the value of the ring normal optical depth for which the SNR drops to about 1. For a 70% confidence level, τ_{TH} is given by Marouf et al. (1986, Eqs. 23–26)

$$\tau_{TH} \cong -\mu_0 \ln\left[1.205 P_N\right]$$

where PN is the average power of the additive noise in watts (W). An estimate $\hat{P}_N$ can be calculated from

$$\hat{P}_N = \frac{1}{SNR_0} \frac{\dot{\rho}_0}{\Delta R}$$

where ΔR is either $\Delta\rho_0$ (the sample spacing of the diffraction-limited data) or ΔR_{eff} (the effective resolution of the reconstructed profiles), $\dot{\rho}_0$ is the ring plane radial velocity of the ring intercept point, and SNR_0 is the free-space SNR in 1-s integration time or 1-Hz bandwidth. The latter can be estimated from the recorded I/Q samples. Alternatively, typical X-band values of about 54 and 48 dB-Hz) for data collected at the 70- and 34-m ground stations, respectively, may be used.

Figure 2.13 illustrates the trade-off discussed above. The plots show the attenuation of the Titan Ringlet located within the Colombo Gap in inner Ring C at three

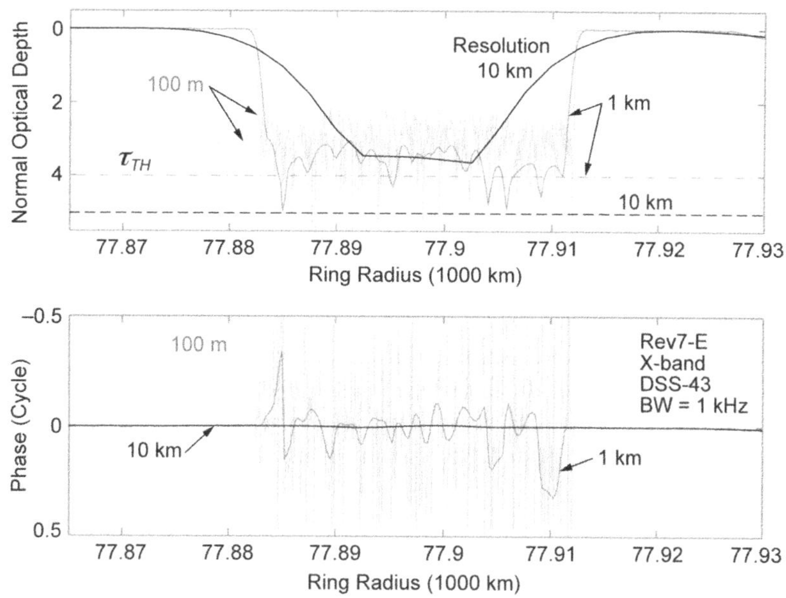

Figure 2.13 Example trade-off between reconstruction resolution and measurement reliability for the Titan ringlet. The horizontal stippled lines identify the threshold optical depth (SNR ~1) for each indicated resolution.

progressively coarser resolutions of 100 m, 1 km, and 10 km. Also shown are the corresponding τ_{TH} levels, horizontal lines of the same color as the corresponding profile. The τ_{TH} levels progressively increase from ~3.1 to 4.0 to 5.0 as the resolution ΔR_{eff} changes from 100 m, to 1 km, to 10 km, respectively.

At 100 m resolution (green profiles in Figure 2.13), almost all structures within the ringlet is obscured by noise; hence, the reconstruction is effectively noise limited, and the phase profile is essentially random. The ringlet edge location is recovered with high accuracy, however. The absence of any profile overshoot in the vicinity of transition to a zero-value free-space optical depth indicates reliable reconstruction of the diffraction effects.

At fine resolution, diffraction-limited data over a relatively large radial interval W are required to compute the inverse Fresnel transform. Any inaccuracies exceeding a small fraction of a cycle in the computed phase of the kernel of the inverse transform can cause undesirable artifacts in the reconstructed profiles.

At the 1-km resolution (blue profiles), more reliable profiling of ringlet structure is possible only over the fraction of its width where the recovered optical depth exceeds the corresponding threshold level (the horizontal blue line). The phase profile is also well behaved when $\tau < \tau_{TH}$. The edges remain relatively sharp at this resolution. At the 10-km resolution (red profiles), reliable measurement of an average optical depth of the ringlet is achieved but the fine structure is lost—for example, the edges are highly smoothed. Relatively large-threshold optical depths, such as the τ_{TH} ~ 5.0 here, are needed for profiling optically thick ring regions such as parts of Ring B.

2.5 Surface Scattering

Bistatic radar (BSR) is the active probing of planetary surfaces by oblique reflection and scattering of microwave signals. Such observations provide statistical information on root-mean-square (rms) surface slopes and on dielectric constant and density at scales comparable to the radio wavelength (centimeters to meters). An unmodulated radio signal, transmitted from the spacecraft, is scattered by the surface, and the echo is received at a ground station on Earth. A direct signal from the spacecraft is often received at the same time, providing a frequency reference for the echo. Fjeldbo (1964) utilized physical optics theory to explain the specular component of the echo as an aggregation of mirror-like reflections from properly oriented elements of the undulating surface. For typical spacecraft transmitters of 1–100 W, surface echoes as small as one zeptowatt (10^{-21} W) can be detected. BSR experiments have been performed on the Moon, Venus, Mars, and Titan using Soviet, US, and European spacecraft (Simpson 2007).

Differential Doppler effects between the direct and reflected paths separate the received echo from the carrier. Dispersion of the echo itself, caused by differential Doppler within the reflecting region, provides a measure of the rms slope of the undulations. Where echo dispersion is not large, the Doppler difference between the two primary paths can be used to infer large-scale topography of the surface. An incident signal with known polarization will generate a reflected signal with both the original component as well as the orthogonal component; application of Fresnel reflection theory to absolutely calibrated echoes in one polarization or to relatively calibrated echoes in orthogonal pairs can be used to recover the dielectric constant of the surface material. A simple model relating dielectric constant to porosity then allows estimation of material density.

The quantities of interest in BSR are the amplitude, frequency, and polarization as well as the time of the echo signal. Incremental echo power from an unresolved surface element is given by the radar equation:

$$dP_R = \left(P_T G_T / 4\pi R_T^2\right)\sigma\left(A_R / 4\pi R_R^2\right), \tag{2.23}$$

where P_T is the transmitted power, G_T is the transmitting antenna gain in the direction of the surface element, R_T is the distance from the transmitter to the surface element, A_R is the effective area of the receiving antenna aperture, and R_R is the distance from the surface element to the receiver. The BSR cross-section of the surface element, σ, is

$$\sigma = \sigma_0 dS, \tag{2.24}$$

where dS (the area of the surface element) and σ_0 is the specific radar cross-section, or radar cross-section per unit area, which is a function of the directions of both the incoming and outgoing wave vectors. For quasi-specular conditions, these reduce to a single angle $\phi = \phi_i = \phi_r$ measured with respect to the local mean surface. Specific radar cross-section is assumed to be proportional to reflectivity, ρ, which is derived from the Fresnel voltage reflection coefficients for horizontally and vertically polarized electromagnetic waves at a planar interface between free space and the planetary surface. So

$$R_{HH} = \left(\cos\phi - \left(\varepsilon - \sin^2\phi\right)^{1/2}\right) / \left(\cos\phi + \left(\varepsilon - \sin^2\phi\right)^{1/2}\right)$$
$$R_{VV} = \left(\varepsilon\cos\phi - \left(\varepsilon - \sin^2\phi\right)^{1/2}\right) / \left(\varepsilon\cos\phi + \left(\varepsilon - \sin^2\phi\right)^{1/2}\right), \tag{2.25}$$

where ε is the dielectric constant of the surface material. The reflection coefficients for circularly polarized waves can be derived from RHH and RVV

$$R_{SC} = \left(R_{VV} + R_{HH}\right)/2$$
$$R_{OC} = \left(R_{VV} - R_{HH}\right)/2, \tag{2.26}$$

where R_{SC} is the voltage reflection coefficient for same sense circular polarization, for example, right-hand circular polarization (RCP) transmitted and RCP received, and ROC is the coefficient for the opposite sense, that is, RCP transmitted and left-hand circular polarization (LCP) received.

The power reflection coefficients r_i are

$$r_{SC} = \left|R_{SC}\right|^2 \text{ and } r_{OC} = \left|R_{OC}\right|^2. \tag{2.27}$$

The ratio ρ_{SC} / ρ_{OC} is sometimes easier to measure than either component by itself because many terms in (2.28) are common to both signal paths (Simpson 2007).

The Doppler shift for a spacecraft transmitter at r_T and receiver at r_R is the rate of change of the path length measured in wavelengths λ (Figure 2.14):

$$f_c = (d/dt)\left|r_R - r_T\right|/\lambda. \tag{2.28}$$

The vectors must be defined in an inertial frame, and light-time adjustments must be included, but these are not shown explicitly. The Doppler shift for a signal reflected by the specular point rP on a smooth target is, similarly

$$f_r = (d/dt)\left|r_R - r_p\right|/\lambda + (d/dt)\left|r_p - r_T\right|/\lambda. \tag{2.29}$$

For a surface on which the roughness is not too severe, the total echo power is the same as the power from the single specular point on the equivalent perfectly smooth sphere. As roughness increases, the spatial distribution of the reflecting facets becomes broad enough that some of the glints fall outside the illumination pattern of the spacecraft antenna, and the observed echo power is reduced accordingly. In the extreme case, only a small fraction of the possible glints are illuminated, and the distribution of incremental echo power reflects the antenna pattern rather than the surface scattering function. Broader echoes result from rougher surfaces and/or more rapidly moving spacecraft. For experiments that are not beam limited, the half-power width of the echo B and the rms roughness of the surface ς are related to V, the velocity of the specular point across the surface by

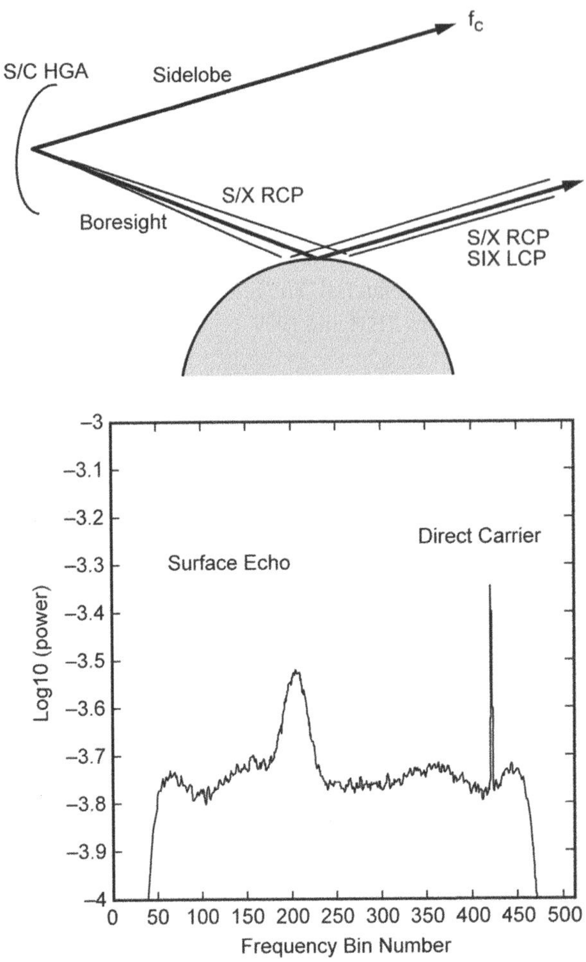

Figure 2.14 Representation of a bistatic radar experiment and resulting spectrum. The directly propagating signal is at frequency bin ~420; the surface echo extends over bins ~175–230. (HGA is high-gain antenna, and SC is spacecraft).

$$B = 4(\ln 2)^{1/2}(V/\lambda)\cos\phi. \tag{2.30}$$

The Cassini Radio Science team carried out a number of bistatic experiments near the Brewster angle at Titan to determine reflectivity and dielectric constant estimates at Cassini's three radio wavelengths.

The first Cassini detectable X-band RS bistatic echoes from Titan's surface were observed on flyby T14 on May 20, 2006 (DOY 140), making Titan the most distant Solar System object for which bistatic echoes were successfully detected. The echoes were weak but clearly detectable. An RCP sinusoidal signal was transmitted by Cassini and both the RCP and LCP surface reflected components were observed at the 70-m station Deep Space Station of the DSN at Madrid, Spain (DSS-63).

Cassini was maneuvered continuously to track the region of Titan's surface where mirror-like (quasi-specular) reflected signals in the direction of Earth could be observed. The ground receivers were tuned to compensate for the large Doppler shift due to relative spacecraft–Titan–Earth motion, allowing detection of the echoes near the center of the recording bandwidth (Figures 2.15 and 2.16).

The experiment was designed to probe incidence angles θ close to the Brewster, or polarization, angle of likely surface compositions. Careful measurements of the system noise temperature allowed determination of the absolute power in each polarized echo component and hence their ratio. The polarization ratio, the known observation geometry, and Fresnel reflectivity were then used to determine the dielectric constant ε.

The T14 surface echoes were detectable on both the inbound and outbound sides, but were intermittent along the ground tracks, suggesting mostly rough terrain occasionally interrupted by patches of relatively flat areas. Polarization ratio measurements were obtained for two localized surface regions of similar latitude ~15 S, well-separated longitudes ~14 and ~140 W, and incidence angles for θ of ~56 deg and ~64 deg, respectively. For both regions, the measurements implied surface dielectric constant ε ~1.6, suggesting liquid hydrocarbons, although alternative interpretations are possible (Marouf et al. 2006).

In Figure 2.15, panels (a) and (b) depict selected geometry parameters of the Cassini T14 bistatic scattering observations. The parameters are plotted versus time relative to closest approach. Inbound (blue) and outbound (red) bistatic observations were implemented before and after the T14 ionospheric/atmospheric occultation (the empty gap in both panels). Panel (c) depicts the corresponding ground tracks. The Huygens Probe

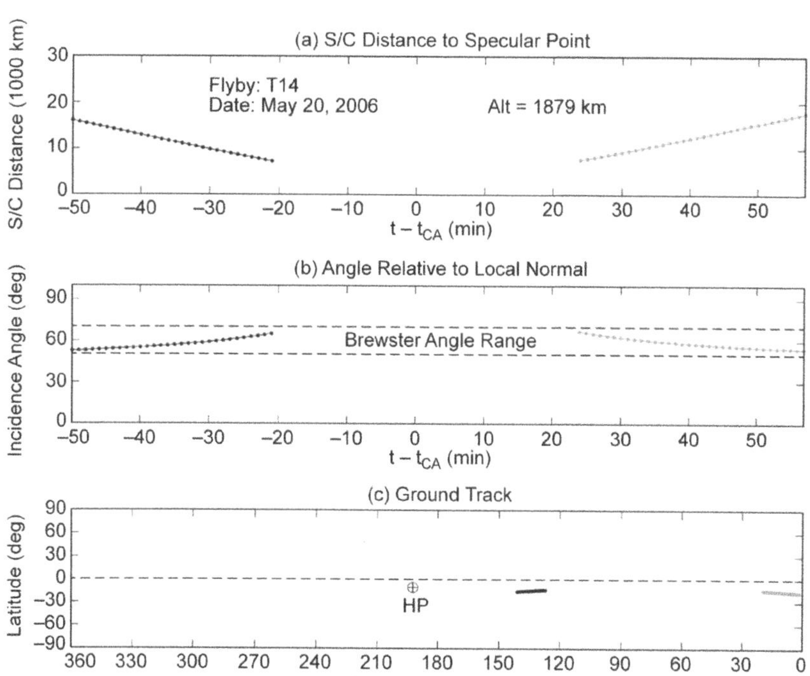

Figure 2.15 Panels (a) and (b) depict selected geometry parameters of the Cassini T14 observations and panel (c) depicts the corresponding ground tracks. The Huygens Probe landing site is shown for reference purposes.

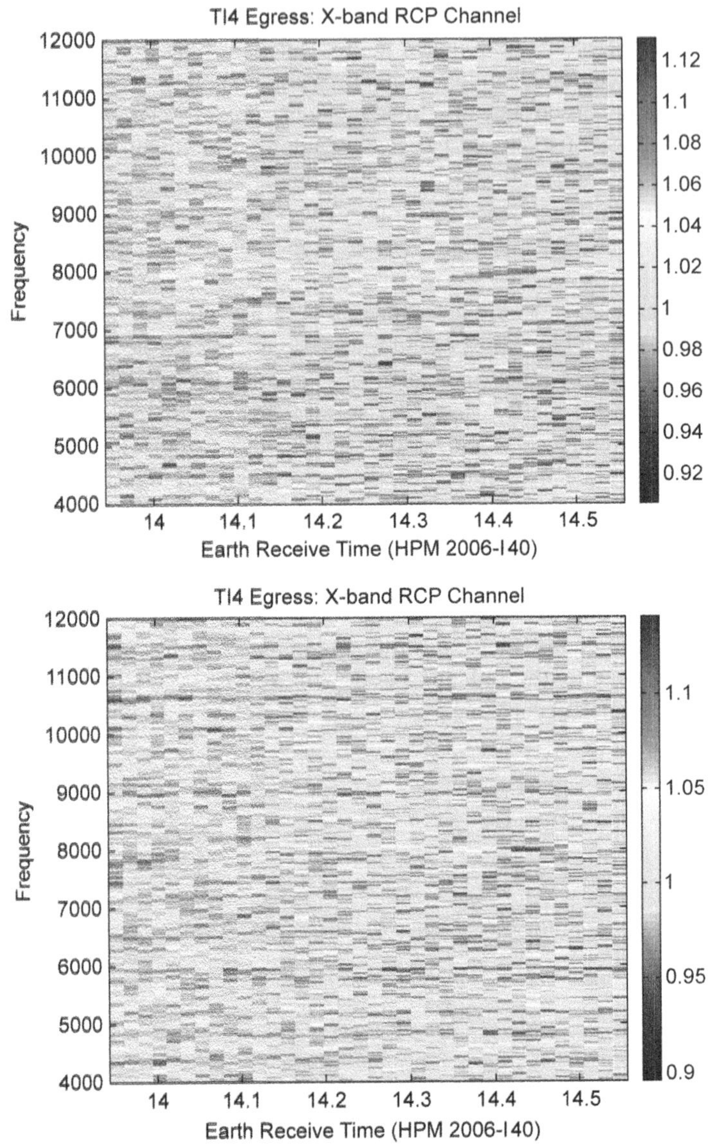

Figure 2.16 Uncalibrated spectrograms of the first positive detection of bistatic echoes from Titan surface observed on the T14 flyby on May 20, 2006 (day of year [DOY] 140, hours past midnight [HPM]). The weak echoes were simultaneously detectable in both the RCP and LCP 16-kHz bandwidth X-band channels of the 70-m ground station of the Madrid, Spain, DSN complex (DSS-63).

landing site is shown for reference purposes. The polarization ratio measurements were obtained for two localized surface regions of similar latitude ~15 S, well-separated longitudes ~14 and ~140 W, and incidence angle θ of ~56 deg and ~64 deg, respectively.

3

Gravity Science and Planetary Interiors

3.1 Overview

Planetary bodies have influence inherent to their mass that extends into space as gravitational fields that is detectable by objects in the vicinity such as spacecraft on scientific missions. One of the earliest applications of radio science (RS) techniques was in the study of planetary gravitational fields for the purpose of deducing planetary properties, such as bulk mass (radio scientists weigh the planets), mass/density distribution, internal density gradients, and interior structures and composition. This research is referred to in the scientific literature as gravity science, planetary gravimetry, planetary geodesy, or celestial mechanics investigations. Table 3.1 lists completed and selected planned gravity experiments from spacecraft radiometric data.

Planetary gravity measurements are grouped into three categories:

1. Static gravity: the gravity field inherent to the body.
2. Time-variable gravity: measurements that change due:
   - Tidal forces
   - Subsurface fluid layers
   - Atmospheric mass motion including thermal tide
   - Atmospheric deposition of polar caps
   - Movement of surface volatiles
3. Rotational dynamics: The effects of planetary motion such as rotational nutation and precession, for the purpose of constraining the physical characteristics of planetary cores.

The field of gravity science is an intersection of two branches of physics: mechanics and electromagnetism. The planetary gravitational forces perturb the motion of a proximate spacecraft by acting on its center of mass (CM) and proportionally alter its trajectory, which is then manifested as a Doppler shift of the frequency of a radio signal, an electromagnetic wave propagating between the spacecraft and a receiving station, or between two spacecraft.

Because gravity acts on the spacecraft's CM, the spacecraft itself is the instrument sensing the planetary gravity. To maximize the experiment sensitivity to gravitational forces acting on the spacecraft, it is important to characterize all nongravitational forces in the measurement system in order to separate them from the gravitational observable. It is also important to account for the primary noise sources that can corrupt the signal along the path of propagation and limit the overall experiment data quality.

Radio Science Techniques for Deep Space Exploration, First Edition. Sami W. Asmar.

Table 3.1 Summary of gravity experiments from spacecraft radiometric data (and selected planned future experiments).

Planetary body	Mission	Year
Mercury	Mariner 10	1974
	MESSENGER	2011–2015
	BepiColombo	Planned for 2025
Venus	Mariner 2	1962
	Mariner 10	1974
	Pioneer Venus	1978
	Magellan	1991
Moon	Luna 10	1966
	Lunar Orbiters I-V	1966–1967
	Apollo 15/16 subsatellites	1971–1972
	Clementine	1994
	Lunar Prospector	1998–1999
	SELENE	2007–2009
	Chang'E-1	2007–2008
	GRAIL	2011–2012
Mars	Mariner 9	1971
	Viking 1, 2	1976
	Mars Global Surveyor	1997–2006
	Mars Pathfinder	1997
	Mars Express (targets)	2003–present
	Mars Recon. Orbiter	2005–present
	Mars Odyssey	2001–present
Phobos	Mars Express (flybys)	2003–present
Deimos	Viking 1, 2 (flybys)	1977
Asteroids Ida	Galileo	1993
Mathilde	NEAR	1997
Eros	NEAR	1999–2001
Itokawa	Hayabusa	2005
Vesta	Dawn	2011–2012
Ceres	Dawn	2015–2016
Lutetia	Rosetta	2010
Ryugu	Hayabusa-2	2018–2019
Bennu	OSIRIS-Rex	Planned for ~2019
Psyche	Psyche	Planned for ~2026
Comets Halley	Giotto	1986

(Continued)

Table 3.1 (continued)

Planetary body	Mission	Year
Grigg-Skjellerup	Giotto	1992
Churyumov-Gerasimenko	Rosetta	2014
Wild 2	Stardust	2004
Jupiter	Pioneer 10, 11	1973–1974
	Voyager 1, 2	1979
	Juno	2016–present
Amalthea	Galileo	2002
Io	Galileo	1995–2003
Ganymede	Galileo	1995–2003
	Planned JUICE mission	~2030
Callisto	Galileo	1995–2003
Europa	Galileo	1995–2003
	Planned Europa Clipper	~2025
Saturn	Pioneer 11	1976
	Voyager 1, 2	1980
	Cassini	2004–2017
Titan	Cassini flybys	2004–2017
Enceladus	Cassini flybys	2004–2017
Rhea	Cassini flybys	2004–2017
Hyperion	Cassini flybys	2004–2017
Dione	Cassini flybys	2004–2017
Iapetus	Cassini flybys	2004–2017
Phoebe	Cassini flybys	2004–2017
Tethys	Cassini flybys	2004–2017
Uranus	Voyager 2	1986
Neptune	Voyager 2	1989
Pluto	New Horizons	2015

Prior to Solar System exploration by robotic spacecraft, scientists calculated planetary orbits and deduced relevant physical properties via optical astrometry and kinematic analysis. The ability to make accurate spacecraft distance and velocity measurements, commonly called range and range-rate radiometric observables, resulted in direct detection of spacecraft trajectory perturbations caused by gravitational fields of planetary bodies, which dramatically improved our understanding of planetary structures.

Range and range-rate measurements are also the radiometric techniques used for navigating robotic missions across the Solar System and beyond and for producing planetary and satellite ephemerides this underscores the important synergy between

RS and navigation (flight dynamics). Additional data types refine knowledge of the planet's location and pole orientation and aid in the interpretation of gravity science data. These data types include:

- Differential very long baseline interferometry (VLBI) (Ichikawa and Kato 1999).
- Delta differential one-way ranging (DOR) (Thornton and Border 2003).
- Optical landmark tracking (Cheng et al. 2003; Yeomans et al. 2000).
- Altimeter crossover measurements (Mazarico et al. 2010; Verma and Margot 2018).
- Dual one-way range (DOWR), radio links between two spacecraft (crosslinks), as in the case of the Gravity Recovery and Climate Experiment (GRACE) and Gravity Recovery and Interior Laboratory (GRAIL) missions (see Figure 3.1).
- Laser links between two spacecraft, introduced by the GRACE Follow-On Earth mission and Laser Interferometer Space Antenna (LISA) Pathfinder and likely to be used in future planetary missions.

Note that gravity measurements produce solutions and models that are not unique since different configurations of material density distribution can sometimes produce indistinguishable gravity signals. To address this, investigators use other available physical, chemical, or geological data to differentiate between feasible interior models.

In addition to the effects of Newtonian gravitational forces acting on spacecraft, there are important effects resulting from planetary rotation, pole nutation, and precession that are derived via Doppler tracking of spacecraft based on their relative velocity to receiving stations. Radiometric tracking of landed spacecraft (stationary on planetary surfaces) such as the Interior exploration using Seismic Investigations, Geodesy and Heat Transport (InSight) mission on Mars enables modeling of interior

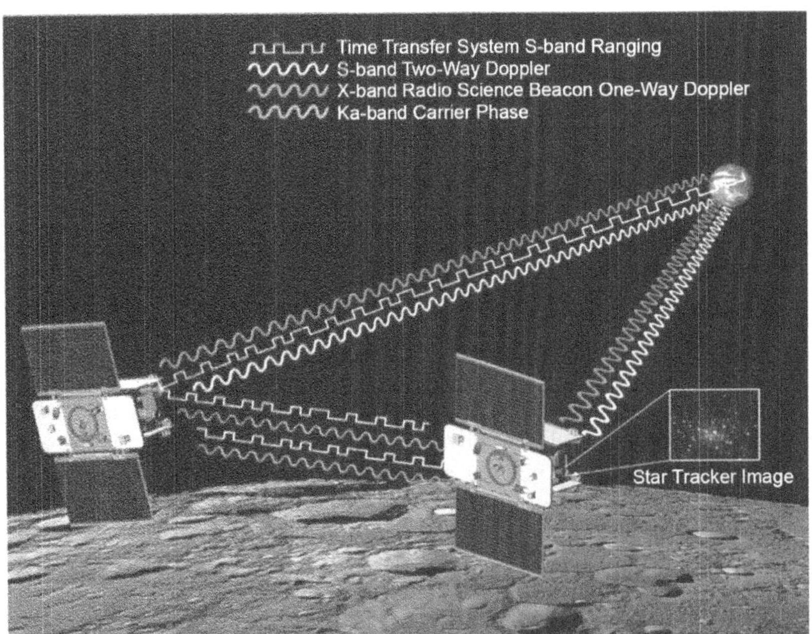

Figure 3.1 The GRAIL spacecraft-to-spacecraft gravity science experiment configuration.

structures and constraining of the size and density of the planetary core through estimation of the precession, the change in the orientation of the rotational axis, and nutation, the nodding motion in that axis.

Relativistic effects also perturb spacecraft trajectories and RS experiments can investigate aspects of the theories of general relativity using high precision radiometric measurements. As detailed in Chapter 4, spacecraft with quiet dynamics from internal systems and instrumented for the calibration techniques described in Chapter 5 are used for investigations of relativistic gravity.

The fundamental formulations of the gravitational potential are detailed in Appendix 3A. A gravitational field can be mathematically represented by two-dimensional spherical harmonic series of degree n and order m, scaled by the gravitational constant multiplied by the planetary mass (GM); this representation utilizes the normalized associated Legendre functions. Figure 3.2 illustrates harmonic contributions to the acceleration at the surface of a rocky planetary body. For any degree at order zero, $\overline{J}_n$ are the longitude-independent coefficients given by $\overline{J}_n = -\overline{C}_{n0}$, where $\overline{J}_2$ is related to the flattening and $\overline{J}_3$ describes the pear shape of the body. The corresponding harmonics are called zonal harmonics because they vanish at small circles and divide a sphere into latitudinal zones; they vary with latitude only and are invariant under rotation.

On the other hand, when the degree and order are equal ($m = n$) the harmonics divide a sphere into sectors and are called sectoral harmonics; they vary with longitude only and vanish at 2 m lines of longitudinal lines the great circles, or meridians. The checkered tesseral spherical harmonics develop when the degree and order are not equal ($m \neq n$); there are ($n - m$) lines of latitude that are zero and 2 m lines of longitude that vanishover the entire sphere.

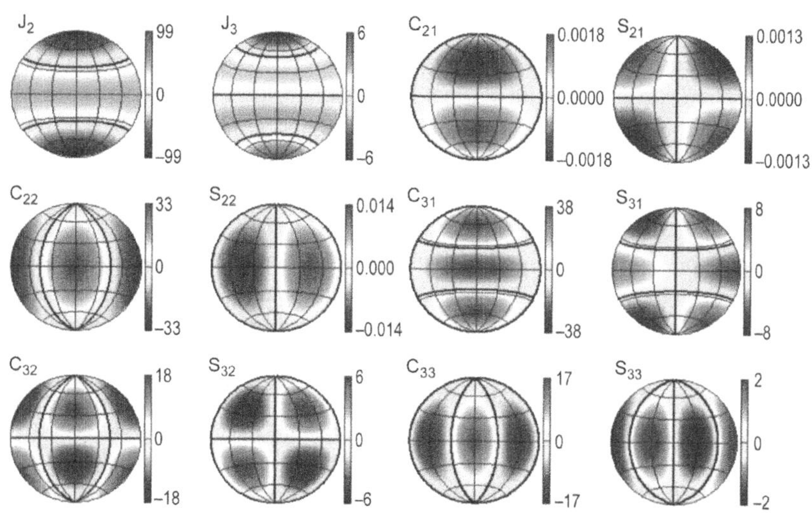

Figure 3.2 Illustration of gravity field spherical harmonic contributions to the acceleration at the surface of the Moon near side in units of milligals (mGal). The zero-contour thick black lines represent the zeros of that harmonic, and the thinner black lines are the latitude and longitude every 30 deg. The gravity amplitudes between the positive and negative regions are not necessarily equal. The positive amplitude for J_2 at the equator is one-half (49 mGal) the amplitude at the poles (99 mGal). After the completion of the GRAIL mission to the Moon, a global degree-and-order 1500 (or higher) field is available.

The resolution of a gravity field model is determined by the highest degree of the spherical harmonic expansion. The gravity field models of the planets have resolutions ranging from zero, only the GM is known as a point mass, to a few degrees for several planetary bodies, and up to many degrees for the more accessible bodies such as the degree-and-order 1800 field recovered for the Moon (available from the Planetary Data System data archives) after the completion of GRAIL mission (Zuber and Smith 2018).

For planetary geodesy, the lower degree terms spherical harmonics of the gravitational field describe global planetary features and large-scale effects, such as dynamic oblateness, and place constraints on the overall deep interior structure such as the existence of a high-density core. Higher degree terms describe smaller scale and shallower regional such as local gravity perturbations referred to as gravity anomalies.

The unit of measuring planetary gravitational acceleration is the Galileo (Gal), in honor of the Italian scientist Galileo Galilei who pioneered early investigations of Earth's gravity. One Gal equals 1 cm/s^2, with the milligal (mGal) commonly used to report observations in planetary science.

3.2 Gravity Observables and Formulations

Planetary gravity field measurements are typically carried out via observations of spacecraft flying near planetary targets or orbiting them while transmitting radio links to the ground stations, as listed in Table 3.1 and also illustrated in Figures 1.1 and 1.6. Radio links can also be transmitted from one spacecraft to another, as shown in Figure 3.1. The Doppler observable is the primary tracking data type; it is derived from two-way signals between spacecraft and ground stations and coherently referenced to highly stable atomic clocks at the stations.

The quality of the Doppler data is characterized by the Doppler noise (see Chapter 5). A key contributor to this noise is the radio link's sensitivity to the charged particles of the interplanetary plasma and Earth's ionosphere. This type of noise is inversely proportional to the square of the frequency of the signal carrier. Of the bands in use in deep space (see Table 1.2), the lower frequency S-band has a typical statistical Doppler noise, in units of velocity, of ~1 millimeter per second (mm/s) at integration times between 10 and 100 s, while the higher frequency X-band links have a typical performance of as much as one order of magnitude better than S-band. Coherent Ka-band links, demonstrated for gravity science on the Cassini, Juno, and BepiColombo spacecraft, further improve the Doppler noise by as much as another order of magnitude as described in Asmar et al. (2005b) and shown in Figure 1.8.

The Sun–Earth–probe (SEP) angle, shown in Figure 3.3, is key to determining the data quality for RS investigations. When the SEP is near 180 deg, or solar opposition, the noise effect on the link from the solar corona or interplanetary plasma is at a minimum. The effect is at its maximum near 0-deg SEP, or solar conjunction, when the ground station is pointed towards the Sun to track the spacecraft signal. Figure 1.8 shows the effect on the Doppler quality as a function of SEP angle in terms of the power spectral density of one-way (for normalization) plasma phase scintillation for the three common deep space links. Figure 1.8 also demonstrates that the higher frequencies are less susceptible to solar plasma noise.

The characteristics of spacecraft orbits are important for Doppler-based investigations. For example, a polar orbit is optimum for acquiring global planetary gravity

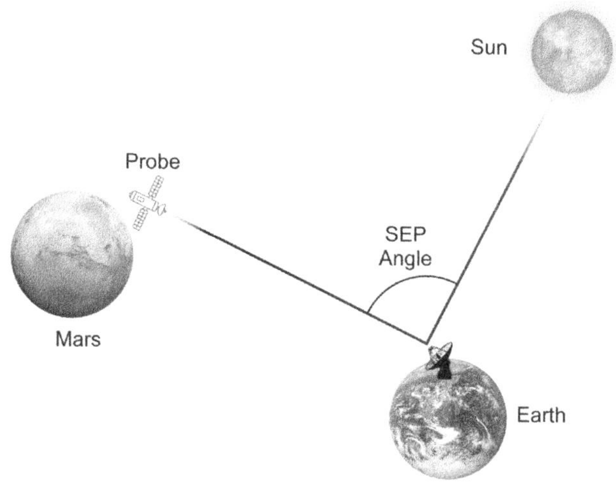

Figure 3.3 Sun–Earth–probe (SEP) is the angle between the Sun, ground station, and the spacecraft and is near 0 at solar conjunctions, where the solar corona impacts the Doppler data quality the most, and 180 deg at solar oppositions, where the solar corona has the minimum impact on the Doppler data quality.

data. For planetary orbiters, the gravity sensitivity depends on whether the orbit is near "face-on" or "edge-on" as seen from Earth, as illustrated in Figure 3.4. During a long orbital mission, the relative geometries of three objects, the Earth, the planet, and the spacecraft, change such that there are times when the spacecraft is always visible from Earth, a face-on geometry, and times when the spacecraft is occulted by the planet at varying angles, an edge-on geometry. Since any Doppler shift is due to relative motion between a transmitter and a receiver along a line-of-sight (LOS), or radial velocity, the sensitivity is maximized during edge-on and minimized during face-on configurations. However, from experience with long-duration Mars orbiters, for example, gravity science teams still acquire valuable data during face-on periods as there often is a residual vector component projection along the LOS.

For gravity investigations, range rate is the primary data type and absolute range is secondary. Spacecraft range data have statistical accuracy of 1 m (1 sigma) at X-band and can be used to improve planetary ephemerides as well as estimating the masses of asteroids, as discussed in Konopliv et al. (2006) and Kuchynka et al. (2014). Deep space ranging at Ka-band has been developed for the first time to be compatible with the BepiColombo mission, reaching accuracy of a few centimeters (see Chapter 5). The data product resulting from crosslink gravity measurements is the DOWR data pioneered by GRACE and GRAIL, also discussed in Chapter 5. GRAIL demonstrated a link quality of 0.03 μm/s with five samples (Konopliv et al. 2014b).

3.2.1 Alternative Basis and Methods

Although the spherical harmonic expansion detailed in Appendix 3A is the most common method for describing planetary gravitational fields when sufficient data are available, there are nonspherical small planetary bodies more suitable for the use of alternate basis functions. Garmier and Barriot (2001) presented a mathematical theory for an ellipsoidal function. Garmier et al. (2002) modeled the elongated Eros gravity

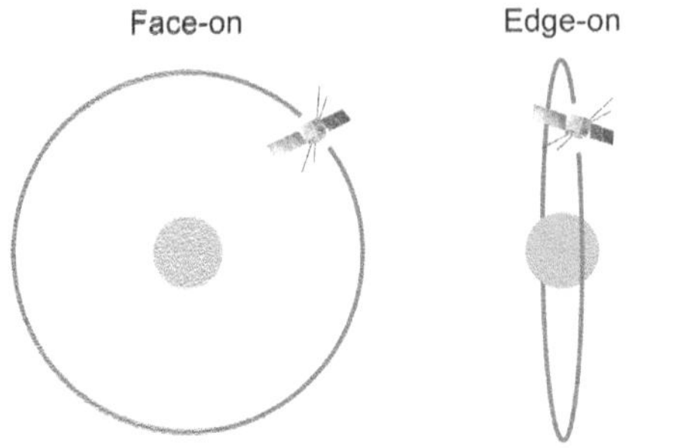

Figure 3.4 Graphical illustration of the face-on versus edge-on spacecraft planetary orbits as viewed from Earth.

field as an ellipsoidal harmonic expansion from the near-Earth asteroid rendezvous (NEAR) spacecraft Doppler tracking data up to degree and order 12.

Park et al. (2014) represented Vesta's gravity field using ellipsoidal harmonics and obtained a 20th degree harmonic field from processing radiometric Doppler and range data from the Dawn mission; gravity anomalies were mapped to a Brillouin ellipsoid (the circumscribing sphere of the body, named after physicist Léon Brillouin) and found to be a substantially closer fit to the surface than the conventional spherical harmonic series, especially near the poles.

In addition to global expansions, there are alternate methods in use for gravity analysis. Barriot and Balmino (1992) describe the LOS technique as a less computationally demanding approach for modeling static local gravity anomalies. The Doppler effect is most sensitive along the LOS between the spacecraft and ground station and least sensitive in the perpendicular plane. Instead of using the full dynamical content of the gravity signal, the technique extracts the residual gravity signal from lateral variations and analyzes its time derivative (Sjogren et al. 1980). This accounts for the effects of the spacecraft altitude and assumes the residual Doppler signal is strongly correlated with local gravity features.

Williams et al. (2008), for example, applied the LOS acceleration profiles for the Mars Global Surveyor (MGS) along with topography data from the Mars Orbiter Laser Altimeter (MOLA) on MGS to estimate the effective elastic thickness for Tharsis and determined the admittance (defined below), with results implying the majority of the topographic rise, was emplaced within the Noachian period.

3.2.2 Tidal Forces and Time Variable Gravity

Global planetary gravitational fields represent static mass distribution. The Newtonian potential theory does not have a time parameter in the equations. However, when gravity measurements are repeated at the same locations and show temporal variability, additional aspects of the body's internal structural dynamics are revealed. Tidal forces on a planetary body from a larger nearby body trigger a response that varies with time as the relative orbital distance changes.

Figures 3.5 and 3.6 show how gravitational parameters for planetary bodies (in this case Saturn's satellites Enceladus and Titan) are derived at various segments of their elliptical orbits. Figure 3.5 depicts a Doppler shift on the radio link transmitted during

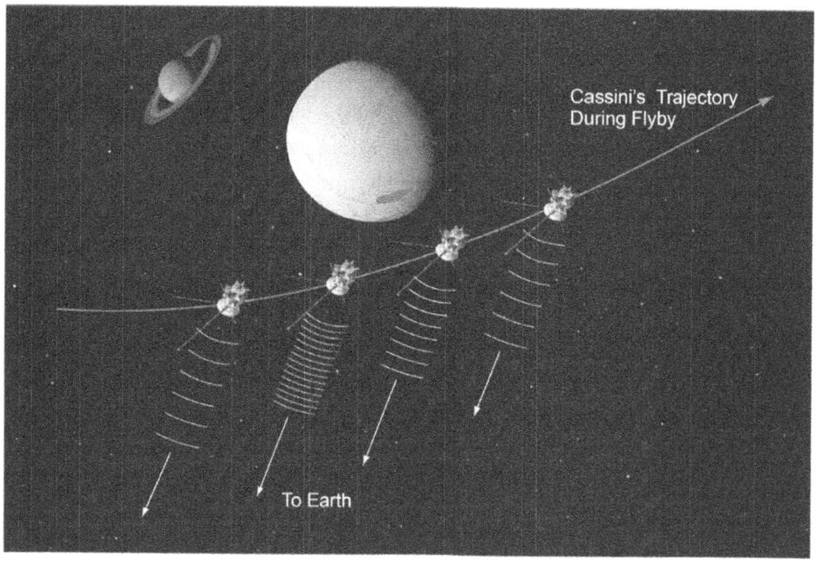

Figure 3.5 Exaggerated graphic illustration of the Doppler shift on the Cassini radio link transmitted to Earth resulting from trajectory perturbations which, in turn, resulted from gravity anomalies at Saturn's moon Enceladus.

Figure 3.6 Illustration of the tidal forces from Saturn acting on Titan at various segments of Titan's elliptical orbit around Saturn. The gravity science team on the Cassini mission measured the gravity field and found it consistent with evidence suggesting the presence of a global subsurface ocean beneath an outer ice shell.

a flyby from the Cassini spacecraft to Earth resulting from trajectory perturbations due to gravity anomalies at Enceladus. Figure 3.6 illustrates the tidal forces from Saturn acting on Titan at various segments of its elliptical orbit around Saturn. Measurements at multiple flybys provided evidence suggesting the presence of subsurface ocean beneath outer ice shells for both satellites (Iess et al. 2012, 2014).

When examining the body's response to external gravitational forces, the dimensionless Love numbers h, k, and l, introduced by Augustus Edward Hough Love for Earth studies, are used to characterize the elastic planetary response to tidal forces (Love 1911; Dehant and Mathews 2015). They characterize the rigidity of a planetary body, which is its susceptibility to deform its shape in response to a tidal potential. For sufficiently large planetary bodies, h describes the elastic radial displacements of the crustal layer relative to an equilibrium pre-tidal ellipsoidal shape, k describes the gravitational modification caused by the deformation (the gravitational potential changes as the body deforms in response to the tidal force), and l, introduced in Shida (1912), describes the elastic tangential displacement.

Examining the interior structure of Enceladus, Hemingway et al. (2018) best describe the equilibrium figure theory and how much of the structure can be deduced from examining tidal and rotational asymmetries in the shape and gravity affecting rotational dynamics. They state:

> ... high internal pressures and low internal viscosities result in relaxation to a figure that approaches the expectation for a strengthless fluid body, in which the inward acceleration due to gravity is everywhere balanced by the gradient in fluid pressure, a condition referred to as hydrostatic equilibrium ... under the influence of self-gravitation alone, the hydrostatic equilibrium figure would be a sphere. Because of their rotation, planetary bodies also experience centrifugal flattening ... satellites in synchronous rotation with their parent bodies also experience permanent elongation along the static tidal axis ... The tidal and rotational deformation results in an asymmetry in the equilibrium figure and consequently the gravitational field ... cause planetary bodies to deviate from spherical symmetry described in terms of disturbances in the potential ... its equilibrium figure will conform to an equipotential. The effect of the disturbing potential is that the surface of the initially spherically symmetric body will no longer be equipotential ... In response to the disturbing potential, the body's figure will relax toward the new equipotential surface. However, the body's mass distribution is altered, effectively causing a small additional disturbance in the gravitational potential.

For an infinitely rigid body, $h = k = l = 0$. For a perfectly homogeneous fluid body, $h = 5/2$ and is lower for bodies with internal density stratification (Hemingway et al. 2018). Furthermore, for perfectly fluid bodies, $h = 1 + k$. For the elastic Earth, the ranges of Love numbers are (Poulsen 2009): $0.304 \leq k \leq 0.312$, $0.616 \leq h \leq 0.624$, and $0.088 \leq l \leq 0.084$.

Since the disturbing gravitational potential is a degree-2 spherical harmonic function, the dominant degree-2 fluid Love numbers are of prime interest. The k_2 Love number is a constant of proportionality between the applied second-degree potential field and the resulting field that it induces at the surface of the planet. Table 3.2 lists k_2 estimates from RS investigations for selected planetary bodies. Note that if there are no lateral, as opposed to radial, variations in the body's properties, the Love numbers of different order are the same for a given degree.

According to Kramm et al. (2011), k_2 measures the level of central condensation of a planet and, the more homogeneous the planet in mass distribution, the bigger the k_2 and maximum homogeneity represented by a planet of constant density, yielding the maximum value of k_2 of 1.5. For a more centrally condensed planet, the Love number decreases and the presence of a core leads to an even smaller Love number. For a homogeneous elastic body,

$$k_2 = 1.5 / \left(1 + \frac{19\mu}{\rho g R}\right),$$

Table 3.2 Love number (k_2) values for selected planetary bodies from RS investigations (subject to update with additional analysis).

Body	k_2	Reference
Mercury	0.53 ± 0.03	Konopliv et al. (2020a)
Venus	0.295 ± 0.066	Konopliv and Yoder (1996)
Moon	0.02405 ± 0.000176 0.02427 ± 0.00026	Konopliv et al. (2013) Lemoine et al. (2013)
Mars	0.169 ± 0.006 0.1697 ± 0.0027	Konopliv et al. (2016) Genova et al. (2016)
Titan	0.616 ± 0.067	Durante et al. (2019)
Saturn	0.390 ± 0.024	Lainey et al. (2017)
Jupiter	0.625 ± 0.063	Iess et al. (2018)

where μ is the body's rigidity, ρ is the mean density, g is the surface gravity, and R is the radius.

The fluid Love numbers are different from the tidal Love numbers. The latter characterize the body's viscoelastic response in shape as well as gravitational potential due to eccentricity tides, which vary the tidal potential as a function of proximity to its parent body. Ignoring these short-timescale variations in shape and gravity, a synchronous satellite in hydrostatic equilibrium thus has a defined mean figure, as described in Hemingway et al. (2018).

Love numbers have a frequency dependence and the fluid Love number is the Love number at zero frequency, or infinite period, that encodes the response of the body to an external potential, allowing an infinite amount of time to respond. Therefore, the fluid Love number is a parametrization of the hydrostatic response of the body to rotation and tidal potential from the main body (planet), and it can be related to hydrostatic values of J_2 and C_{22}. The relationship between the J_2 and the fluid Love is linear; so, with no forcing (no rotation or tides), J_2 will be zero but the fluid Love number does not vanish.

There is also a response of the body to external forcing during the orbital period due to eccentricity of the orbit and, therefore, changing distance to the planet (or other periods). This response is typically smaller than that during the infinite period since the body does not have time to adjust to constantly varying external potential. For fast rotating giant planets, the fluid Love number and the Love number at orbital frequencies of their satellites are nearly the same since the bulk of the gas giants is fluid (see also Wahl et al. 2017).

Ragozzine and Wolf (2009) discuss the formulations if planets were purely spherical masses, they would obey a r^{-2} force law and execute closed elliptical orbits. The presence of external potential causes nonspherical mass effects. They explain that the centrifugal potential of spinning bodies causes rotational flattening and the tidal potential of a nearby mass raises tidal bulges and rotational and tidal bulges create gravitational quadrupole fields (r^{-3}) that lead to orbital precession. Due to the orthogonality of the Legendre polynomials shown in Appendix 3A, if the planet is responding to a second-degree harmonic field, the second-degree harmonic of the planet's gravity field is altered, to first-order. Ragozzine and Wolf (2009) add that

> k_2 is a measure of how the redistribution of mass caused by the applied potential actually affects the external gravity field of the planet ... By examining the variations in k_2 for giant planets we can gain a feel for its expected values and how sensitive it is to internal structure. Even though Jupiter may have a 10 Earth mass core, it is small in comparison to Jupiter's total mass, and thus it has minor effect on the value of k_2. Saturn, on the other hand, has a roughly 20 Earth mass core and is less than 1/3 of Jupiter's mass.

Wieczorek (2007) describes planetary time-variable rotation and the k_2 Love number. The former is measured to investigate the moment of inertia factor when combined with low-degree gravity measurements. Constraints on the density and viscosity profile can be obtained from k_2. The second-degree gravity coefficients determine the location of the principal axes relative to the spin axis and can lead to detection of a possible wobble. The detection of the size of a tidal bulge, for example, can imply a distinction between liquid or solid cores.

Measurement of the precession of the spin axis together with the second-degree gravity field parameter can constrain core size and composition. For detection of a possible Mars wobble, see, for example, Konopliv et al. (2020 and 2006). Solar tide detection for both Mars (Yoder et al. 2003) and Venus (Konopliv and Yoder 1996) were examined to determine the state of the planetary core. For constraining the Mars core size, see Folkner et al. (1997a, 1997b). Modeling of the Galilean satellites using Love numbers is discussed in Section 3.5.5. Modeling of the Chandler Wobble is discussed in Section 3.5.4.

An accurately determined gravity field is also necessary for estimating planetary atmospheric density, as has been done for Venus and Mars, the orientation of the Venus and Mars principal axes of inertia (by the harmonic coefficients of degree-2 of the gravity field), as well as the determination of the masses of the Phobos and Deimos satellites. For the Moon, see the analysis in Williams et al. (2014) of the spherical harmonic second-degree gravity coefficients as well as the Love number describing the lunar elastic response to tidal forces. Combined with determination of the principal moments of inertia from lunar laser ranging data, the density, moment, and Love number resulted in calculated models that have a possible fluid outer core.

Tidal effects can lead to inferences about the body's moment of inertia (MoI). Helled et al. (2011) posit that the moment of inertia of a giant planet reveals important information about the planet's internal density structure and this information is not identical to that contained in the gravitational moments. They examined Jupiter's normalized moment of inertia (NMoI), equal to C/MR^2 (C is the polar moment of inertia, M is Jupiter's mass, and R is Jupiter's mean radius), which can be obtained by measuring Jupiter's pole precession and the Lense–Thirring acceleration of the spacecraft.

Jupiter's moment of inertia together with its mass, shape, size, and gravitational coefficients can provide key information about its current interior structure, leading to an improved understanding of Jupiter's formation and evolution. Detailed interior models using theoretical hydrogen and helium equations of state (EOS) have been used together with Jupiter's measured gravitational field (J_2, J_4, J_6) and other measured physical parameters (e.g., mass, radius, rotation rate) to better constrain the internal structure of Jupiter.

Fortes (2012) estimated Titan's MoI from Cassini's quadrupole gravity measurement to be 0.342 (the MoI of a sphere of uniform density is 2/5), consistent with a differentiated internal structure that includes a shell of water ice overlying a low-density silicate core. Durante et al. (2019) reported a MoI for Titan from a new solution close to 0.341 and note that such significant nonhydrostatic topography is a reminder that hydrostatic equilibrium is not guaranteed, "precluding a definitive determination of the moment of inertia." They compared the nonhydrostatic parts of the gravitational potential with the topography and observed that Titan's topography is highly compensated, which is expected if the ice shell is thin, permitting effective compensation, or the ice shell is stiff, permitting overcompensation that remains effective even for a thicker ice shell.

3.2.3 Covariance Analysis

Geophysicists use the error spectra as the leading evaluation criteria of data quality for gravity experiment and the optimization of their systemic configurations. This is typically shown as the root mean square (RMS) of the magnitude versus degree harmonic of the gravity field solution. For example, in Konopliv et al. (2018), the Ceres gravity field RMS magnitude spectrum of the spherical harmonic coefficients shows the improvement at low-altitude mapping orbit (LAMO) as compared to measurements at high-altitude mapping orbit (HAMO), along with the uncertainties given by the dashed lines.

Furthermore, a comparison is made with a gravity field derived from the shape of the body assuming uniform density and other cases, see Figure 3.7. The field error spectra for different cases are shown in Figure 3.8 along with the nonhydrostatic spectrum showing the removed J_2 and J_4 coefficients, and the Kaula power law (Kaula 1966) that fits the higher degree part of the spectrum. Table 3.3 lists the power law, commonly known as the Kaula law, scale factors for the inner planets and the Moon.

Covariance analysis is a necessary step to producing the error spectra. This analysis provides estimates of how the errors in the parameter produced by the software depend upon the various input information, where the observations are related to such parameters via an information matrix of partial derivatives of observations with respect to parameters, including:

Table 3.3 Kaula power law scale factors for the inner planets and Moon.

Body	Earth scaled power law (K_P/n^2)	Observed power law
Mercury	$6.9 \times 10^{-5}/n^2$	$5 \times 10^{-5}/n^2$ (Konopliv et al. 2020a)
Venus	$1.2 \times 10^{-5}/n^2$	$1.2 \times 10^{-5}/n^2$ (Konopliv et al. 1999)
Moon	$36.4 \times 10^{-5}/n^2$	$25 \times 10^{-5}/n^2$ (Konopliv et al. 2014b)
Mars	$6.9 \times 10^{-5}/n^2$	$122 \times 10^{-5}/n^{2.57}$ (Konopliv et al. 2016)

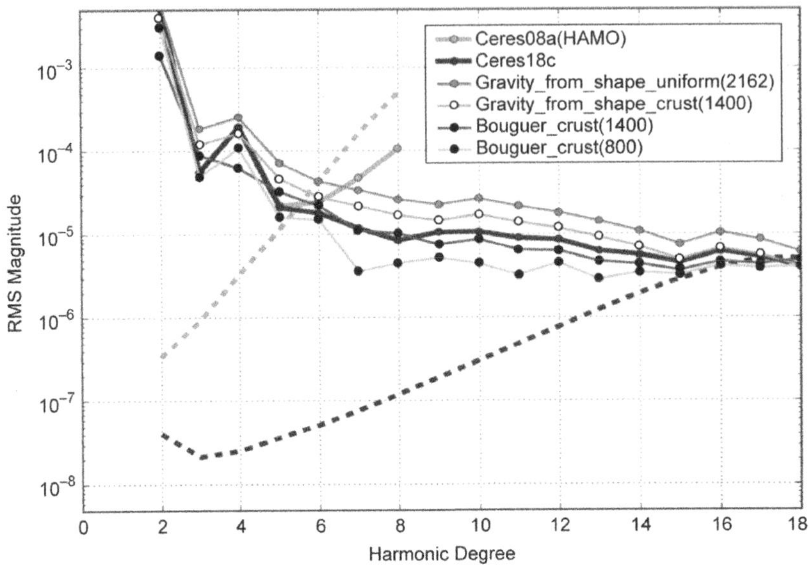

Figure 3.7 Ceres gravity field RMS magnitude spectrum of the spherical harmonic coefficients from Konopliv et al. (2018), showing improvement from the high-altitude (Ceres08a) to the low-altitude phase (Ceres18c), with uncertainties shown by the dashed lines. The gravity amplitudes (blue) are noticeably reduced relative to that derived from shape model (gold) assuming uniform density and for a two-layer model (cyan).

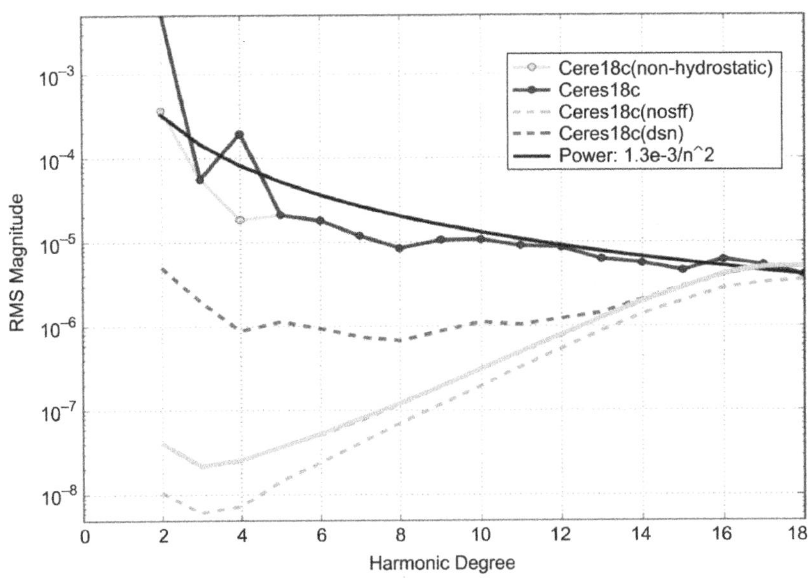

Figure 3.8 Ceres gravity field error spectra for two different cases from Konopliv et al. (2018) where one (labeled nosff) assumes no disruptive attitude control maneuvers and one (labeled dsn) assumes no optical landmark tracking, which shows the significant contribution of the optical landmark data, especially for the low harmonic coefficient solutions. Also shown are the nonhydrostatic spectrum (removed J_2 and J_4 coefficients) and the Kaula power law that fits the higher degree part of the spectrum.

- Orbit geometry
- Measurement accuracy/data quality (Doppler noise)
- Nongravitational forces
- Arc length
- Mission duration

Ermakov et al. (2018) built on the knowledge that, for the terrestrial planets, the power spectrum of the gravity field follows a power law, the Kaula rule, and a similar rule was derived for topography (Vening Meinesz, 1951) and sought a generalization of the power law dependence of the gravity and topography spectra for solid surface Solar System bodies. They found a good fit, for regularly shaped bodies, using a general scaling based on the body's radius and mean density.

Then, Bills and Ermakov (2019) derived analytic expressions for estimating gravity field error spectra for the purpose of experiment planning or mission proposal activities. Simple analytic expressions were derived and yielded estimates of the resulting error spectra, with reasonable accuracy. An application to Vesta was used as a validation exercise and showed possible models for Mars and Venus.

The examined cases are as follows:

1. The traditional approach of a single spacecraft orbiting a target body and tracked from Earth.
2. Two co-orbiting spacecraft, with measurements of the range-rate variations between them, as used by GRACE and GRAIL (see Section 3.3).
3. The gradient of the gravitational acceleration vector measured via onboard gradiometers as in the example of the Gravity field and steady-state Ocean Circulation Explorer (GOCE) mission.

For a two-dimensional representation of the gravitational potential of a target body, covariance analyses are applied for each of the three configurations; then analytical forms for the resulting error spectra are obtained. Figures 3.9–3.12 (contributed by B. Bills) show illustrative cases for hypothetical missions to Earth, Moon, Venus, and Mars.

A pattern appears where, in each case, the satellite pair outperforms the single satellite case due to a lower velocity error for the intersatellite tracking at the Ka-band compared to the interplanetary tracking at the X-band. The gradiometer performance is better than the other two, though it suffers more lack of accuracy from increasing orbital altitude than do the other approaches. Furthermore, according to Bills and Ermakov (2019), smaller orbits yield better results, not considering the important atmospheric drag effects.

3.3 Earth and Moon Gravity Measurements and the Development of Crosslinks

Many concepts related to the internal structure of rocky planetary bodies are based on models initially developed for Earth using various methods, such as seismic wave analysis and the subsequent advances in precision measurements of plate tectonics. Milsom (2018) best summarizes the progress in the hunt for Earth's gravity as a century of work by Copernicus, Tycho Brahe, and Kepler to "destroy Ptolemy's geocentric universe of spheres and circles and replace it with a heliocentric universe of elliptical orbits."

In Milsom's narrative, Galileo, a contemporary of Kepler, introduced the idea of acceleration and established the relationship between the time of the fall of a body and the time of the swing of a pendulum, laying the ground work for others after him to estimate the value of g, the gravitational acceleration near the Earth's surface, and other generations, including Newton, to connect it with G, the universal gravitational constant.

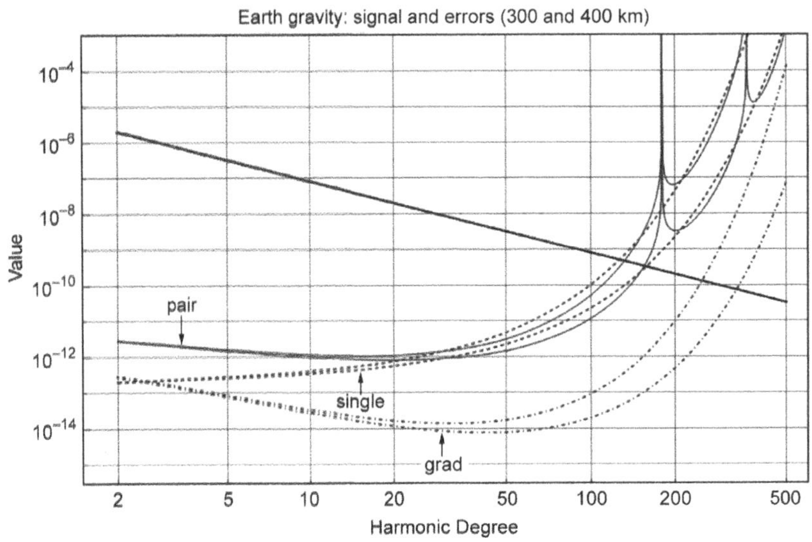

Figure 3.9 Gravity signal and error plots showing the signal amplitude spectrum, or Kaula's power law, and error spectra at two different orbital altitudes for single satellite tracking, co-orbital pair, and gradiometer for Earth at altitudes of 300 and 400 km.

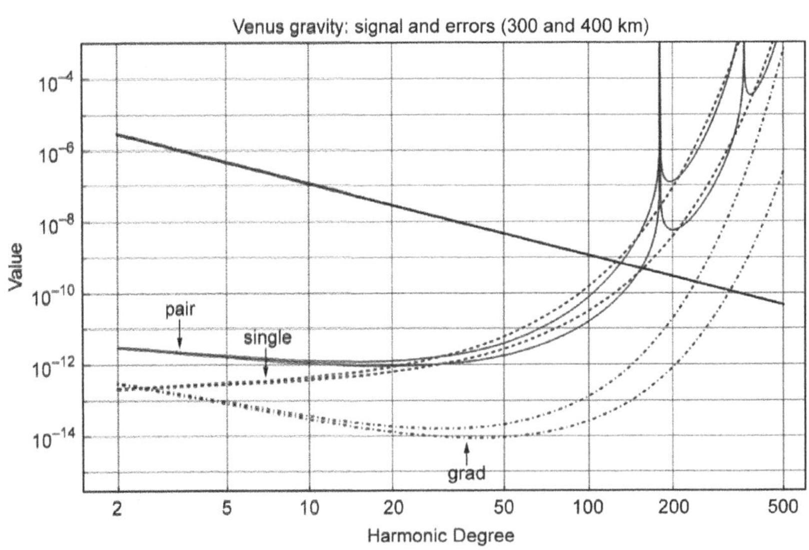

Figure 3.10 Gravity signal and error plots showing the signal amplitude spectrum and error spectra at two different orbital altitudes for single satellite tracking, co-orbital pair, and gradiometer for Venus at altitudes are 300 and 400 km.

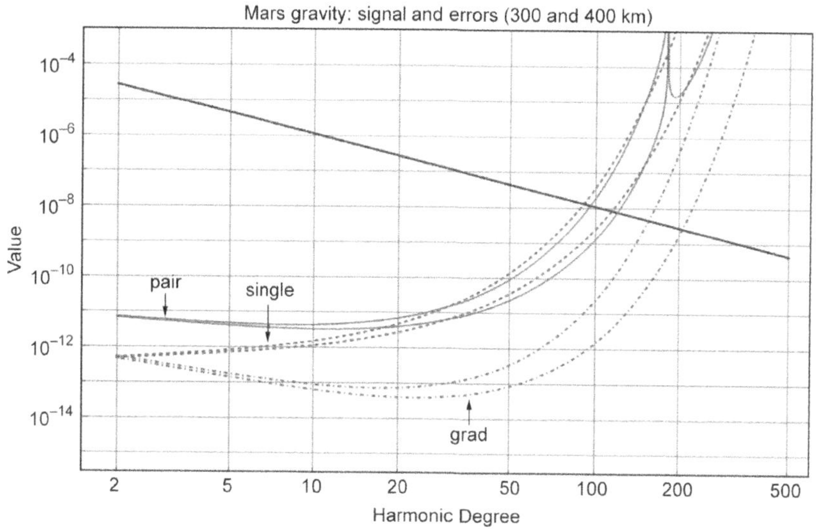

Figure 3.11 Gravity signal and error plots showing the signal amplitude spectrum and error spectra at two different orbital altitudes for single satellite tracking, co-orbital pair, and gradiometer for Mars at altitudes are 300 and 400 km.

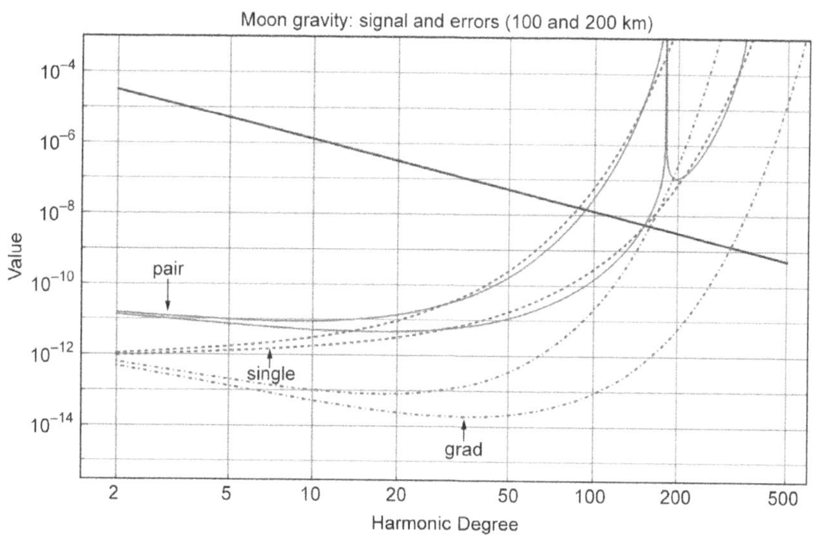

Figure 3.12 Gravity signal and error plots showing the signal amplitude spectrum and error spectra at two different orbital altitudes for single satellite tracking, co-orbital pair, and gradiometer for the Mars at altitudes of 100 and 200 km.

Another concept from Earth science applicable to rocky planets is isostatic equilibrium, the gravitational equilibrium between the crust and mantle with three principal models (Turcotte and Schubert 2014; Watts 2001): Airy isostasy, where topographic heights, called gravity loads, are compensated by variation in the thickness of a crust of constant density via lateral variations in the crustal thickness; Pratt isostasy, where topographic heights are compensated by lateral variations in crustal density;

and flexural isostasy, where the lithosphere acts as an elastic plate with rigidity that distributes the localized loads over a broader region by flexing.

The lunar gravity measurement method is typical for one spacecraft with radio link(s) to ground stations. There are two planetary bodies where this method has significant practical limitations, the Moon and the Earth. The method is limited at the Moon because a LOS is required from a spacecraft to an Earth station to observe the Doppler effect and this can only work from the nearside of the Moon and not from the farside that never faces Earth. The method is limited for studying the Earth because any one tracking station on Earth would have a LOS to an orbiting spacecraft overflying the station for only a few minutes and, thus, a very large number of globally distributed and time-synchronized stations would be required. If there were a Deep Space Network (DSN) station on the Moon or Mars, for example, tracking an Earth-orbiting spacecraft, an Earth gravitational field would be measured via a method equivalent to tracking a lunar spacecraft from Earth.

These challenges led to innovative development; the links from a spacecraft to a ground station were replaced by links from one spacecraft to another, typically identical spacecraft and in formation-flying configuration. The radiometric properties of the links from one spacecraft to another, called crosslinks, are examined in this section.

Pioneering the use of spacecraft-to-spacecraft radiometric tracking as the primary measurement method, an accelerometer for calibrating nongravitational forces, and the global positioning system (GPS) for timing synchronization, the GRACE mission has generated a very high-resolution gravitational field of Earth (Tapley et al. 2004; Watkins et al. 1995). One of the most notable results from the long period of observations with GRACE is the study of the movement of Earth's water and ice. Using the concept of time-varying gravity, a high-resolution gravity map is produced monthly and the comparison from month to month shows short-term gravitational fluctuations due to the movement of water, the effect of sea-level rise, polar ice-cap mass loss, or changes in large-scale water storage, among other effects (Ivins et al. 2011; Han et al. 2004).

The GRAIL mission has taken the GRACE measurement concept, without the accelerometer and GPS timing system, and successfully applied it to the Moon. Park et al. (2012) and Asmar et al. (2013b) describe the method and design of the GRAIL system for high-resolution gravity field measurements. In this method, illustrated in Figure 3.1, two nearly identical (sometimes called mirror identical) spacecraft operate in precision formation-flying in order to traverse the same ground tracks while maintaining radio links between them. The DOWR measurements are generated on both spacecraft for post-processing to obtain fully calibrated range-rate observables. The accuracy of the GRAIL measurements, driven by one ultra-stable oscillator (USO) on each spacecraft, has RMS range-rate residuals from the Ka-band instrumentation of 0.02–0.05 μm/s (Zuber et al. 2013b).

Another example of spacecraft-to-spacecraft precision links, in this case laser links, is the planned LISA European mission designed to measure gravitational waves (Danzmann 2000). The SELenological and ENgineering Explorer (SELENE; also known as Kaguya) mission (Kato et al. 2008) also used a form of spacecraft-to-spacecraft links to investigate global lunar gravity (see Section 3.5.1).

A different method of measuring planetary gravity fields is the use of gravity gradiometers, which have been flown on board Earth missions to measure the variations in gravitational acceleration, specifically the rate of change of the gravitational acceleration, or gradient, in the spatial domain. The variations are caused by subsurface differences in the density of material as well centripetal and tidal forces, among other factors.

A recent example of a mission utilizing gravity gradiometers is the European Space Agency's (ESA) GOCE mission, which carried a set of six 3-axis accelerometers mounted in a diamond configuration in an ultra-stable structure. According to Drinkwater et al. (2006), each accelerometer pair formed a gradiometer arm 50-cm long with the difference in gravitational pull measured between the two ends. Three arms were mounted orthogonally along-track, cross-track, and vertically. An onboard GPS receiver used as a spacecraft-to-spacecraft tracking instrument supplemented the gradiometer measurements. A comparison of the performance and data quality between spacecraft-to-ground tracking, crosslinks, and gravity gradiometers is shown in Section 3.2.

3.4 Shape and Topography Data for Interpretation of Gravity Measurements

Once a gravity model of a planetary body is produced, a key step in the geophysical interpretation process is a comparison of the gravity anomalies with known surface or subsurface features from available topography, imaging, or other data. Accurate knowledge of the body's shape is needed to properly account for the gravity created by topographic masses. In addition, the body's shape itself bears information about planetary internal structure.

The terms shape and topography are sometimes used interchangeably in the literature. Here, shape refers to a three-dimensional representation of the body surface, and topography is defined as the elevation difference between the shape and a reference surface. The reference surface could be defined by a hypothetical surface of the body if it were in a state of hydrostatic equilibrium. The reference surface can also be defined as an equipotential surface.

For the Earth, this is the geoid, which is often described in terms of the global mean sea level, and is irregular due to the uneven distribution of mass. It is possible to make inferences about the internal structure of a planetary body based on its shape alone (e.g., Ermakov et al. 2018). For example, the body's rotationally induced oblateness depends on its differentiation state, that is, the increase of density with depth; assuming all other parameters being equal, a more differentiated body would be less oblate (see Dermott (1979) for a two-layer model).

Topography data can also be described in the same mathematical field representation as gravity using spherical harmonic expansion (Esposito et al. 1992; Rappaport and Plaut 1994; Rappaport et al. 1999; Wieczorek 2007). Correlation of a gravitational field with a topography field in the spherical harmonic domain is a powerful tool for deducing information about the internal structure of the planet (e.g., McGovern et al. 2004; Wieczorek 2007). This information can be further combined with optical or radar maps of surface features, as well as other available geological, geochemical, or geophysical information, to place constraints on models of the differentiation of interior layers and the chemical composition and physical state of the interior.

Knowledge of planetary shape helps predict the magnitude of the perturbations in the motion of the spacecraft due to any non-sphericity of a body's gravity field (Ermakov et al. 2018). For the terrestrial planets, the gravity field's and topography's power spectra follow a power law that scales as g^{-2} (g is the surface gravity) known as the Kaula rule (Kaula 1963).

Ermakov et al. (2018) found that for minor bodies, the higher the density of a minor body, the higher the amplitude of topography it is expected to have. However, it is expected that the topography of minor bodies will be supported mostly by their frictional strength (Melosh 2011). Since the coefficient of friction is nearly independent of density, Ermakov et al. (2018) concluded not only frictional strength supports the topography at least for the selected minor bodies.

On the other hand, for planets, whose topography is in the gravity-dominated regime, higher density leads to an increase of surface gravity, which limits the amplitude of topography. Therefore, as expected, amplitude of planetary topography was found to be negatively correlated with their mean densities. In addition, Ermakov et al. (2018) found the wavelength dependence is a little steeper for minor bodies than for planets, which implies that simple rescaling of the power law from the terrestrial planets to asteroids will not necessarily give an accurate estimate of the asteroids' topography RMS spectra. Finally, Ermakov et al. (2018) found that for irregularly shaped bodies, the gravity power spectrum is no longer a power law even if their topography spectrum is a power law due to the nonlinearity of the gravity–topography relationship.

Gravity–topography admittance is a useful quantity as it depends on the mechanism of topographic support (Nimmo 2002). Gravity–topography admittance is defined as the wavelength-dependent ratio of powers of gravity and topography. Admittance can be used to study whether or not the topography is isostatically compensated (see Park et al. 2016; Ermakov et al. 2017; Sori et al. 2018). In addition, admittance can be used to constrain the density profile (Besserer et al. 2014).

To study the subsurface density heterogeneity in the spatial domain, several topography-related corrections need to be applied to the observed gravity field. First, the measurements need to be reduced to a common reference surface. This correction is known as the free-air correction. Second, the effect of topography masses (e.g., Heiskanen and Moritz 1967; Wieczorek and Phillips 1998) needs to be subtracted from the observed gravity field. This step is known as the Bouguer correction and involves choosing appropriate density from the crust, which may or may not be known.

Therefore, the Bouguer anomaly is a gravity anomaly corrected for the altitude above the surface and for the effect of the gravity due to the topography. If the crustal density is not known, it can be chosen empirically by minimizing the correlation between the Bouguer anomaly and topography as the Bouguer correction is proportional to crustal density. This, however, involves an assumption that the subsurface heterogeneities are also not correlated with topography.

A classic example of a subsurface structure inference using Bouguer anomalies is mass concentrations, or mascons, on the Moon. A mascon is a region or structure of the crust where subsurface material denser than its surroundings causes an increase in the gravity but has no observed effect in the topography or imagery. This discovery resulted from early radiometric gravity investigation of the Moon by Muller and Sjogren (1968) and can have possible implications to other rocky bodies (O'Leary et al. 1969; Sjogren 1979).

Figures 3.13(a–c) illustrates how gravity instrumentation such as that of GRAIL can detect topographic and subsurface features and density anomalies using Doppler

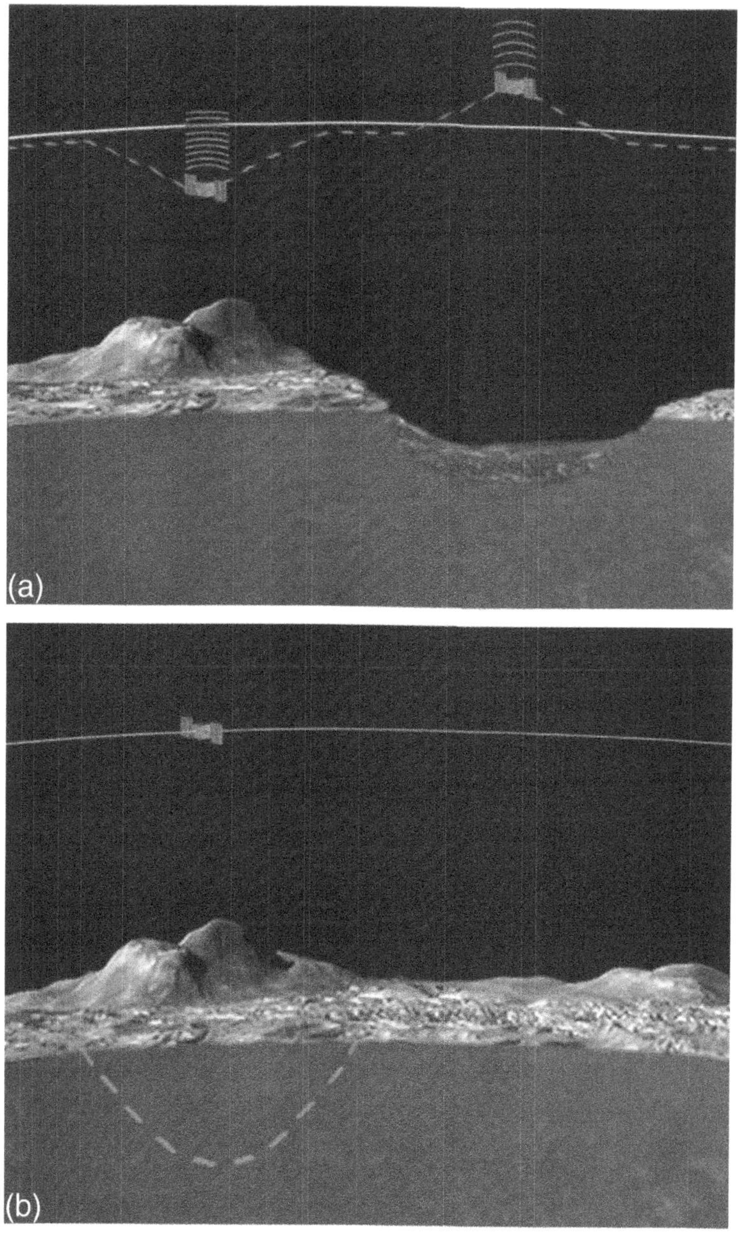

Figure 3.13 Illustration of gravity instrumentation detecting topographic and/or subsurface features and density anomalies using Doppler variations. Panel (a) shows a spacecraft flying over positive and negative gravity anomalies also observed in surface topography and imagery (mountain and crater) resulting in proportional Doppler variations. Panel (b) shows a spacecraft flying over a topographic feature that has been compensated via sub-surface density variation with no changes in the Doppler effect due to the feature. Panel (c) shows a spacecraft flying over a mascon, a subsurface mass concentration that has no representation or detection in the topography or surface imagery, with a detectable variation in the Doppler signal.

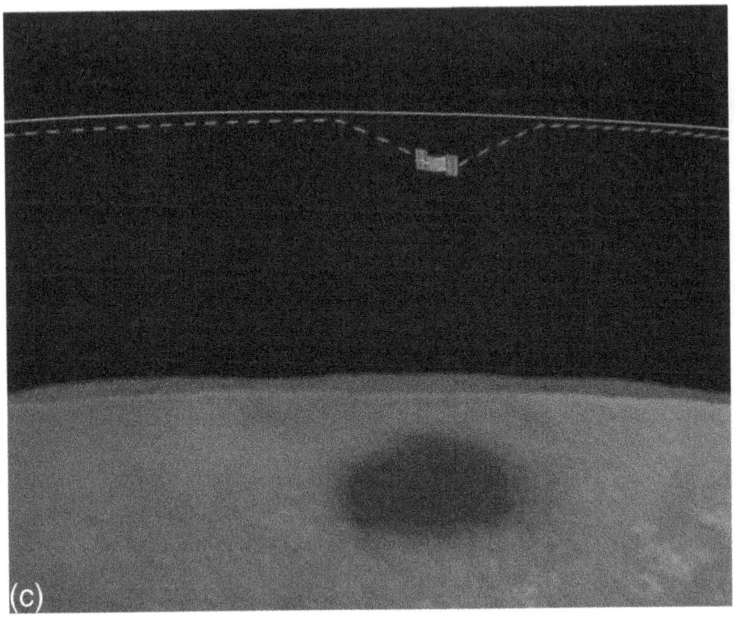

Figure 3.13 (Cont'd)

variations. Figure 3.14(a–d) illustrates gravity instrumentation sensitivity to nongravitational effects at the Moon due to solar radiation pressure (SRP), the release of gas spacecraft equipment in the vacuum of space, lunar albedo, and infrared radiation (see details in Asmar et al. 2013b). Figure 3.15 illustrates how gravity and topography data correlated for the Moon and rocky planets to deduce internal structure models.

Lunar mascons are characterized by the presence of a central positive Bouguer anomaly surrounded by a ring of a negative gravity anomaly. Such gravity patterns are commonly observed in large basins where the low topography would imply a low gravity signal but the opposite result, a positive gravity anomaly, uncovers a density anomaly within the crust or upper mantle, supported by the lithosphere.

A team from the GRAIL mission attempted to explain the origins of the lunar mascons. Melosh et al. (2013) proposed that the mascon gravity pattern results from impact basin excavation and collapse, which is followed by isostatic adjustment, cooling, and contraction of a melt pool. Melosh et al. (2013) found that the key parameters that control the mascon formation are the impactor energy, the thermal gradient within the crust at the time of impact, the crustal thickness, and the extent of volcanic fill. This phenomenon further shows the power of gravity science for deducing internal structures since the hidden density anomalies are not otherwise detectable on optical or topographic maps.

The fact that Earth is geologically active makes finding mascons on Earth less likely. Karimi and Lewis (2018) demonstrated that the Moon's relatively cold thermal state, anhydrous rheology, and low gravitational acceleration inhibit the efficient relaxation of surface/subsurface topography, making it well suited to the formation of mascon, while large craters on Mars, Venus, and Mercury undergo much faster relaxation than the Moon and exhibit smaller likelihood of mascon formation.

Shape and topography can be determined from various methods, including imagery, altimetry, ground-based radar, or radio occultations, which are described in the following sections.

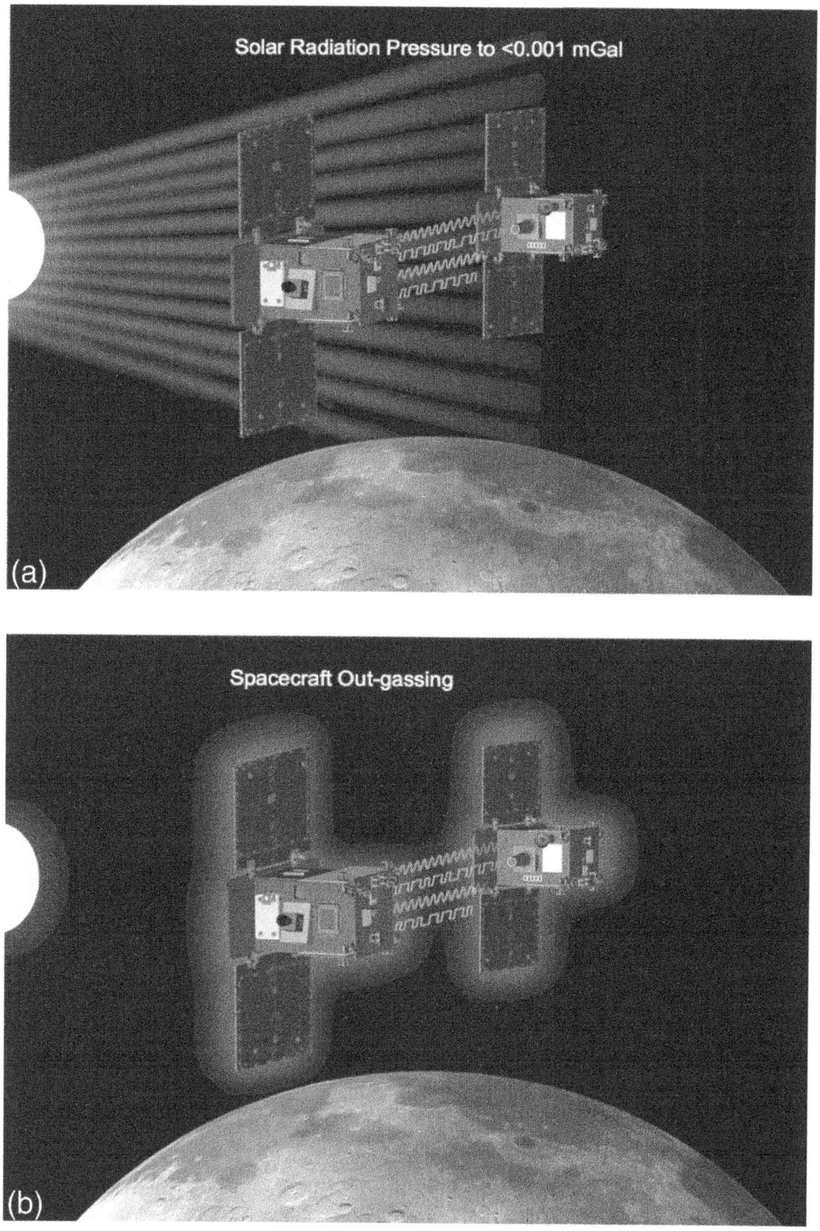

Figure 3.14 Illustrations of gravity instrumentation sensitivity to nongravitational effects at the Moon. Panel (a) shows the forces on the spacecraft surface due to solar radiation pressure (SRP). Panel (b) shows the forces on the spacecraft due to the release of gas from spacecraft equipment in the vacuum of space. Panel (c) shows the forces on the spacecraft due to lunar albedo. Panel (d) shows the forces on the spacecraft due to infrared radiation.

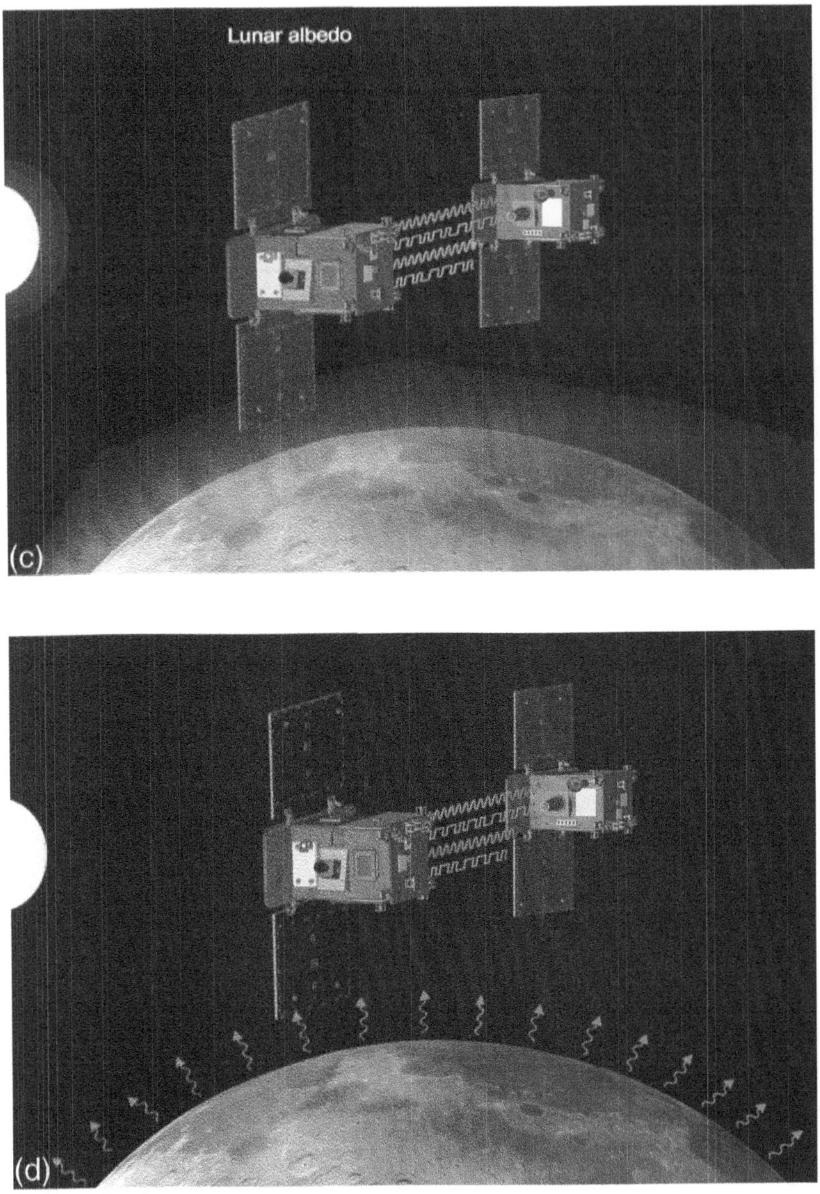

Figure 3.14 (Cont'd)

3.4.1 Imagery

Gaskell et al. (2008) discuss advances in the imagery and characterization of small body surfaces with stereo-photo-clinometry (SPC). Landmark maps, along with topography maps of resolution comparable to that of the images, can accurately be located on a global reference frame for scientific and navigation applications. Jaumann et al. (2012) applied imagery techniques to examine Vesta's shape and morphology from data acquired by the Dawn mission.

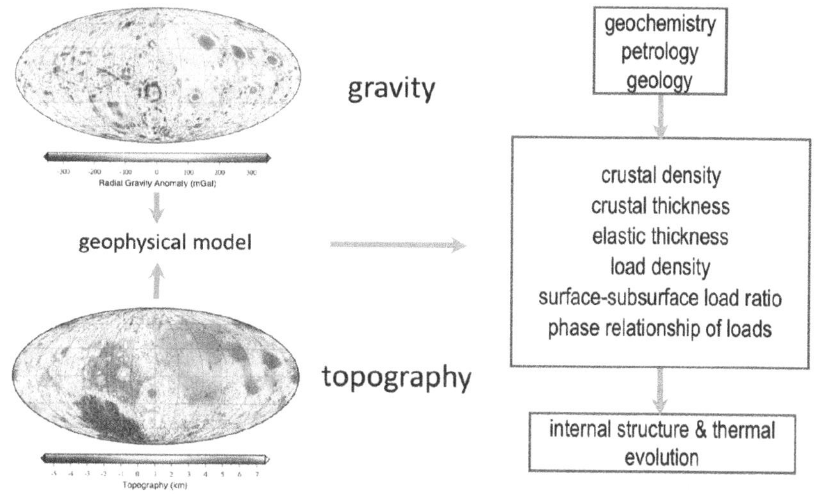

Figure 3.15 Illustrations of lunar gravity, the correlation of the gravity, and topography data, when available in comparable resolution using geophysical models to deduce important parameters for the internal structure.

Observations confirmed the large impact basin, Rheasilvia, and revealed evidence for an earlier, underlying large basin, Veneneia. Prior to the Dawn mission to Vesta and Ceres, Thomas et al. (2005) investigated Hubble telescope images to look for evidence of the differentiation of Ceres as revealed by its shape. Castillo-Rogez and McCord (2010) modeled Ceres' thermo-physical-chemical evolution based on shape measurements.

Stereo imaging techniques have also led to the production of two-dimensional long-wavelength topography, or limb profiles, of the satellites of the outer planets. Nimmo et al. (2010) found these limb profiles important because (1) the spherical harmonic degree-2 topography may be used to infer the moment of inertia of the satellites; (2) they may provide evidence of other physical processes, such as tidal heating, impact cratering, flexure, and/or relaxation processes; and (3) uncompensated long-wavelength topography can complicate the interpretation of degree-2 gravity anomalies and their implications for internal structures. In Nimmo et al. (2011), limb profiles were applied to produce the degree-2 shapes of Mimas, Enceladus, and Tethys.

3.4.2 Altimetry

The MOLA (Zuber et al. 1992; Smith et al. 1999) is an example of a superior data set available to planetary scientists. The instrument produced global high-resolution topography of Mars from elevation data at a level suitable for detailed scientific investigations. The literature related to data analysis and scientific interpretation based on the MOLA data is very large, especially when combined with gravity data for interior structure analysis.

Zuber et al. (2012) reported that the laser altimetry on board the MErcury Surface, Space ENvironment, GEochemistry, and Ranging (MESSENGER) spacecraft had yielded a topographic model of the northern hemisphere of Mercury with the most prominent feature being an extensive lowland at high northern latitudes that hosts the volcanic northern plains.

3.4.3 Space-based Radar

Shape can also be reconstructed using space-based radar assets. Corlies et al. (2017) used the synthetic aperture radar (SAR) technique to create the shape model of Titan using Cassini radar data (e.g., Elachi et al. 2004). This model was subsequently used to compute gravity anomalies (Durante et al. 2019). For Titan, the correlation between the low-degree shape and low-degree gravity also observed by Cassini is close to zero indicating the importance of subsurface density inhomogeneities responsible for gravity anomalies.

3.4.4 Radio Occultations

Like the stellar occultation timing techniques mentioned in Chapter 2, radio occultations can be used for precision timing of ingress and egress events to produce shape models. Cain et al. (1972) produced a Mars shape model from the Mariner 9 occultation data, one of the earliest missions to perform RS experiments, and later using Viking orbiter data (Lindal et al. 1979). The analysis used the assumption that the disk of Mars acts as a knife edge in the ray path, and Fresnel limb diffraction theory dictates that the received power at grazing incidence will be 6 dB below the level in free space; the small atmospheric effect was ignored.

Asmar et al. (1999) applied the same technique to the Moon to improve on the high-altitude topography of the Moon at the time, using Lunar Prospector occultation data. Perry et al. (2011, 2015) also applied MESSENGER radio occultation timing to Mercury to better than 0.1 s. The local topography was evaluated using images to identify the high-elevation feature that defines the radio frequency (RF) path or using altimeter data to quantify surface roughness. Radius measurements are accurate to 150 m.

3.4.5 Ground-based Radar

The DSN-based Goldstone Solar System Radar (Slade et al. 2010; Ostro 1993) has been used to determine the shapes of planetary bodies. Anderson et al. (1996) describe an investigation of the shape and orientation of Mercury from radar ranging data. In another application, Ostro et al. (2000) describe 2380-MHz radar observations of main-belt M-class asteroid 216 Kleopatra that revealed a dumbbell-shaped object. Interestingly, stellar occultation timing analysis had also suggested an elongated shape (Dunham 1981).

For an investigation of Mars, Esposito (Esposito et al. 1992) describes the method and limitations as the range component of the radar echo provides an elevation measurement for a patch of the ground located within the resolution cell or radar footprint; however, there is some uncertainty with this process that would have to be accounted for. In addition, the global accuracy depends on the accuracy of the planetary ephemeris.

3.4.6 Examples of Results of Gravity–Topography Analysis

In their study of Mars, Esposito et al. (1992) list a set of basic output of the comparison between the gravity and shape (or topography) to include:

- Global figure from estimates of the polar and equatorial radii from a shape model as well as oblateness estimates from a lower degrees gravity field.

- Mean density from the ratio of the bulk mass to the total volume derived from a shape model—what they call the degree zero harmonics of gravity and topography.
- Rotational period, which can be obtained from radio tracking of stationary landers or landmark tracking when possible.
- Center of figure versus CM, where the former is derived from the first-degree topography harmonics and the latter from the equivalent gravity parameter.

Examples of results of correlating gravity and topography/shape models include:

- Maps of crustal thickness for the Moon. The GRAIL team used the coherence function between the observed lunar gravity and the gravity predicted from unit density topography to show the high percentage of the gravitational signature associated with topography; the remaining small amount represents fine details of subsurface structure not previously resolved (Zuber et al. 2013b).
- Maps of crustal thickness for Mars (Neumann et al. 2004).
- The use of the admittance function to estimate the mean crustal thickness and density of the Martian hemispheric dichotomy (Nimmo et al. 2010). Admittance estimates from LOS acceleration profiles of the MGS spacecraft are used to constrain the mean crustal thickness and surface density centered on the hemispheric dichotomy.
- Lithospheric thickness estimates for Venus using spatiospectral localization of harmonic data (Simons et al. 1997).
- Evidence of large empty lava tubes on the Moon using GRAIL gravity (Chappaz et al. 2017).
- Mechanisms for support of topographic loads, such as Airy or Pratt isostatic compensation and dynamic or thermal support and internal composition for Europa (Anderson et al. 1998).
- Iess et al. (2010) used Cassini tracking data to determine Titan's mass and gravity harmonics to degree 3 and found the quadrupole field is consistent with a hydrostatically relaxed body shaped by tidal and rotational effects.
- Bills et al. (2014) examined the gravity and topography of Vesta based on Dawn data and observed that the gravity is highly correlated with the observed topography, suggesting little lateral variation in density. The variance spectra of both gravity and topography follow power laws similar to those seen for the Moon, Mars, Venus, and Earth, but different in that both gravity and topography are significantly anisotropic, with more north–south variation than east–west variation.

3.5 Application to Solar System Bodies

Table 3.1 lists planetary gravity experiments carried out by or planned for Solar System missions. These experiments began in 1962 with the Mariner 2 mission to Venus, the first successful planetary flyby mission. Using the Doppler profiles from the Mariner radio signal generated at the ground station, Anderson and Warner (1966) made order-of-magnitude improvements over the previous Earth-based determinations of the masses of the Moon and Venus. An estimate was also made of the offset between the lunar center of gravity and center of figure. Subsequently, Mariner 4 at Mars, Mariner 5 at Venus, and Mariner 9 at Mars provided the masses and first estimates of low-order gravitational harmonics for Venus and Mars (Anderson et al. 1974).

3.5.1 Moon

The Moon's gravitational field has been studied extensively since those early missions. The discovery of mascons is discussed in Section 3.4. The Lunar Prospector mission (1998–1999) produced a degree and order 165 gravitational field when combined with tracking data from previous missions to the Moon without aliasing when evaluated up to degree and order 110 (Konopliv et al. 2001). This was the last mission to examine lunar gravity using the LOS method of a radio link from a single spacecraft to a station on Earth. As discussed in Section 3.3, the absence of a direct measurement of farside gravity anomalies made the global solution dominated by high-quality data from the nearside.

A new method was needed to peek into the farside, while still using Doppler tracking. The Japanese Space Agency's (JAXA) SELENE mission led to a significant improvement in the farside gravity field of the Moon via Doppler tracking of a relay subsatellite placed in view of the farside, producing a data type called four-way Doppler (defined in Chapter 5). For the first time, mascons were fully resolved on the lunar farside for several basins (Namiki et al. 2009).

Matsumoto et al. (2010) produced a spherical harmonic solution of the lunar gravity field to degree and order 100 with gravity anomaly errors predicted from the error covariance, of 26 mGal and 35 mGal for the nearside and the farside, respectively. Goossens et al. (2011) produced another degree-100 field with an emphasis on using same-beam S-band differential VLBI data obtained via the SELENE mission; improved orbit determination (OD) also led to improved gravity field models.

The GRAIL mission to the Moon used another method developed for the GRACE Earth mission to measure the signal at the lunar farside. GRAIL used two identical spacecraft that tracked each other, and the resulting spacecraft-to-spacecraft radiometric data were used to investigate the structure of the lunar interior by producing a detailed map of the gravity field; the resulting model of the interior was used to address outstanding questions regarding the Moon's thermal evolution.

Each GRAIL orbiter contained a Lunar Gravity Ranging System instrument that conducted dual-one-way ranging measurements to precisely measure the relative motions between the spacecraft. These measurements were then used to develop the lunar gravity field map (Zuber 2013a and 2013b). The short mission, 3 months of a prime mission at an average altitude of 50 km and 3 months of an extended mission at an average altitude of 25 km, had a declared mission objective of reaching a 180-deg gravitational field. By the end of the science analysis phase of the mission, after publishing several intermediate fields, the GRAIL team produced a field of degree and order 1500, an order of magnitude higher resolution than the declared objective.

The GRAIL field revealed previously unresolved features, including tectonic structures, volcanic landforms, basin rings, crater central peaks, and numerous simple craters (Lemoine et al. 2013; Zuber et al. 2013b, 2016; Konopliv et al. 2014b). Figure 1.5 (in Chapter 1) shows a high-resolution lunar gravity field from the GRAIL mission. The tracking data from each GRAIL orbiter to the ground stations were also incorporated in the processing.

3.5.2 Mercury

Mercury was observed by two Mariner 10 flybys in 1974 and 1975, and analysis yielded the mass and dynamical oblateness of Mercury (Howard et al. 1974b; Esposito et al.

1976). Further analysis produced a mass accurate to 0.4%, second-order coefficients of the gravity field, and identification of a gravity anomaly in the vicinity of the spacecraft's closest approach point (Anderson et al. 1987b). Radio tracking of the MESSENGER spacecraft for more than 4 years provided a model of Mercury's gravity field and revealed several large Northern Hemisphere gravity anomalies. The low-degree gravity field, combined with planetary spin parameters, yielded the moment of inertia $C/MR^2 = 0.353 \pm 0.017$. A model for Mercury's radial density distribution includes a solid silicate crust, a mantle overlying a solid iron-sulfide layer, an iron-rich liquid outer core, and perhaps a solid inner core (Smith et al. 2012).

According to Mazarico et al. (2016), the gravity and topography data provided important geophysical constraints to better understand Mercury and its evolution, from its deep internal structure, when combined with ground-based radar observations, to its crust. Lower periapsis altitudes in the final year greatly enhanced the sensitivity to shorter wavelength gravity anomalies. A field solution to degree and order 100 was obtained. The team computed a degree strength map from the full covariance matrix obtained during the inversion; the degree strength at each location is the degree at which the computed error spectrum intersects the Kaula constraint (Kaula 1966).

Genova and Iess (2013) used the altimetric and radio observables to retrieve the topography and the gravity fields of Mercury to test the OD procedures developed in view of a similar and more accurate investigation on ESA's BepiColombo mission to Mercury. Launched in 2018, ESA's BepiColombo mission to Mercury will carry out precise range and range rate measurements via the Mercury Orbiter Radio Experiment (MORE). MORE addresses the mission's scientific goals in geodesy, geophysics, and fundamental physics. The key instrument is a Ka/Ka band digital transponder enabling high-phase coherence between uplink and downlink carriers and supporting a wideband ranging tone. This is complemented by a ground system based on the simultaneous transmission and reception of multiple frequencies at X- and Ka-band. The new wideband ranging system is designed for an end-to-end accuracy of ~20 cm using integration times of a few seconds (Iess et al. 2009).

3.5.3 Venus

Since the first Mariner missions to Venus, two long-term orbiters, Pioneer Venus Orbiter (1979–1992) and Magellan (1990–1994), performed detailed gravity experiments. The combined data from these two missions produced a degree and order 180 spherical harmonic gravity model, benefitting from lowering Magellan's altitude for specific gravity campaign periods (Konopliv et al. 1999). The Magellan radar data also resulted in a degree and order 360 spherical harmonic topography model (Rappaport et al. 1999). Due to comparable resolution in the available gravity and topography fields, various interesting features were examined by correlating the two fields, including the Ishtar Terra plateau (Sjogren et al. 1984), the Beta and Atla Regions (Sjogren et al. 1983), Aphrodite Terra (Black et al. 1991), and others (Smrekar and Phillips 1991).

3.5.4 Mars

Mars has had the highest number of orbiting spacecraft that have contributed to gravity measurements starting with Mariner 9 in 1971 and the two Viking orbiters arriving in

1976. The Mars Global Surveyor orbited for nearly a decade (1997–2006) with gravity being a prime science objective. The long-duration Mars Odyssey and Mars Reconnaissance Orbiter (MRO) also contributed significantly by providing radiometric data ingested into a multi-spacecraft gravity field solution. ESA's Mars Express mission has also carried out gravity experiments.

The post-Viking gravity models reached degree and order 50 (Smith et al. 1993; Konopliv and Sjogren 1995). Studies of specific features included Olympus Mons (which is the largest known gravity anomaly in the Solar System), the mascons in the Isidis basin (Sjogren 1979), the Hellas Planitia low gravity region (Sjogren and Wimberly 1981), and the Tharsis region (Janle and Erkul 1991).

By the end of the MGS mission, which had a low altitude (378 km), nearly circular orbit as well as polar inclination, a degree and order 95 global field was obtained (Konopliv et al. 2006). In addition to the first global high-resolution gravity field, the measurement of the solar tide from the MGS tracking data indicated an outer liquid Mars core (Yoder et al. 2003). A discussion of gravity results from previous Mars missions, as well as predicted and initial MGS contributions to Mars gravity field knowledge, appear in Rosborough and Lemoine (1991), Tyler et al. (1992), Smith et al. (1999), Lemoine et al. (2001), and Yuan et al. (2001). In 2006, MRO began collecting tracking data for the RS gravity investigation (Zuber et al. 2007a, 2007b) at a low altitude (255 km periapsis) that enabled a gravity field model solution to degree and order 110 (Konopliv et al. 2011b).

An additional 4 years of range data improved the Mars ephemeris (Folkner et al. 2014). Konopliv et al. (2011b) determined 21 asteroid masses and bounded the solar mass loss as $dGM_{\odot} / dt < 1.6 \times 10^{-13}$ per year. Furthermore, combination of MGS and Mars Odyssey tracking data have led to improved measurement of the seasonal changes in the gravity coefficients caused by the mass exchange between the polar ice caps and atmosphere, and the free wobble of the Mars' spin axis has been constrained to be less than 10 mas by examining the temporal history of C21 and S21 terms; the tidal Love number solution was consistent with the liquid outer core determination from MGS tracking data (Yoder et al. 2003), giving a combined solution of $k_2 = 0.152 \pm 0.009$ (Konopliv et al. 2011b). Figure 3.16 shows a degree and order 120 gravitational field of Mars from Konopliv et al. (2016) with a similar 120-deg field also generated by Genova et al. (2016).

For NASA's InSight mission, Folkner et al. (2018) describe the onboard Rotation and Interior Structure Experiment (RISE) using the lander's X-band radio system and DSN stations to determine the rotation of Mars. RISE measures the nutation of the planet spin axis for possible detection of a liquid core and constraining the core radius and density. It will also measure changes in the rotation rate of Mars on seasonal time-scales in order to constrain the atmospheric angular momentum budget. Furthermore, RISE will provide a tie between the cartographic and inertial reference frames; see recent results in Kahan et al. (2021).

Similarly, for ESA's planned ExoMars 2020 mission, Dehant et al. (2009) describe the onboard Lander Radio Science (LaRa) experiment using the lander's X-band transponder for two-way Doppler measurements between the Mars lander and Earth stations to obtain Mars' orientation in space and precession, nutation, and length-of-day variations. Precise measurement of variations in the orientation of Mars' spin axis will enable the determination of the size and moment of inertia of the core. For a liquid core, the free core nutation resonance induces a change, as compared to a solid core,

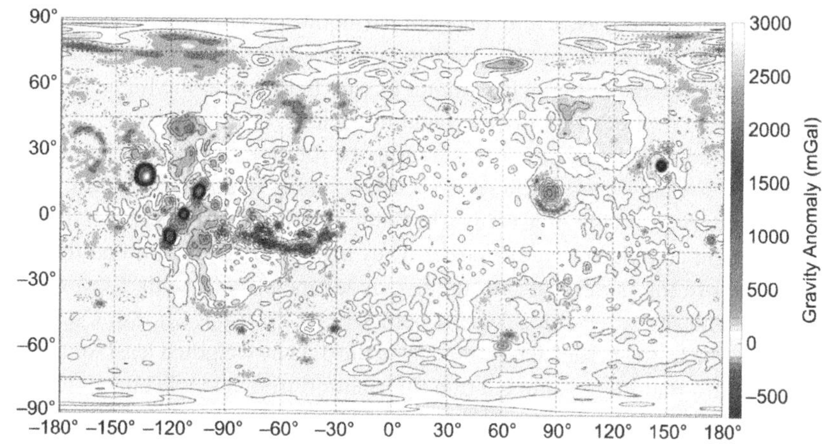

Figure 3.16 A 120-deg-and-order gravitational field map of Mars (Konopliv et al. 2016).

at the percent level in the semi-annual prograde nutation amplitude and even more for the retrograde terannual nutation amplitude. The resonance amplification depends on the size, moment of inertia, and flattening of the core.

Le Maistre et al. (2012) explain that Mars orientation and rotation parameters—precession, nutation, polar motion, and length-of-day variations—are related to the interior of the planet as well as to the dynamics of its atmosphere; their analysis showed that the latitude of the landing site must be higher than 20 deg to detect the Chandler Wobble component of the polar motion. Le Maistre et al. (2018) discuss simulations for a case study involving a moving rover on Mars and showed that after one Martian year of operation, the polar motion could be determined with 5 mas of precision from a link between a rover and a Mars orbiter due to improvements in position accuracy. These improvements, especially when combined with historical and anticipated future lander data, could enable the first direct detection of the Chandler wobble amplitude in the polar motion of Mars.

Konopliv et al. (2020b) announced that for the first time for any planetary body other than the Earth, the free wobble of the pole of Mars called the Chandler wobble, which constrains the rheology of the Martian mantle and in particular its long-term frequency dependence, has been detected with a period of 206.9 ± 0.5 days and amplitude of 10 cm from radio-tracking observations of Mars Odyssey, MRO, and MGS missions. The motion of the rotation pole location on the surface of Mars was observed using joint global estimates of Mars' orientation and its gravity field and time series solutions of C_{21} and S_{21}. For Mars interior models, the Chandler wobble period is combined with other measurements including the moments of inertia from our estimated precession rate of -7603.9 ± 1.3 mas/year and tidal Love number $k_2 = 0.169 \pm 0.006$.

3.5.5 Jupiter

Prior to the Galileo mission, Jupiter had flybys by Pioneers 10 and 11 and Voyagers 1 and 2, and data were combined from the four experiments to produce mass calculations for the four Galilean satellites but not any harmonic terms (Campbell and Synnott 1985). The mass of Jupiter was estimated along with a few low even harmonic coefficients (Anderson et al. 1992); note that odd harmonics, if detected, would correspond to north–south asymmetries. The Galileo Orbiter (1995–2003) did not significantly improve our understanding of the Jupiter gravity field due to the high-altitude orbit (required due to Jupiter's very intense radiation belts).

Galileo contributed significantly to understanding the rotational and tidal characteristics of the four Jovian large satellites (Io, Europa, Ganymede, and Callisto) by measuring their external gravitational fields. The Galileo team inferred Io's gravitational quadrupole moments (Anderson et al. 2001b). They explain that they limited the Io gravity parameters to the monopole GM, the five quadrupole coefficients J_2 (=$-C_{20}$), C_{21}, S_{21}, C_{22}, S_{22}, and the seven third-degree coefficients J_3 (=$-C_{30}$), C_{31}, S_{31}, C_{32}, S_{32}, C_{33}, and S_{33}, all in the fitting model determined from the data by weighted least squares. They further explain that there is not enough information from a single encounter to determine all the coefficients and imposed the constraint that J_2 is 10/3 C_{22}.

Anderson et al. (2001b) concluded that gravitational and other data strongly suggest that Io is in hydrostatic equilibrium and has a metallic core. The team used the Doppler data from four encounters of Europa to refine a model of its interior that is likely differentiated into a metallic core surrounded by a rock mantle and a water ice–liquid outer shell, but the data could not eliminate the possibility of a uniform mixture of dense silicate and metal beneath the water ice–liquid shell (Anderson et al. 1998). Moore and Schubert (2003) further discuss how Io, the Galilean satellite nearest to Jupiter, is subject to the most intense tidal forces that deform it in a way determined by the properties of Io's interior. They incorporate Io's rotational and tidal deformation measured by imaging (Thomas et al. 1998) and posit that, on the basis of density alone, it can be inferred that Io has an iron core of considerable size.

Doppler data from five Galileo encounters with Callisto were used to determine the GM and unnormalized quadrupole gravity coefficients in Callisto's external gravitational field; possible two- and three-layer interior models that satisfy the given constraints for radius, density, and moment of inertia show that Callisto cannot be entirely differentiated (Anderson et al. 2001b). Four encounters of Ganymede provided data of sufficient quality to yield a mass estimate and conclude the presence of gravity anomalies (Anderson et al. 1996, 2001b). Two surface mass anomalies, one a positive mass at high latitude and the other a negative mass at low latitude, can explain the data, but there are no obvious geological features that can be identified with the anomalies (Anderson et al. 2004b).

Schubert et al. (1996) discuss the role of the magnetic field in the understanding of the interior structure of Ganymede; Moore and Schubert (2003) discuss the tidal response of Ganymede and Callisto; and Palguta et al. (2006) discuss the mass anomalies on Ganymede from the gravity investigation. Jacobson (2001) provides a summary of the gravity field of the Jovian system and the orbits of the Jovian satellites.

The Juno mission has led to a better understanding of Jupiter's interior structure and dynamical properties by mapping its gravitational and magnetic fields (Bolton 2010). The Juno radio instrumentation is based on highly stable two-way (uplink and downlink) radio links at X- (8.4 GHz downlink) and Ka-bands (32.5 GHz downlink), providing range-rate data accurate to 3 μm/s at 1000 s integration time.

In 2016, Juno was inserted into a highly eccentric polar orbit ($e = 0.9466$) with a period of 53.5 days. The low pericenter of about 3500–5000 km altitude makes the orbit especially sensitive to the zonal gravity field. The Juno spacecraft is also exposed to significant relativistic effects, such as the acceleration due to the Lense-Thirring precession, which can contribute to understanding the angular momentum of the planet (Finocchiaro et al. 2011).

Parisi et al. (2021) describe constraining the depth of Jupiter's Great Red Spot with the Juno gravity overflights. Folkner et al. (2017) describe the initial improvement in estimating the Jovian gravity field from the first two orbits with emphasis on the low-degree spherical harmonic coefficients. Durante et al. (2017) proposed to exploit Juno's extremely accurate RS data to examine the gravity perturbations produced by Jupiter's acoustic modes and concluded that the oscillations yield effects comparable to those expected from shallow zonal winds.

Iess et al. (2018) reported Jupiter's zonal harmonics from the first two perijove passes dedicated to the gravity experiment, pointing out that the gravity harmonics of a fluid rotating planet can be decomposed into static components arising from solid-body rotation and dynamic components arising from flows. Furthermore, Iess et al. (2018) report

> in the absence of internal dynamics, the gravity field is axially and hemispherically symmetric and is dominated by even zonal gravity harmonics that are approximately proportional to the ratio between centrifugal acceleration and gravity at the planet's equator ... any asymmetry in the gravity field is attributed to differential rotation and deep atmospheric flows. The odd harmonics are a measure of the depth of the winds in the different zones of the atmosphere.

This research detected north–south asymmetry, a signature of atmospheric and interior flows.

Durante et al. (2019) exploited a larger data set for a combined multi-arc least square estimation filter by interpreting the results from first two gravity passes to suggest the presence of a diluted core and north–south asymmetry in Jupiter's gravity field due to nonzero odd harmonics, which have been explained by the atmospheric wind dynamics. Using this data set, Durante et al. obtained a new solution and concluded that the new solution was consistent with the results from Iess et al. (2018). The new solution also examined Jupiter's spin axis motion and found the estimated positions to be close to the International Astronomical Union (IAU) 2006 model predictions.

Notaro et al. (2019) reported on the possibility of determining Jupiter's frequency-dependent Love numbers using the Juno gravity data. Jupiter's fast rotation introduces a dependency of the tidal response of the planet on the distance of the perturber, which is small but detectable, and would provide information on the interior structure of the planet, especially in case of a dynamical tide detection. Dynamical amplification of the frequency-dependent tidal response can arise from resonances between the forcing frequency (Io, Europa, Ganymede, or Callisto's orbital frequency) and Jupiter's natural oscillation eigenfrequencies, providing a way to sample the oscillation spectrum of the planet. See also Juno results in Buccino et al. (2020), Durante et al. (2020), and Notaro et al. (2021).

The ESA's Jupiter Icy Moons Explorer (JUICE) mission will fly by and observe the Galilean satellites Europa, Ganymede, and Callisto and characterize the Jovian system in a lengthy Jupiter-orbit phase. The mission will then orbit Ganymede for in-depth studies of habitability, evolution, and the local environment. One of the JUICE mission objectives is the characterization of the internal mass distribution, dynamics, and evolution of the interiors (Grasset et al. 2013; Plaut et al. 2014).

Asmar et al. (2014c) describe the advanced instrumentation planned for JUICE and other future planetary missions. The advanced instrumentation centers on high-precision range and range-rate observables simultaneously at two coherent wavelengths, X- and Ka-bands, and full calibration of the intervening media: interplanetary plasma, Earth's ionosphere, and Earth's troposphere. Special instrumentation on board the spacecraft and at the ground stations would also be used.

3.5.6 Saturn

Voyager 1 and Voyager 2 (1980 and 1981), along with Pioneer 11, also made flybys of Saturn. Though Pioneer 11 flew closer to the planet, proximity of the ray path to the Sun introduced a large amount of plasma noise in the signal; therefore, data from the two Voyagers were an improvement (Null et al. 1981) over the Pioneer results. The Voyagers also obtained closer looks at some of the larger Saturnian moons.

A study combining the Voyager and Pioneer data was produced by Campbell and Anderson (1989) and contained descriptions of previous results for Saturn gravitational characteristics. Improved masses for the moons Rhea, Titan, Tethys, and Iapetus, as well as several harmonic coefficients of Saturn's gravity field, were part of this study. In addition, upper limits on second-order coefficients of Titan's gravity field were also calculated. One small moon, 1981 S13, was first detected with the Voyager 2 Network-generated Doppler data. After gravitational perturbations were noticed in the spacecraft's motion, optical images of candidate body locations revealed the new satellite (Showalter 1991).

Iess et al. (2019) reported the interior structure of Saturn, the depth of its winds, and the mass and age of its rings. When the spacecraft finally traveled between the planet and its innermost ring at altitudes of 2600–3900 km above the cloud tops, the radio signal was received during six crossings and the gravitational field of the planet as well as the mass of its rings were measured. Saturn's gravity deviated from theoretical expectations, requiring differential rotation of the atmosphere to a depth of at ~9000 km. Iess et al. (2019) reported the total mass of the rings to be $(1.54 \pm 0.49) \times 10^{19}$ kg, which is 0.41 times that of the moon Mimas, indicating that the rings may have formed 10^7 to 108 years ago.

Data from the Cassini orbiter have been used to study Saturn, and the Saturnian gravity field has been determined to a degree and order 2 so far. As Cassini looped through the Saturnian system, encounters with several of the other large moons filled in information on their masses and mean densities and better characterize the interdynamics of the planet and its satellites (Kliore et al. 2004).

3.5.6.1 Titan

Iess et al. (2010) describe the determination of Titan's mass and gravity harmonics to degree 3 and deduce that a quadrupole field is consistent with a hydrostatically relaxed body shaped by tidal and rotational effects, and an inferred moment of inertia factor of ~0.34 implies incomplete differentiation.

Iess et al. (2012) used precise measurements of the spacecraft acceleration during six flybys (2006–2011) to show that Titan responds to the variable tidal field exerted by Saturn with periodic changes of its quadrupole gravity, at about 4% of the static value. They also reported detecting the signature of the periodic tidal stresses within Titan, driven by the eccentricity ($e = 0.028$) of its 16-day orbit around Saturn; furthermore, the 2σ degree-2 Love number ($k_2 = 0.589 \pm 0.150$ and $k_2 = 0.637 \pm 0.224$) requires that Titan's interior be deformable in a manner consistent with a global ocean at depth.

Durante et al. (2019) reported updates to the gravity solution based on the final data set including data where the closest approach occurred at a high latitude (75° N), over an area not previously sampled. The magnitude of the response quadrupole field, expressed in the tidal Love number k_2, used to infer the existence of an internal ocean, was updated to 0.62, accurate to a level of a few percent. The interpretation includes possibilities of a high-density ocean, a partially viscous response of the deeper region, or a dynamic contribution to the tidal response. The new solution includes higher degree and order harmonic coefficients, up to 5. They report the geoid is poorly correlated with the topography, implying strong compensation.

3.5.6.2 Enceladus

For the small and active Saturnian moon Enceladus, Iess et al. (2014) used three Cassini flybys and reported indications of the presence of a negative mass anomaly in the South Polar region, largely compensated by a positive subsurface anomaly compatible with the presence of a regional subsurface sea at depths of 30–40 km and extending up to south latitudes of about 50°; the reported moment of inertia of ~0.335 MR^2 suggests a differentiated body with a low-density core.

3.5.6.3 Rhea

After two close flybys of Saturn's second largest moon Rhea, Tortora et al. (2016) reported the results of gravity quadrupole unnormalized coefficients of $J_2 \times 10^6 = 946.0 \pm 13.9$ and $C_{22} \times 10^6 = 242.1 \pm 4.0$. Their ratio is statistically not compatible with the theoretical value of 10/3, making it not possible to infer the moment of inertia factor directly using the Radau–Darwin approximation. A range of moments of inertia was modeled but low ones (more differentiation) require greater magnitudes of excess core topography to satisfy the observations.

Anderson and Schubert (2007) reported a mean density of 1233 ± 5 kg/m^3 and an axial moment of inertia consistent with an undifferentiated interior and implication of a homogeneous mix of ice and rock. Further examination by Anderson and Schubert (2010) of the effects of thermal emission confirmed, under the assumption that Rhea is in hydrostatic equilibrium, that Rhea is an undifferentiated satellite made up of about 25% rock-metal and 75% water ice by mass.

Iess et al. (2007a) and Mackenzie et al. (2008) reported J_2 and C_{22} values consistent with hydrostatic equilibrium and a moment of inertia that favors an almost undifferentiated Rhea where a very large uniform core is surrounded by a relatively thin ice shell containing no rock at all. Nimmo et al. (2010) incorporate information from gravity science to examine the geophysical implications of the long-wavelength topography of Rhea; Jacobson et al. (2006, 2007) summarize results from the Saturnian satellites.

3.5.7 Uranus

The best information on Uranus came from the Voyager 2 flyby of the planet between late 1985 and early 1986. The combination of Doppler information and visual identification of mutual perturbations between the satellites allowed greatly improved mass determinations for the five major satellites and therefore produced mean densities and indications of the moons' composition. The compositional information, in fact, led to the conclusion that the satellites could not be captured comets, as had been suggested (Anderson et al. 1987b). More accurate values were gained for both the Uranian mass and the solar-to-Uranian system mass ratio, though significant information for other gravitational field parameters was not obtained. A model of Uranian gravity field would be particularly interesting as Uranus is the only planet whose axis of rotation appears nearly parallel to the ecliptic plane.

3.5.8 Neptune

Neptune was the last encounter for Voyager 2, in mid-1989. Because of the close approach and the past difficulties in obtaining information about such a distant planet using only Earth-bound techniques, the Voyager flyby increased the accuracy of the solar-to-Neptunian-system-mass ratio by a factor of 1000, using both navigation and Doppler data (Tyler et al. 1989). From the data, the masses of Neptune and Triton were estimated to 0.0003% and 0.3%, respectively. Even the small satellites Nereid and 1989N1 were observable in the spacecraft Doppler signature (Tyler et al. 1989). Estimates of a few low-order coefficients of the Neptunian gravity field were also obtained. A discussion of the interior structure of Neptune appears in Hubbard et al. (1991).

3.5.9 Pluto

The New Horizons mission acquired radiometric data at Pluto during a close flyby in 2015 (Stern et al. 2015). The mass determination of Pluto and Charon from the Radio Science Experiment (REX) was a prime objective. During the flyby, the spacecraft velocity was slightly perturbed and the resulting shift in the Doppler profile on the uplink X-band signal was recorded, as described in Linscott et al. (2011). For a detailed description of the investigation and instrumentation design and implementation, see Tyler et al. (2009).

3.5.10 Asteroids and Comets

Most asteroids are too small to produce noticeable effects on spacecraft. Galileo's 1991 encounter with the asteroid Gaspra produced only images. In 1993, Galileo flew by the asteroid 243 Ida, during which it returned images of a second object, 1993(243)1 Dactyl, the first confirmed satellite of an asteroid. Belton et al. (1994) applied statistical and dynamical arguments to constrain the range of possible orbits, and hence the mass of Ida, and yielded a bulk density of 2.6 ± 0.5 g/cm^3.

The NEAR spacecraft had a close flyby of asteroid 253 Mathilde and the gravitational perturbation on the spacecraft was observed in the tracking data. As a result of the accurate targeting, data were used to determine Mathilde's mass as 1.033 ($\pm$0.044) $\times 10^{20}$ g. Coupled with a volume estimate provided by the NEAR imaging team, this suggests a low bulk density for Mathilde of 1.3 g/cm^3 (Yeomans et al. 1997).

The mission, later named NEAR Shoemaker, orbited 433 Eros and then landed on the latter asteroid. When it flew within 3830 km of the asteroid on December 23, 1998, the gravitational perturbation on NEAR was evident in the spacecraft tracking data, and the mass of Eros was determined to be $(7.2 \pm 1.8) \times 10^{18}$ g and, coupled with a volume estimate provided by the NEAR imaging team, the bulk density was 2.5 ± 0.8 g/cm^3 (Yeomans et al. 1999). As noted in Section 3.2.1, Garmier et al. (2002) modeled the Eros gravity field from spacecraft Doppler tracking data up to degree and order 12.

The Vesta gravity field and related physical parameters have been precisely measured using 10 months of radiometric Doppler and range data as well as optical landmark tracking from the Dawn spacecraft. Konopliv et al. (2014a) estimated Vesta's gravity field, orientation parameters, landmark locations, and orbit and reported the resulting spherical harmonic gravity field has a half-wavelength resolution of 42 km (degree 20). The inertial spin pole location was determined to better than 0.0001°, and the uncertainty in the rotation period was reduced by nearly a factor of 100 from previously known results. The combined precession and nutation of the pole of Vesta has been detected with angular rates about 70% of expected values, but not well enough to constrain the moment of inertia. The implications for interior structure were also addressed in Asmar et al. (2012b).

Park et al. (2016) reported that the gravity and shape measurements of Ceres obtained from the Dawn spacecraft show that Ceres is in hydrostatic equilibrium with its inferred normalized mean moment of inertia of 0.37. These data show that Ceres is a partially differentiated body, with a rocky core overlaid by a volatile-rich shell, and the gravity signal is strongly suppressed compared to that predicted by the topographic variation. This indicates that Ceres is isostatically compensated, such that topographic highs are supported by displacement of a denser interior. Konopliv et al. (2018) determined the Ceres' gravity field to degree 18, indicating the gravity field is mostly compensated, consistent with the Airy isostatic model. Ermakov et al. (2017) and Fu et al. (2017) placed constraints on Ceres' internal structure and evolution from its shape and gravity measured.

The Rosetta mission observation of 67P/Churyumov-Gerasimenko included an RS investigation. Pätzold et al. (2016) reported the precise mass, bulk density, porosity, and internal structure of the nucleus of comet 67P/Churyumov-Gerasimenko based on its gravity field as derived from measured spacecraft velocity perturbations at flyby distances between 10 and 100 km GM was measured to be 666.2 ± 0.2 cubic meters per second squared, giving a mass $M = (9982 \pm 3) \times 10^9$ kg. The average bulk density of the nucleus is 533 ± 6 kg per cubic meter (kg/m^3). The team concluded that the most likely composition mix has approximately four times more dust than ice by mass and two times more dust than ice by volume, and that the interior of the nucleus is homogeneous and constant in density on a global scale without large voids.

JAXA's Hayabusa-2 mission (Tsuda et al. 2019) to asteroid Ryugu has resulted in determining the mass (GM) of the body (Takeuchi and Saito 1972).

A gravity experiment for the Didymos system (see Tortora et al. 2019) is the planned by the National Aeronautics and Space Administration (NASA) Double Asteroid Redirection Test (DART), Agenzia Spaziale Italiana (ASI; Italian Space Agency) Light Italian Cubesat for Imaging of Asteroid (LICIACube), and ESA's Hera missions.

3.5.11 Pioneer and Earth Flyby Anomalies

During the many decades of Solar System exploration, scientists have identified two anomalies based on RS data that received considerable attention in the literature. The first is the Pioneer anomaly, which is an observed deviation from predicted accelerations of the Pioneer 10 and Pioneer 11 spacecraft, launched in the early 1970s, after they reached ~20 AU (Anderson et al. 2002; Nieto and Turyshev 2004). It was expected that both spacecraft would slow down as they moved further away from the Sun, but radiometric observables revealed that they were slowing slightly more than expected by a very small acceleration towards the Sun of $(8.74 \pm 1.33) \times 10^{-10}$ m/s^2.

The anomaly led to numerous proposed explanations, including new physics, from professional scientists as well as the public worldwide at international conferences and collaboration societies. The currently accepted explanation of the cause of this anomaly is an anisotropic radiation pressure due to the spacecraft's heat loss, as explained in Turyshev and Toth (2010) and Modenini and Tortora (2014), after recovering and analyzing about three decades of historical data, with some support from the Planetary Society. One positive outcome of the Pioneer anomaly is the wide interest in the type of data as well as the detail required to calibrate all possible effects that make up the error budget of a planetary gravity science experiment.

The other phenomenon observed from the same type of data observables is called the Earth flyby anomaly. Many planetary missions targeting various parts of the Solar System use a trajectory that brings them back to Earth for a gravitational flyby. Called a gravity assist, this maneuver saves on energy needed by the spacecraft to reach its target by utilizing the gravity of Earth or another planet providing the gravity assist (Van Allen 2003).

When the Galileo mission carried out an Earth flyby in 1990, a discrepancy was observed in Doppler data where 66 mHz, or a velocity of under 4 mm/s, was unaccounted for. Subsequently, similar discrepancies were observed with the NEAR, Cassini, and Rosetta Earth flybys (Anderson et al. 2008). As with the Pioneer anomaly, numerous explanations have been proposed. To date, no widely accepted explanation of the Earth flyby anomaly has yet been published.

3.6 A User's Guide

As explained in Section 1.4, the success of RS experiments relies on the thorough understanding, modeling, and calibration of error sources. Asmar et al. (2005a) (with significant input from P. Tortora, M. Zannoni, L. Iess, M. Parisi) detail a noise budget for gravity experiments when accounting for propagation and instrumental noise, in addition to systematic errors. Asmar et al. (2013b) provides quantification of errors for precision Doppler in spacecraft-to-spacecraft links, such as the GRAIL mission where errors are grouped into dynamic or kinematic sources; the former are forces acting on the spacecraft and perturb their orbit or motion and the latter are local effects, such as temperature and timing.

Figure 3.17 is a representative diagram of gravity processing flowchart. As noted in Asmar et al. (2014b), all non-negligible forces acting on the spacecraft must be modeled, including:

- Point-mass Newtonian gravitational acceleration due to the Sun, planets, and their satellites, as well as asteroids and comets.
- Newtonian acceleration due to the known gravity fields of the planets, their satellites, and small bodies in the Solar System, including gravitational acceleration due to planetary rings and mascons.

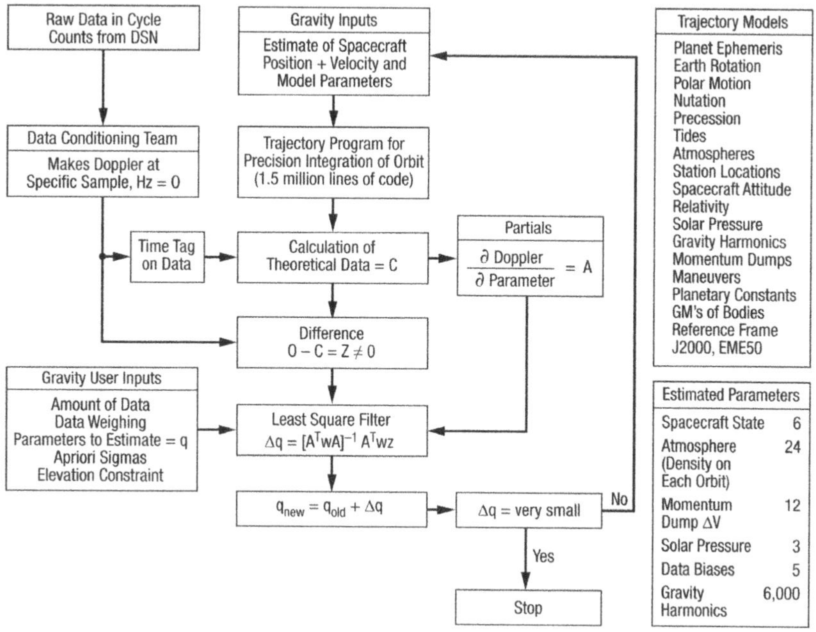

Figure 3.17 A diagram of representative process of data analysis of planetary gravity field data from spacecraft Doppler observables.

- Point-mass relativistic perturbing gravitational acceleration due to the Sun, the planets, and their satellites.
- Relativistic acceleration observed for missions near Earth (Turyshev and Toth 2010, JPL IOM).
- Acceleration due to geodesic precession.
- Acceleration due to Lense-Thirring precession.
- Acceleration caused by tidal effects on the central planetary body.
- Acceleration due to time-varying gravity effects, such as atmospheric motion and atmospheric transport of volatiles.
- Solar Radiation Pressure.
- Planetary radiation pressure: acceleration due to the radiation emitted from the surface of a planet, such as reflected visible light (albedo) and thermal emission.
- Thermal imbalance: acceleration due to nonuniform spacecraft surface heating.
- Gas leakage: acceleration due to spacecraft control jet leakage.
- Planetary atmospheric drag.
- Perturbations due to spacecraft maneuvers.

In addition, the following kinematic error must be modeled:

- Timing errors such as those caused by differences in the stability and drift of spacecraft and ground clocks.
- Errors in station location.
- Earth tides; first order displacement is ~50 cm, second order is ~13 mm
- Ocean loading displacements (typically less than 4 mm).

- Polar motion, which modifies the centrifugal potential of the Earth and causes a solid tide that produces a displacement of the tracking station (under 2 cm/year).
- Plate motion on the order of 10 cm/year.
- Offset between Earth's CM and the origin of the considered terrestrial reference frame.
- Offset between the ground station's center reference point and electrical phase center.
- Offset between the spacecraft center of gravity point and antenna electrical phase center.
- Spacecraft charging when passing through planetary magnetic fields (if needed).

Asmar et al. (2014b) provide detailed procedures for deriving gravity field models by examining spacecraft OD, an iterative estimation based on the comparison between measured observables and computations from the mathematical models described here and in Moyer (2003). The goal of the OD process is to estimate a set of "solve-for" parameters that unambiguously define the spacecraft trajectory along with the formal uncertainty of the solution. These estimated parameters minimize the frequency residuals of interest to the gravity analysis in a least square manner.

Calibrations are applied to the observables in order to correct for the signal group delay in the ground and spacecraft electronics (see Chapter 5), as well as group and phase delay due to the troposphere, ionosphere, and interplanetary plasma. The major steps of the OD process are summarized in the following paragraphs.

3.6.1 Calculation of Observables and Partials

The observables are computed using mathematical models of all non-negligible physical effects that depend on the numerical values of three types of parameters:

- **Solve-for parameters** are a set of model parameters that include the initial position and velocity of the spacecraft where an estimation filter computes a differential correction of their central values along with uncertainties in the form of a covariance matrix. They are imperfectly known and their estimate can be improved through the OD process. The covariance matrix is obtained by inverting the information matrix built from the partial derivatives of the observables with respect to the solve-for parameters and the residuals. For the first iteration, solve-for parameter a priori values and uncertainties are obtained from the best measurements or theoretical estimation; subsequent iterations build on the first iteration.
- **Consider parameters** are the best available estimates used in the mathematical models and are not improved through the OD process. Their uncertainties increase the overall solution uncertainty but do not change the estimates of the solve-for parameters.
- **Exact parameters** are OD parameters with uncertainties that do not affect the estimation because they have negligible influence on observables.

The partial derivatives of the observables with respect to the solve-for and consider parameters are computed and used to obtain new estimates of the values and their covariance matrices. Calculation of observables and their partials requires reconstructing the spacecraft trajectory by integrating the equations of motion with respect to a nonrotating reference frame, which includes the coordinate system origin or center of integration (COI), from an initial state.

It may be necessary to update the planetary or satellite ephemerides since the measured observables are sensitive to the relative positions and velocities of the spacecraft, and the planet and errors in the satellite ephemeris may alias into estimation errors on the spacecraft state and/or the gravity coefficients. Updated ephemerides are generated by integrating the full relativistic equations of motion of the planets and their satellites from their updated initial states. To compute the partials and update the ephemerides, the variational equations are integrated—see Appendix 3A.

Note that a conversion of the station location from the terrestrial reference frame to the celestial reference frame consists of a series of rotations: polar motion, Earth rotation, nutation, and precession. The Jet Propulsion Laboratory (JPL) defines these rotations according to the IAU 1980 theory of nutation and the IAU 1976 precession model. The geocentric state of the tracking station must be converted to the Solar System barycentric frame of reference through proper Lorentz transformations. Obtaining the light-time solution in the Solar System barycentric frame, necessary to find the transmission and the reflection times, is based upon the known speed of light and travel time from the transmitter to the receiver, which has two components:

1. Newtonian light time that represents the time for light to travel along a straight-line path at the speed of light in a vacuum, and
2. Relativistic light-time delay, also called the Shapiro effect, which accounts for the reduction of the coordinate velocity of light and bending of the light path due to presence of a gravity field that curves the space-time.

3.6.2 Estimation Filter

The estimation filter computes parameter values that minimize the frequency residual differences between the measured and computed observables. These are weighted by their measurement uncertainties, which depend on the noise in the link; a priori estimate based on available theoretical error budgets and experience is used. In the next iteration, the noise is estimated from the standard deviation of the residuals. It is held constant in subsequent iterations.

The filter also minimizes differences between the solve-for parameters and their a priori values that are weighted using the uncertainty of the a priori information. The estimation filter produces a differential correction to a priori values of the solve-for parameters. The filter gives the updated estimate of the solve-for parameters and the covariance matrix of the solve-for parameters that represents the solution uncertainty. The a priori covariance matrix is updated based on the influences of measurement noises and consider parameter uncertainties on the solve-for parameters.

3.6.3 Solution Analysis

The solution obtained from the estimation filter is verified using these criteria:

- Convergence, reached if two successive iterations produce statistically equivalent solutions.
- Residual analysis checks if the solution residuals represent only measurement noise by inspecting a zero mean, noise spectrum of residuals, and standard deviations should be compatible with the residual weights used in the estimation filter.

- Covariance matrix comparison between a priori and computed formal uncertainties. A parameter is fully derivable from data if the computed uncertainties are much less than the a priori ones. Typically, a factor 10 can be expected; otherwise, the estimation critically depends on a priori information and relatively less on measurements. Analysts also examine the cross-correlation between solve-for parameters to determine if any solve-for parameters are very strongly correlated.
- Comparison between a priori values and the new estimates for statistical compatibility.
- Solution stability, which is checked by executing another estimation process using the solution as a priori values, with increased a priori uncertainties. If the solution is stable, the new estimation will be statistically compatible with the testing solution.

Appendix 3A

Planetary Geodesy

Contributed by A. S. Konopliv

3A.1 Planetary Geodesy: Gravitational Potentials and Fields

Following Kaula (1966), Heiskanen and Moritz (1967), and Konopliv et al. (2011b), the gravitational potential is modeled by a spherical harmonic expansion in the body-fixed reference frame with normalized coefficients ($\bar{C}_{nm}$, $\bar{S}_{nm}$) and is given by (the symbols n and m are used here for harmonic degree and order; other literature uses other symbols)

$$U = \frac{GM}{r} + \frac{GM}{r} \sum_{n=1}^{\infty} \sum_{m=0}^{n} \left(\frac{R_e}{r} \right)^n \bar{P}_{nm}(\sin \varphi_{\text{lat}}) \left[\bar{C}_{nm} \cos(m\lambda) + \bar{S}_{nm} \sin(m\lambda) \right], \quad \text{(3A.1)}$$

where GM is the gravitational constant times the mass of the central body, n is the degree, m is the order, $\bar{P}_{nm}$ are the fully normalized associated Legendre polynomials, R_e is the reference radius of the body, ϕ_{lat} is the latitude, and λ is the longitude. The gravity coefficients are normalized such that the integral of the harmonic squared equals the area of a unit sphere and are related to the unnormalized coefficients as (Kaula 1966; Lambeck 1988)

$$\begin{pmatrix} C_{nm} \\ S_{nm} \end{pmatrix} = \left[\frac{(n-m)!(2n+1)(2-\delta_{0m})}{(n+m)!} \right]^{1/2} \begin{pmatrix} \bar{C}_{nm} \\ \bar{S}_{nm} \end{pmatrix} = f_{nm} \begin{pmatrix} \bar{C}_{nm} \\ \bar{S}_{nm} \end{pmatrix}. \quad \text{(3A.2)}$$

The corresponding normalized Legendre polynomials $\bar{P}_{nm}$ are thus related to the unnormalized polynomials P_{nm} by

$$P_{nm} = \frac{1}{f_{nm}} \bar{P}_{nm}. \quad \text{(3A.3)}$$

The factors f_{nm} and unnormalized Legendre polynomials P_{nm} through degree and order 3 are listed in Table 3A.1.

$\bar{J}_n$ are the longitude independent zonal coefficients and are given by $\bar{J}_n = -\bar{C}_{n0}$, where $\bar{J}_2$ is related to the flattening and $\bar{J}_3$ describes the pear shape of the body. The corresponding zonal harmonics have n zeros with each zero following a band of latitude. The tesseral ($m \neq n$) and sectoral harmonics ($m = n$) have ($n - m$) lines of latitude that are zero and 2 m lines of longitude that are zero over the entire sphere as illustrated in Figure 3.2.

Table 3A.1 Gravity coefficient normalization factors and unnormalized Legendre polynomials. $y = \sqrt{1 - x^2}$, or in terms of latitude $x = cos\varphi_{lat}$ and $y = sin\varphi_{lat}$.

n	m	f_{nm}	$P_{nm}(x)$
1	0	$\sqrt{3}$	x
1	1	$\sqrt{3}$	y
2	0	$\sqrt{5}$	$-\frac{1}{2} + \frac{3}{2}x^2$
2	1	$\sqrt{5/3}$	$3xy$
2	2	$\sqrt{5/12}$	$3y^2$
3	0	$\sqrt{7}$	$-\frac{3}{2}x + \frac{5}{2}x^3$
3	1	$\sqrt{7/6}$	$-\frac{3}{2}y + \frac{15}{2}x^2y$
3	2	$\sqrt{7/60}$	$15xy^2$
3	3	$\sqrt{7/360}$	$15y^3$
4	0	3	$\frac{3}{8} - \frac{15}{4}x^2 + \frac{35}{8}x^4$
5	0	$\sqrt{11}$	$\frac{15}{8}x - \frac{35}{4}x^3 + \frac{63}{8}x^5$

The unnormalized degree-1 coefficients are related to the body's center of mass (CM) by

$$\begin{aligned} C_{10} &= Z_{cm} / R_e \\ C_{11} &= X_{cm} / R_e \\ S_{11} &= Y_{cm} / R_e. \end{aligned} \tag{3A.4}$$

Since the origin of the coordinate system is most often chosen to be the CM, the degree-1 coefficients are usually zero. However, the topography of a planet is often determined using the CM as the origin since the spacecraft position is determined in that frame. The center of figure of the planet relative to the CM is also given by similar spherical harmonic representation of the topography.

The unnormalized second-degree coefficients of the gravity field are related to the moments of inertia (A, B, and C for x- and y-axes) as given by (Lambeck 1988; Yoder 1995a)

$$\begin{aligned} J_2 &= \frac{1}{MR_e^2}\left(C - \frac{1}{2}(A+B)\right) \\ C_{21} &= \frac{I_{xz}}{MR_e^2} \qquad S_{21} = \frac{I_{xy}}{MR_e^2} \\ C_{22} &= \frac{1}{4MR_e^2}(B-A) \qquad S_{22} = \frac{I_{xy}}{2MR_e^2}. \end{aligned} \tag{3A.5}$$

If the rotational axis of the central body is equal to the principal axis, then the C_{21} and S_{21} gravity coefficients vanish. For $J_2 \gg C_{22}$ and S_{22}, the offset of the rotation axis ($\delta\phi$) from the principal axis is given by $\delta\varphi^2 = \left(C_{21}^2 + S_{21}^2\right) / J_2^2$ and indicates a wobble. The free Eulerian wobble period P_W (motion of the spin axis relative to the body surface or polar motion) for a rigid triaxial body not in spin-orbit resonance (e.g., Mars, Venus) is (Yoder 1995b)

$$P_W = \frac{2\pi}{\omega}\left(\frac{AB}{(C-A)(C-B)}\right)^{1/2}, \tag{3A.6}$$

where ω is the spin rate of the body. The precession or drift of the body's spin axis in inertial space due to the Sun is given by (Yoder 1995a)

$$\frac{d\psi}{dt} = \frac{-3MR_e^2}{2C\omega} J_2 n^2, \tag{3A.7}$$

where n is the mean motion of the body around the Sun. In the case of Mars, the measured precession and J_2 determine the polar moment of inertia (Folkner et al. 1997b; Konopliv et al. 2006).

The gravitational potential also includes modifications due to tides caused by a perturbing body. The tidal potential acting on a satellite with position $\vec{r}$ (magnitude r and unit vector $\hat{r}$) from the central body and perturbing body (e.g., the Sun) at position $\vec{r}_p$ (magnitude r_p and unit vector $\hat{r}_p$) relative to the central body is given by

$$U = k_2 \frac{GM_p}{R} \frac{R^6}{r^3 r_p^3}\left[\frac{3}{2}\left(\hat{r}\cdot\hat{r}_p\right) - \frac{1}{2}\right], \tag{3A.8}$$

where k_2 is the Love number, M_p is the mass of the perturbing body causing the tide, and R is the equatorial radius of the central body. The tides cause the degree and order 2 coefficients to have time-varying components as follows for normalized values (e.g., Eanes et al. 1983; McCarthy 1992):

$$\begin{aligned}
\Delta\bar{J}_2 &= -k_{20}\sqrt{\frac{1}{5}}\frac{GM_p R^3}{GMr_p^3}\left[\frac{3}{2}\sin^2\varphi_p - \frac{1}{2}\right] \\
\Delta\bar{C}_{21} &= k_{21}\sqrt{\frac{3}{5}}\frac{GM_p R^3}{GMr_p^3}\sin\varphi_p\cos\varphi_p\cos\lambda_p \\
\Delta\bar{S}_{21} &= k_{21}\sqrt{\frac{3}{5}}\frac{GM_p R^3}{GMr_p^3}\sin\varphi_p\cos\varphi_p\sin\lambda_p \\
\Delta\bar{C}_{22} &= k_{22}\sqrt{\frac{3}{20}}\frac{GM_p R^3}{GMr_p^3}\cos^2\varphi_p\cos 2\lambda_p \\
\Delta\bar{S}_{22} &= k_{22}\sqrt{\frac{3}{20}}\frac{GM_p R^3}{GMr_p^3}\cos^2\varphi_p\sin 2\lambda_p.
\end{aligned} \tag{3A.9}$$

ϕ_p and λ_p are the latitude and longitude of the perturbing body on the surface of the central body; separate Love numbers have been used for each order. The expressions in Equation (3A.8) consist of a constant plus periodic part. The constant part of the tide is typically not included in the published gravity coefficient, in which case the complete expression of Equation (3A.8) is added to the input gravity coefficient (see McCarthy (1992) for a discussion on the permanent tide for the Earth). For Venus, the latitude of the Sun is near zero, and the largest period term comes from $\bar{C}_{22}$ and $\bar{S}_{22}$ that changes due to the solar tide.

The acceleration due to the gravitational potential must be rotated from the body-fixed frame to the inertial International Celestial Reference Frame (ICRF) using the corresponding model for the body (Seidelmann et al. 2002). Using the Legendre polynomials as a function of latitude results in singularities at the pole in the partials of the gravity acceleration with respect to the spacecraft position. Thus, the Orbit Determination Program (ODP) uses a nonsingular formulation including recursion relations given by Pines (1973) for calculating acceleration and partials.

3A.2 Gravity Determination Technique

Since the early 1970s, spacecraft tracking data via the Deep Space Network (DSN) were processed using the Jet Propulsion Laboratory's (JPL) ODP (Moyer 2003), the institutional software for navigating planetary spacecraft. The estimation of the gravity field follows the same steps as the orbit determination process in navigation but involves more parameters and methods to constrain the gravity field. The ODP consists of three major subsets of programs that integrate the spacecraft motion, process the spacecraft observations, and filter or estimate the spacecraft state and related parameters using the observations.

The ODP has been replaced by a new software called Mission analysis, Operations, and Navigation Toolkit Environment (MONTE), which has similar functionalities. The Gravity Recovery and Interior Laboratory (GRAIL) science team at JPL used a version of the ODP called the Multiple Interferometric Ranging Analysis using GPS Ensemble (MIRAGE).

3A.3 Dynamical Integration

To determine the spacecraft dynamical path, the ODP numerically integrates the spacecraft Cartesian state by including all known forces acting on the spacecraft, such as gravity, solar pressure, spacecraft thrusting, and atmospheric drag. The spacecraft state and the force model partial derivatives (e.g., gravity harmonics) that are later estimated are integrated using the variable order Adams method (Krogh 1973). The inertial coordinate system is centered at the planet, moon, asteroid, or comet of interest and defined by the nonrotating ICRF that is nearly equal to the Earth's mean equator and equinox at the epoch of J2000.

After Konopliv et al. (2013), with $\vec{r}$ and $\vec{v}$ being the position and velocity vectors of the spacecraft relative to the central body, the ODP integrates the second-order differential equations:

$$\ddot{\vec{r}} = \vec{f}(\vec{r},\vec{v},\vec{q}) = \nabla U(\vec{r}) + \vec{f}_{pm} + \vec{f}_{in-pm} + \vec{f}_{in-obl} + \vec{f}_{srp} + \vec{f}_{att} + \vec{f}_{drag} + \vec{f}_{rel} + \cdots. \quad (3A.10)$$

The above equation is integrated in inertial Cartesian coordinates. For the acceleration due to gravity, which has longitude and latitude dependence, the acceleration is computed in the planet's body-fixed frame and then rotated to the ICRF frame. The same is true for any other forces requiring the body-fixed frame such as atmospheric drag.

$\vec{f}(\vec{r},\vec{v},\vec{q})$ is the total acceleration on the spacecraft, and $\vec{q}$ are all the constant ($\dot{\vec{q}} = 0$) model parameters that will be estimated, such as the gravity harmonic coefficients. Contributions to the total acceleration include spacecraft acceleration relative to the central body due to the gravitational potential of the central body $\nabla U(\vec{r})$; spacecraft acceleration due to other solar system bodies treated as point masses $\vec{f}_{pm}$; indirect point mass acceleration of the central body in the solar system barycentric frame due to the other planets and natural satellites $\vec{f}_{in-pm}$; indirect oblateness acceleration, acceleration of the central body (e.g., the Moon) due to another body's oblateness (e.g., the Earth) $\vec{f}_{in-obl}$; acceleration of the spacecraft due to solar radiation pressure $\vec{f}_{srp}$; acceleration due to spacecraft cold gas thrusting for attitude control maneuvers (usually for de-spinning angular momentum wheels) $\vec{f}_{att}$; acceleration due to atmospheric drag $\vec{f}_{drag}$; and acceleration due to general relativity corrections $\vec{f}_{rel}$. Other accelerations may include spacecraft thermal forces, planetary albedo, and infrared radiation, tides, and empirical (usually periodic) acceleration models.

In addition to integrating the spacecraft position and velocity, the ODP also integrates the variational equations to estimate the epoch spacecraft state and constant parameters. Following the nomenclature of Tapley et al. (2004), the nominal orbit or trajectory is given by

$$X^*(t) = \begin{pmatrix} \vec{r}^*(t) \\ \vec{v}^*(t) \\ \vec{q}^* \end{pmatrix}. \quad (3A.11)$$

The asterisk refers to the nominal orbit. The interest is in linear changes relative to the nominal orbit, i.e., the first term in a Taylor series. The first-order differential equation to integrate to determine the nominal orbit is

$$\dot{X}^*(t) = \begin{pmatrix} \vec{v}^* \\ \vec{f}(\vec{r}^*,\vec{v}^*,\vec{q}^*) \\ 0 \end{pmatrix} = F(X^*,t). \quad (3A.12)$$

The variation of the trajectory from its nominal path is given by $x(t) = X(t) - X^*(t)$ and the linearized equations

$$\dot{x}(t) = A(t)x(t) = \left(\frac{\partial F(t)}{\partial X(t)}\right)^* x(t). \quad (3A.13)$$

The integrated solution is the state transition matrix $\Phi(t,t_0)$

$$x(t) = \Phi(t,t_0)x(t_0) = \begin{bmatrix} U_{6x6} & V_{6xp} \\ 0_{px6} & I_{pxp} \end{bmatrix} x(t_0), \quad (3A.14)$$

which relates the deviation from the nominal path at epoch t_0 to the deviation from the nominal path at time t for the six-position and velocity epoch parameters (U_{6x6}) and the p constant model parameters (V_{6xp}).

The ODP integrates a number of second-order differential equations, including 3 position variables, 18 variables representing the changes in position and velocity due to small changes in epoch position and velocity to get the matrix U_{6x6}, and 3 equations for each dynamic parameter or constant in Equation (3A.13) being estimated. For a complete gravity field of degree and order n, the total number of parameters is given by $(n-1)(n-3)$ or, for example, 14,637 parameters for a degree and order 120 field.

3A.4 Processing of Observations

The second major step is to process the Doppler and range observations of the spacecraft. Following Tapley et al. (2004), the general form of the observation equation is

$$Y = G(X,t) + \varepsilon, \tag{3A.15}$$

where Y is the actual observation, $G(X,t)$ is the expression for the modeled observation value, and ε is the observation error. The Doppler data are not an instantaneous velocity measurement but instead are processed similar to a range observable and given by a differenced range measurement for two-way Doppler as

$$G(X,t) = \left((r_{12} + r_{23})_e - (r_{12} + r_{23})_s\right) / \Delta t + \rightleftharpoons ..., \tag{3A.16}$$

where r_{12} is the uplink range between the Earth station at the time of signal transmission and the spacecraft at the time the signal is received, and r_{23} is the downlink range from the spacecraft at the time the signal was transmitted to the location the Earth station will be when the signal arrives, with subscripts denoting the end and start of the Doppler count interval (Δt). One of the main tasks in processing the observation is to solve the light-time equation in a solar system barycentric frame, i.e., find the original transmit time at the first station and the receive time at the spacecraft using an iterative procedure.

Other corrections are also required in Equation (3A.16) for Doppler or the similar expression for range, such as relativistic propagation delay due to the Sun and planets (Moyer 2003), media corrections for the Earth ionosphere and troposphere refraction using calibrations provided by the DSN (Chao 1972; Callahan 1973; Mannucci et al. 1998), and solar plasma delays due to the solar corona of the Sun and any measurement biases.

The linearized form of Equation (3A.15) is also computed in processing the observables for the purpose of estimating the gravity field. The observation residual y is given by the difference of the actual observable Y and computed observable using the nominal orbit $Y^*(t)$ as given by

$$y = Y - Y^*(t). \tag{3A.17}$$

Using the state transition matrix to map to the epoch time, Equation (3A.15) is then written as

$$y = \left(\frac{\partial G}{\partial X}\right)^* \Phi(t,t_0)x_0 + \varepsilon = Hx_0 + \varepsilon. \tag{3A.18}$$

The last equation is then minimized using the vector of residuals y and partials matrix H. The calculation of the computed observable and related partials involves the precise determination of the Earth station in a solar system barycentric ICRF frame (e.g., see Yuan et al. 2001). The Earth-fixed coordinate system is consistent with the International Earth Rotation and Reference Systems Service (IERS) terrestrial reference frame (ITRF) labeled ITRF93 (Boucher et al. 1994).

Rotation of the Earth-fixed coordinates of the DSN station locations to the Earth-centered inertial system requires a series of coordinate transformations due to several factors:

- Precession (according to the International Astronomical Union (IAU) 1976 model (Lieske 1977)).
- Nutation of the mean pole (according to the IAU 1980 nutation theory (Wahr 1981; Seidelmann 1982), with daily corrections to the model from the JPL Earth Orientation Platform (EOP) product of Folkner et al. (1993)).
- Rotation of the Earth (according to Aoki et al. (1982) and Aoki and Kinoshita (1983), and the Coordinated Universal Time (UTC) minus a smoothed a priori measurement of the Earth's orientation called UT1R, or UTC-UT1R corrections of the JPL EOP file).
- Polar motion of the rotation axis (according to the JPL EOP file).

The JPL EOP product is derived from very long baseline interferometry (VLBI) and lunar laser ranging (LLR) observations and includes Earth rotation and polar motion calibrations, as well as nutation correction parameters necessary to determine inertial station locations to the level of a few centimeters. Note that in timing systems, UT0 is universal time measured using the daily motions of the stars and UT1 is UT0 corrected for polar motion and is the same everywhere on Earth; UT1R is UT1 corrected for tidal effects.

The body-fixed ITRF93 DSN station locations have been determined with VLBI measurements and conventional and global positioning system (GPS) surveying. The coordinate uncertainties are about 4 cm for DSN stations that have participated in regular VLBI experiments and about 10 cm for other stations (Folkner 1996). Folkner (1996) also provides the antenna phase center offset vector for each DSN station. The "no-net-rotation relative velocities new global plate motion model" or NNR-NUVEL1 (Argus and Gordon 1991) is the plate motion model consistent with the solution of DSN station locations. The variations of DSN station coordinates, caused by solid Earth tides, ocean tide loading, and rotational deformation due to polar motion, are corrected according to the IERS standards of 1992 (McCarthy 1992).

3A.5 Filtering of Observations

Once the observation equations are found, the ODP estimates the spacecraft state and other parameters using a square-root information weighted least-squares filter (SRIF) (Bierman 1977; Lawson and Hanson 1995). This is the most computationally intensive task, and for the larger planetary gravity fields of Mars, Venus, and the Moon, results are obtained with the help of supercomputers or Linux clusters. In normal form, the solution is given by

$$\hat{x} = \left(H^T W H + P_{ap}^{-1}\right)^{-1} H^T W y, \tag{3A.19}$$

where W is the weight matrix for the observations and P_{ap} is the a priori covariance matrix of the parameters being estimated. In the ODP SRIF filter, the solution equation is kept in the form

$$R\hat{x} = z, \tag{3A.20}$$

when processing observations by using Householder transformations. In this equation, R is the upper triangular square-root of the information array, and R and z are related to the normal equations as

$$\begin{aligned} R^T R &= H^T W H + P_{ap}^{-1} \\ z &= \left(R^T\right)^{-1} H^T W y, \end{aligned} \tag{3A.21}$$

and the covariance P of the solution (inverse of the information array) is given by

$$P = R^{-1}\left(R^{-1}\right)^T. \tag{3A.22}$$

The observations for gravity field determination are usually separated into different time spans or data arcs. The parameters estimated in the gravity solutions consist of arc-dependent variables (spacecraft state, atmospheric drag, and other things) that are determined separately for each data arc and global variables (gravity coefficients, orientation parameters, and other things) that are common to all data arcs. The global parameters are determined by merging only the global parameter portion of the square-root information arrays from all the data arcs—equivalent to solving for the global parameters plus arc-dependent parameters of all arcs. This technique is described by Kaula (1966) using partitioned normal matrices and was first used to analyze Earth orbiter data; for the type of filter used in this work (square-root information), the method is outlined by Ellis (1980).

In practice, a separate square-root information upper triangular matrix or SRIF is maintained for the global variables for each spacecraft and mission phase or sections of a mission phase with only a priori constraints on the local variables. The global portion of the SRIF matrices is optimally merged to minimize the data residuals as described by Yuan et al. (2001) and Yuan (1991) and is equivalent to solving for all the local and global variables. For a total of q SRIF equations for just the global variables (z_j and R_j, $j = 1$ to q), the relative weighting factor f_j for each equation is given by

$$f_j^2 = \frac{n_j}{\left(z_j - R_j\hat{x}\right)^T\left(z_j - R_j\hat{x}\right) + \varepsilon_j^T\varepsilon_j}, \tag{3A.23}$$

where $\varepsilon_j^T\varepsilon_j$ is the sum of squares of the residuals for the jth SRIF equation, n_j is the number of observations for the jth equation, and $\hat{x}$ is the nominal solution with weighting factors equal to one (Yuan 1991). In the ODP filter, the square root of the sum of squares (RSOS) of the residuals is the last element of the last column of the upper triangular matrix and the observation vector or right-hand size z_j is the last column. Next, all the SRIF equations are merged using Householder transformations and multiplying SRIF equation R_j and z_j by f_j. Once the global SRIF equations are merged, a gravity a priori constraint is applied.

The gravity field solution is generally estimated to a higher resolution and includes more coefficients than what is inherent in the tracking data (more unknowns than equations). For this reason, the gravity coefficients (or some of them) must be constrained to obtain the most realistic gravity solution. The two different methods to constrain the gravity coefficients for terrestrial planets are a spectral or Kaula power law constraint or a spatial surface acceleration constraint that depends on the local resolution of the gravity field. The normalized gravity coefficients decrease in magnitude M_n versus increasing degree n according to power law as first noted by Kaula (1963a) for the Earth:

$$M_n = \sqrt{\frac{\sum_{m=0}^{n}\left(\bar{C}_{nm}^2 + \bar{S}_{nm}^2\right)}{2n+1}} = \frac{K_E}{n^2} \approx \frac{10^{-5}}{n^2}. \tag{3A.24}$$

The scale factor of the power law of the Earth K_E can be scaled to other planets or moons K_P (Kaula 1963a; Lorell 1969; Nerem et al. 1995) using

$$K_P = K_E \left(\frac{M_E}{M_P}\right)^2 \left(\frac{R_P}{R_E}\right)^4 = K_f K_E. \tag{3A.25}$$

The factors for the power law and the observed power laws are given in Table 3.3. The Kaula power constraint biases the coefficients toward zero based upon the power law (i.e., the a priori value of the coefficient is zero, and the a priori sigma is given by the power law using the degree of the coefficient). If there is a fairly uniform distribution of data over the surface, then it may not be necessary to constrain all the coefficients. For the Mars gravity analysis, only the coefficients of degree 59 and greater are constrained using a Kaula-type power law (Konopliv et al. 2006). This is done to avoid suppressing any gravity amplitudes in the data.

The second type of gravity constraint or surface acceleration constraint was used with the Venus gravity effort (Konopliv et al. 1999) and the Mars gravity efforts prior to Mars Global Surveyor (MGS) (Konopliv and Sjogren 1995). This method is useful for cases where there is not an equal distribution of data over the surface of the body and resolution varies with location. This technique applies less of a constraint where the data strength is greatest and more of a constraint where the data is not as good.

The a priori constraint applied for this spatial method evaluates the radial acceleration and its uncertainty on the reference sphere (i.e., $r = R_e$). At that surface, the radial acceleration (a_n) from all coefficients of degree n is given by

$$a_n = \frac{GM}{R_e^2}(n+1)\sum_{m=0}^{n}\bar{P}_{nm}(\sin\varphi)(\bar{C}_{nm}\cos m\lambda + \bar{S}_{nm}\sin m\lambda). \tag{3A.26}$$

To create a profile of acceleration contributions versus degree, the root mean square (RMS) of the acceleration a_n is obtained over the sphere. As a good approximation, the RMS magnitude spectrum of the gravity coefficients follows the Kaula rule K_p/n^2. The expected acceleration profile at the surface is then given by (for $n \gg 1$)

$$(a_n)_{\text{RMS}} = \frac{GM}{R_e^2}K_p\sqrt{2/n}, \tag{3A.27}$$

which for Venus is in mGals

$$(a_n)_{\text{RMS}} = 15/\sqrt{n} \tag{3A.28}$$

and for Mars is in mGals

$$(a_n)_{\text{RMS}} = 68 / \sqrt{n} \tag{3A.29}$$

and for the Moon is in mGals

$$(a_n)_{\text{RMS}} = 57 / \sqrt{n}. \tag{3A.30}$$

This is the expected "signal" for the acceleration at each point on the surface of the reference sphere. The signal could also be determined empirically by taking the RMS of a given gravity field over different regions. However, in applications of this method, only one signal profile has been used for all latitudes and longitudes.

The next task is to map the acceleration uncertainty at the surface into an uncertainty or "noise" profile showing the error in acceleration versus harmonic degree. The acceleration uncertainty from the summed contributions of all coefficients from degree 2 to degree n is given by

$$\sigma(a_{2,n}) = \frac{\partial a_{2,n}^T}{\partial G_{2,n}} P_{noap(2,n)} \frac{\partial a_{2,n}}{\partial G_{2,n}}, \tag{3A.31}$$

where $G_{2,n}$ is the vector of all normalized gravity coefficients from degree 2 to degree n and $P_{noap(2,n)}$ is the corresponding covariance. The covariance of the coefficients from degree 2 to n is the covariance as if the higher degree coefficients ($>n$) are not estimated. Hence, it is a truncation, or submatrix, of the full degree and order covariance without any constraint applied to the gravity field (*noap* for "no a priori"). The partial derivatives of the acceleration with respect to the coefficients of degree n and order m are functions of latitude and longitude and are given by

$$\begin{aligned} \frac{\partial a_{2,n}}{\partial \bar{C}_{nm}} &= \frac{GM}{R_e^2}(n+1)\bar{P}_{nm}(\sin\phi)\cos m\lambda \\ \frac{\partial a_{2,n}}{\partial S_{nm}} &= \frac{GM}{R_e^2}(n+1)\bar{P}_{nm}(\sin\phi)\sin m\lambda. \end{aligned} \tag{3A.32}$$

The uncertainty for the coefficients of degree n is then given by the difference of the sum total error to degree n and the sum total error to degree $n - 1$ as

$$\sigma(a_n) = \sigma(a_{2,n}) - \sigma(a_{2,n-1}). \tag{3A.33}$$

Both the "signal" or the expected magnitude of the coefficients from the Kaula power rule and the "noise" or uncertainty in the coefficients can be plotted versus degree. The crossing point of the Kaula signal with the acceleration uncertainty or noise is called the degree strength of the gravity field for that particular latitude and longitude. For degrees greater than the degree strength, the noise in the data exceeds the signal.

The purpose of the gravity constraint method is to constrain the noise of the gravity field to zero with some uncertainty when the noise exceeds the signal. The acceleration at the surface from all harmonic coefficients greater than or equal to the degree strength is constrained to zero with an uncertainty approximately equal to the expected signal at the degree strength. This amounts to generating observations over the entire surface of the sphere based upon the degree strength at each latitude and longitude. An observation ($a_{D,N}$) for degree strength D and gravity field estimated to degree N is

$$a_{D,N} = \frac{GM}{R_e^2} \sum_{n=D}^{N} \sum_{m=0}^{n} (n+1)\bar{P}_{nm}(\sin\phi)(\bar{C}_{nm}\cos m\lambda + \bar{S}_{nm}\sin m\lambda). \tag{3A.34}$$

The linearized observation Equation (3A.18) has y as the difference between the observed value and the nominal value of the observation, or accumulated acceleration at the surface for degrees D to N from the a priori gravity, and H is the row vector of observation partials (the partial of the observation with respect to all the gravity parameters being estimated). The partials H to construct the observation equation are

$$H = \frac{\partial a_{D,N}}{\partial G}, \tag{3A.35}$$

where G is the vector of all gravity coefficients. The elements of H for coefficients with degrees less than the degree strength D are zero.

The surface acceleration observations for the spatial constraint are then merged with the unconstrained gravity square root information array using Householder transformations. In normal form, the constrained gravity estimate $\hat{x}$ is written as

$$\hat{x} = (H^T WH + P_{noap}^{-1})^{-1}(H^T Wy + P_{noap}^{-1} x_{noap}), \tag{3A.36}$$

where P_{noap} is the unconstrained covariance of the gravity coefficients, H is the matrix of observation partials with each row an acceleration observation, x_{noap} is the unconstrained gravity estimate, and y is the vector of linearized acceleration observations. The new constrained covariance P is then

$$P = \left(H^T WH + P_{noap}^{-1}\right)^{-1}. \tag{3A.37}$$

The observations should be spaced such that at least two observations are generated over the shortest harmonic wavelength—the Nyquist rate. The weight used for an observation is then proportional to the area between observations and is approximately equal to the signal at the degree strength. The observations are globally distributed on a rectangular grid of latitude and longitude typically with a spacing of 1 degree.

The main advantage of using this spatial constraint instead of a straight Kaula rule on the spectrum appears to be better determination of peak amplitudes. Since the well-determined degrees are not constrained directly, the amplitudes for those degrees are not biased toward zero. It is also flexible in allowing relaxation of selected regions for any reason, such as incorrect data weighting or a region exhibiting greater signal than the power rule. So, one can refine local or regional solutions similar to line-of-sight methods using a global technique that is fully dynamic. For the higher degree terms, this local varying constraint does not affect other more remote regions as a global power rule constraint would. In general, there are only slight differences between the two constraint methods for Venus since it does not have strong local deviations from the Kaula power spectrum. For Mars, prior to MGS, the differences were more pronounced for the Tharsis volcanoes and doubled the peak amplitudes (Konopliv and Sjogren 1995).

4

Solar and Fundamental Physics

In Collaboration with J. W. Armstrong

4.1 Principles of Heliospheric Observations

Radio science (RS) observations can probe the inner heliosphere and the solar wind on spatial scales inaccessible to in-situ observations. The first use of the deep space network (DSN) for heliospheric research was a spectral broadening experiment as the line-of-sight (LOS) of the Mariner 4 spacecraft passed within 0.6 degrees (deg) of the Sun. An open-loop receiver (see Chapter 5) was used to record the pre-detection electric field data. In 1968, Goldstein (1969) duplicated the experiment during the superior conjunction of Pioneer 6. Table 4.1 summarizes observable quantities and their relevance to heliospheric investigation goals. Appendix 4A describes solar corona observation methodology using Mars Express data for illustration.

Overviews of spacecraft radio sounding techniques for solar corona studies have been presented by Bird and Edenhofer (1990) and Woo (1993). There is an extensive literature on propagation theory and how to relate radio observations to properties of the media causing the scintillation, including Tatarski (1961), Lovelace (1970), Jokipii (1973), Coles (1978), Ishimaru (1978), Rickett (1990), and Woo (1993). The solar plasma has a density- and wavelength-dependent index of refraction. The refractivity, which is the difference between refractive index and unity, is proportional to the local electron density and inversely proportional to the square of the radio frequency (RF), as discussed in Chapter 2.

As a wave propagates through the turbulent plasma, it undergoes a geometrical optics phase shift and the integral along the LOS, taken to be the z-direction, of $(2\pi / \lambda)\mu(r)dz$. This two-dimensional (2D) phase function of position in the (x,y) plane perpendicular to the LOS has an associated 2D power spectrum,

$$P_{2\phi}(q_x,q_y) = \text{const}\, P_{3n}(q_x,q_y,q_z = 0), \tag{4.1}$$

where $P_3\text{n}(q)$ is the three-dimensional (3D) power spectrum of electron density and (qx,qy) is the 2D wave vector associated with the (x,y) plane. If the scattering irregularities are frozen in a frame moving with transverse velocity v, assumed in the x direction, then spatial frequencies are translated to temporal frequencies (f) with the resulting observed one-dimensional (1D) phase spectrum (Lovelace 1970):

$$P_{1\phi}(f) = \frac{2\pi}{v}\int_{-\infty}^{\infty} P_{2\phi}\left(q_x = \frac{2\pi f}{v}, q_y\right) dq_y. \tag{4.2}$$

Radio Science Techniques for Deep Space Exploration, First Edition. Sami W. Asmar.

Table 4.1 Radio science observables relevant to heliospheric studies.

Observation	Relevance	Comment
• **Phase scintillation** • Spectrum • Allan variance	• Directly related to electron density spectrum over wide range of spatial scales • Sense changes with solar distance • Sense changes in/out of ecliptic	• Sensitive and w/large dynamic range (does not "saturate" like intensity scintillation) • S-X or X-K isolates downlink plasma w/ excellent control of systematics • ~10^{-4}–10 Hz • Useful over whole sky: Sun–Earth–spacecraft angles 0–180 deg
• **Intensity fluctuations**Spectrum • Statistical moments (variance, third central moment) • Bi-spectrum • Cross-correlation of two-frequency intensity to determine velocity (refraction separates rays at different frequencies) • Cross-correlation of intensity simultaneously received at separated ground stations to determine velocity	• Directly related to electron density spectrum for spatial scales smaller than about 100 km • Sense changes with solar distance • Sense changes in/out of ecliptic • Sensitive to short time-scale nonstationarities/space weather • Measure solar wind speed in inner heliosphere and out-of-ecliptic • Intensity statistics have engineering relevance (telecom links for NASA missions, e.g., MESSENGER and Solar Probe) • Very high time resolution for sensing changes in small scale • Turbulence level • Turbulence spectrum shape • Nonstationarity/space weather	• Sensitive to very small scales (~100 km or smaller) • Fresnel scattering/weak scintillation imposes natural length scale onto data—spectral width then gives a velocity estimate • Monochromatic point source means accurate normalization • Excellent link SNR allows high-time resolution—statistical moments can be computed accurately with ~minute of data • Dynamic range limited due to "saturation" in strong scintillation • Useful close to Sun: Sun–Earth–Spacecraft angles <10 deg

Radio-frequency spectrum broadening	• Directly related to electron density spectrum for spatial scales smaller than about 100 km
	• Very high time resolution for sensing changes in small scale
	  • Turbulence level
	  • Turbulence spectrum shape
	  • Nonstationarity/Space weather
		• Directly related to density structure function via robust 2nd moment theory
		• Excellent link SNR allows high time resolution—can be computed in ~seconds
		• Sensitive to very small scales, ~10 to 100 km
		• Most useful close to Sun or in nonstationary intervals

(continued)

Table 4.1 (Continued)

Observation	Relevance	Comment
Cross and autocorrelation functions of multilink Doppler/phase time series	• Localization of transients and separation of spatial and temporal events • Assessment of nonhomogeneity of turbulence along the line-of-sight • Nonstationarity/"space weather"	Transfer function of one- and two-way Doppler allows space-time localization of plasma disturbances in the inner heliosphere and out-of-ecliptic
Ranging	Complementary to coronal images; ranging column density can give absolute calibration of simultaneous coronal image	• Sensitive to intermediate and large scales • High dynamic range—no saturation • Gives absolute electron column density to spacecraft
Faraday rotation	Gives model-dependent information on magnetic field in regions of heliosphere otherwise inaccessible	Measures product of longitudinal magnetic field and electron density

References for Topics in Table 4.1

- Phase scintillation: Lovelace (1970); Woo and Armstrong (1979); Coles et al. (1991); Woo (1993); Pätzold et al. (1996).
- Intensity fluctuations: Tatarski (1961); Little and Ekers (1971); Armstrong and Coles (1972); Coles (1978); Kaufman (1976); Armstrong and Woo (1981); Scott et al. (1983); Bourgois et al. (1985); Rickett (1990); Armstrong (1993).
- Spectrum broadening: Woo and Armstrong (1979, 1981); Harmon and Coles (1983); Coles and Harmon (1989); Woo (1993).
- Cross and autocorrelation: Armstrong (2006); Bertotti et al. (2003); Richie-Halford et al. (2009). Ranging: Woo (1996).
- Faraday rotation: Stelzried et al. (1970); Pätzold et al. (1987).

In the special case where spectrum is a power law with spectral index α, then $P_{1\phi}$ is also power law with spectral index $\alpha - 1$.

Coherently related dual-frequency downlinks (both links are phase-referenced to the same oscillator) can be used to isolate the dispersive plasma contribution on the downlink. This eliminates confusion with competing systematic effects (described by Woo et al. 1976; Asmar et al. 2005b). The special case of the Cassini multilink system (X-up/X-down, X-up/Ka-down, Ka-up/Ka-down, see Chapter 5) allows isolation of the uplink and downlink plasma separately for plasma studies (Bertotti et al. 2003; Richie-Halford et al. 2009).

Interplanetary intensity scintillation, analogous to the optical twinkling of stars caused by turbulence in the neutral atmosphere, is also observed (Coles 1978). A formula for the intensity spectrum (Kaufman 1976) similar to that for $P_{1\phi}$ applies when intensity scintillations are weak and the root mean square (rms) intensity fluctuations $\ll$ mean intensity:

$$P_{1I}(f) = \frac{2\pi}{v} \int_{-\infty}^{\infty} P_{2I}\left(q_x = \frac{2\pi f}{v}, q_y\right) dq_y, \tag{4.3}$$

where P_2I is proportional to $\sin^2\left(\left(q_x^2 + q_y^2\right)\lambda z / 4\pi\right) P_{3n}\left(q_x, q_y, q_z = 0\right)$ and z is the distance to the scattering medium, here simplified as a thin screen. This multiplicative $\sin^2$ term imposes a purely geometric scale $\sim (\lambda z)^{1/2}$ on the intensity pattern at the Earth for observed solar wind density spectra.

Measurement of the phase spectrum can give information on the electron density spectrum over a wide range of spatial scales; measurement of the intensity spectrum gives information only on smaller scale electron density fluctuations. Since the typical time scale in weak scintillation for intensity fluctuations is proportional to $(\lambda z)^{1/2}/\nu$, information on the pattern velocity is possible from intensity scintillation. Statistical moments of the intensity fluctuations can be accurately normalized, since a spacecraft is effectively a monochromatic point source, and complications associated with natural radio sources, finite diameter and finite bandwidth, are avoided.

Although signals transmitted from a spacecraft are approximately monochromatic, scattering by irregular movement of the media between the spacecraft and the receiver causes the received signal to be broadened into a band of frequencies about the carrier known as RF spectral broadening. The broadening is directly related to the density structure function, which is connected to the density spectrum via an integral transform (Ishimaru 1978; Harmon and Coles 1983; Rickett 1990). Spectral broadening can be interpreted using "second-moment" theory, that is, the interpretation does not require the scintillation to be weak (Ishimaru 1978; Rickett 1990; Woo 1993).

Ranging information (see Chapter 5 for the ranging data type), or the contribution to the apparent range due to solar wind electrons along the LOS, can be obtained from time-of-flight of the modulated radio signals. The effect of plasma on the group velocity of radio waves, expressed as apparent range change, is given, for example, by Koehler (1968). Range measurements as a function of Sun–Earth–spacecraft angle (often called Sun–Earth–Probe, or SEP angle) and solar latitude of the principal scattering region provide a sensitive method for measuring electron density in the inner heliosphere.

Faraday rotation observations give information on the magnetic field in the inner heliosphere. The quasi-longitudinal approximation for Faraday rotation is given by Stelzried et al. (1970). Measurement of coronal Faraday rotation provides information on both electron density and the solar magnetic field (Stelzried et al. 1970; Rusch and Stelzried 1972; Pätzold et al. 1987). Appendix 4B describes Faraday rotation methodology illustrated by Magellan observations.

4.2 Inner Heliospheric Electron Density

Muhleman et al. (1977) used ranging during the solar conjunctions of Mariners 6 and 7 to estimate inner heliospheric electron density using the single S-band frequency. It is difficult to disentangle radial, temporal, latitudinal, and longitudinal effects (these issues were later addressed by Woo 1996), so range variations were initially modeled as purely radial. Dual-frequency ranging observations in 1985 and 1988 using Voyager were made by Anderson et al. (1987c), Krisher et al. (1991), and with Ulysses as described in Bird et al. (1994).

4.3 Density Power Spectrum

The power spectrum is fundamental in the description of stochastic density variations. At frequencies between ~10^{-4} and 10 Hz, the density spectrum can be inferred from the spectrum of a radio signal's phase, with coherent dual-frequency data preferred to isolate the plasma from systematic effects. This inference is made under plausible assumptions including frozen flow, unique spatial-to-temporal frequency conversion, and statistical stationarity of the plasma fluctuations along the LOS, which, in some cases, can be independently verified or refuted using other simultaneously-taken radio data types. The 1976–77 and 1978–79 Viking solar conjunctions were used for scintillation measurements (Tyler et al. 1977; Muhleman and Anderson 1981).

Kinman and Asmar (1988) used the 1985 Voyager 2 solar conjunction to study the two-way coherent Doppler error due to the solar corona inducing phase scintillations on the uplink signal. Woo and Armstrong (1979) studied the density spectrum using dual-frequency Viking data that were taken mostly in the ecliptic plane, over Sun–Earth–spacecraft angles ≈1–175 deg. Those spectra were often well represented by a pure power law, $P_1(f) \sim f^{-\alpha}$, with the average value of α close to the 2.67 value anticipated for Kolmogorov turbulence. Some spectra deviated from a pure Kolmogorov power law, however, showing "bowing" (see Woo and Armstrong 1979) and flattening close to the Sun (which may be related to the solar wind at the higher latitude of the sub-scattering point for the very near-Sun observations). The spectral levels at $f = 1$ milli-Hertz (mHz) as a function of Sun–Earth–spacecraft angle, combining the Viking data with dual frequency X-minus-Ka-band Cassini observations, are summarized in Asmar et al. (2005b).

The high-frequency part of the spatial density spectrum can be determined from spectral broadening observations. Woo and Armstrong (1979) and Armstrong and Woo (1981) showed density spectra often well approximated by a unique power law. Other observations, for example, Harmon and Coles (1983), show clear evidence for an "inner scale" that evolves as a function of distance from the Sun. This is important in the interpretation of Voyager radio observations of the solar wind/interstellar medium boundary—for example, Armstrong et al. (2000).

At very low frequencies ($<10^{-4}$ Hz) ranging variations can be used to study plasma fluctuations. Near-Sun low-frequency range variations are caused at least in part by quasi-stationary density structures having identifiable counterparts in coronal white light, which rotate through the LOS to cause temporal variability (Woo and Habbal 1999; Woo 2010).

4.4 Intermittency, Nonstationarity, and Events

Radio data are LOS averaged; the data sometimes show more variability than would be expected for quiescent background turbulence. Some of these variations are clearly due to "events"—very large disturbances crossing the LOS during the observation and affecting radio-wave phase and amplitude (e.g., Woo and Armstrong 1981; Bird et al. 1985). Sometimes data from LOS to multiple spacecraft are simultaneously available, and timing disturbances as they cross different LOSs can be used to measure speed very accurately over a wide heliocentric distance range. Woo et al. (1985) used this method for shocks observed inside 0.3 astronomical units (AU, 1 AU = 1.5×10^{11} m) and were the first to show that coronal mass ejections (CMEs) unexpectedly *accelerate* close to the Sun before then decelerating farther from the Sun.

Because of the excellent signal-to-noise ratio (SNR) on spacecraft radio links, the RF spectrum can be estimated in short integration times. Spectral broadening can then be used, for example, for high-time resolution studies of density spectrum level and shape before and in the wake of shock waves crossing the LOS. These shock waves can be high signal-to-noise events, with the width of the RF carrier increasing by 10–100 × in ~ minutes and the plasma signature visible in the RF spectrum over bandwidths >100 Hz. The spectral index sharply steepens in the wake of a shock and remains steep for ~hours after the event, with the data consistent with the density spectrum having a (time-dependent) power law shape throughout (Woo and Armstrong 1992).

A lower level of variability—statistically larger than expected from estimation error or LOS integration over a smoothly varying background, however—is seen in some high-time resolution observations, for example, in observations of radio signal intensity variance observed on time scales ~minutes. These may be associated with nonstationarity and intermittency expected in the solar plasma (Aschwanden et al. 2001). Space-time localization of plasma irregularities causing these effects (Richie-Halford et al. 2009) may be able to assess electron density intermittency and nonstationarity quantitatively in the inner heliosphere.

4.5 Faraday Rotation

The first coronal Faraday rotation observations performed using an artificial signal (Pioneer 6) were done at the DSN's Goldstone complex (Stelzried et al. 1970). Goldstone's then-64-m station was equipped with a remotely rotatable linear feed. A closed-loop polarimeter (Ohlson et al. 1974) was used to track the orientation of the received signal polarization as a function of time. Observations during November–December 1968 gave information on solar coronal transient events (Levy et al. 1969) and the background solar corona. The polarization tracking capability was continued with Pioneer and, later, Helios spacecraft observations—see the reviews of coronal Faraday rotation by Volland et al. (1977) and Bird et al. (1977). The Helios Faraday rotation data were also used in studies of the magnetic field in CMEs (Bird et al. 1985) and the average coronal field (Pätzold et al. 1987); see Appendix 4B.

4.6 Spaced-receiver Measurements

Solar wind speed can be measured in and out of the ecliptic and inside the solar wind acceleration region using spaced receiver measurements. The idea is illustrated in Figure 4.1. Solar wind scattering forms an intensity diffraction pattern at the surface of the Earth. This pattern moves (in the simplest case) with a unique velocity. By cross-correlating the intensity at separated stations, one can deduce time lags and thus the pattern velocity (speed and direction) from which the solar wind velocity can be inferred.

This technique was pioneered by astronomers using small-diameter natural radio sources which must be compact so that intensity scintillations are not quenched. For example, in weak scintillation, diffraction imposes the Fresnel scale, ~ $(\lambda z)^{1/2}$, on the

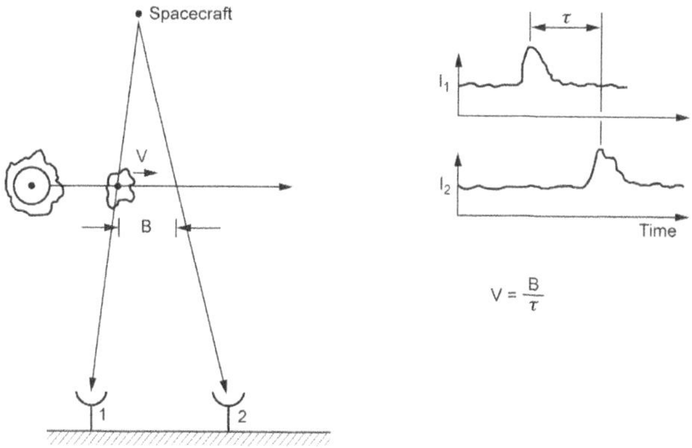

Figure 4.1 Conceptual diagram for measuring solar wind speed via the spaced-receiver method. A spacecraft transmits a signal that is scattered by the solar wind to produce a scintillation diffraction pattern on the Earth. The intensity of the signal is measured at two or more separated stations (I_1 and I_2 in the diagram) as a function of time. Pairwise cross-correlating the intensity time series from the stations gives time offsets due to motion of the pattern. From the geometry, the projected baseline, B, is known; hence, a pattern speed can be computed.

intensity pattern, where λ is the radio wavelength and z is the characteristic distance to the scattering region. Elements of the source brightness distribution separated by angle θ produce diffraction patterns offset on the ground by θz. If $\theta z >> (\lambda z)^{1/2}$ the patterns will overlap and the scintillation will be suppressed ("stars twinkle; planets don't"). Numerically: for $z = 1$ AU, $\lambda = 13$ cm, the angular size of the radio source must be smaller than about a 10th of an arcsecond to scintillate well (Hewish et al. 1964; Dennison and Hewish 1967; Kakinuma 1977; Coles et al. 1980; Bourgois et al. 1985; Armstrong et al. 1986; Coles 1993; Grall et al. 1996). Use of *spacecraft* as the radio source for intensity IPS velocity studies, however, has been limited (Armstrong and Woo 1981) in part because of lack of suitably spaced spacecraft tracking stations. Another method for detecting coronal velocities is to use spaced transmitters as described by Efimov et al. (2009).

A related method is to use a single receiver with multiple-frequency downlinks. The idea is that refraction in the solar wind causes each link to sense a spatially offset part of the plasma at the ray paths' closest points of approach to the Sun. Cross-correlating the intensity fluctuations observed at different frequencies gives a time offset. The "baseline" (ray path separation at closest approach) is given by a refraction model of the average solar wind; the ratio of the baseline to the time-offset then gives the (assumed radial) solar wind speed. This has been done successfully with natural radio sources at meter wavelengths (Scott et al. 1983). At the microwave frequencies used by spacecraft, there is only a narrow range of Sun–Earth–spacecraft angles where the ray path offsets are large enough to measure a significant time offset (Efimov et al. 2005).

4.7 Space-time Localization of Plasma Irregularities

When spacecraft observations are in the two-way mode (downlink is phase-locked to the uplink) plasma fluctuations have a specific, location-dependent transfer function to observed phase scintillation (Estabrook and Wahlquist 1975), which is illustrated schematically in Figure 4.2. The ground station is continuously both transmitting a signal to the spacecraft and receiving a signal from the spacecraft: the difference between the transmitted and received signals is the Doppler shift. A discrete plasma blob shifting the phase on both the up- and downlink, as illustrated, gives rise to "events" in the multilink tracking time series separated in time depending on the blob's location. Echoes in the two-way time series can thus localize position of blobs along the LOS.

With the multilink Cassini system more can be done; the plasma contribution to the up- and downlinks can be separately determined (Bertotti et al. 2003). Cross correlation of the up- and downlinks gives the time delay, and so the distance "x" from the Earth to the blob. The technique allows localization of plasma events in time and three space dimensions (plane of the sky and along the LOS; for example, Armstrong 2006; Richie-Halford et al. 2009; and discussions in Wohlmuth et al. 1997b).

4.8 Utility for Telecommunications Engineering

In addition to their science relevance, radio scintillation data have telecommunications engineering relevance for flight projects involving near-Sun radio communications and relevance for feasibility studies of other radio science investigations taken

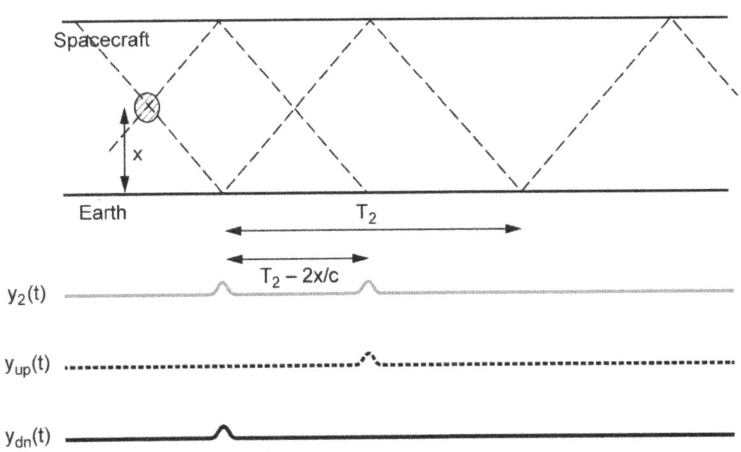

Figure 4.2 Transfer functions of plasma fluctuations to one- and two-way phase/Doppler scintillation allows localization of plasma blobs in space and time. *Upper plot* is a space-time diagram (space vertically, time horizontally). The ground station and the spacecraft are continuously exchanging microwave signals, some of which are shown as dashed lines. If the signals pass through a plasma blob (indicated by the cross-hatched circular area) the phase is perturbed. This perturbation is on both the uplink and downlink signals. The Doppler/phase time series are shown in the *lower plots*. In the two-way data, $y_2(t)$, the phase perturbation is seen first on the downlink, and then later when the perturbation on the uplink is phase-coherently re-transmitted back to the ground. The effect is two positively correlated features in the two-way time series, separated by $T_2 - 2x/c$, where T_2 is the two-way light time between the Earth and the spacecraft and x is the distance of the blob from the Earth. The Cassini five-link system allows separate determination of the up- and downlink plasma perturbations, $y_{up}(t)$ and $y_{dn}(t)$, as shown above. These time series can be pairwise cross-correlated to determine time offset and hence localize plasma disturbances in space and time (Figure 4.1 of Richie-Halford et al. 2009).

when communications paths are close to the Sun (e.g., Woo and Armstrong 1979; Armstrong and Woo 1980; Asmar et al. 2005b; Armstrong 2006). For example, Figure 4.3 shows the scintillation index (rms intensity/mean intensity) for spacecraft data taken close to the Sun mostly in the ecliptic. Figure 4.4 compares the third central moment of the intensity fluctuations to the second moment, with overplots of the predictions of two popular models of scintillation ("Rice-squared" and log-normal, Armstrong 1993) sometimes used to model scintillation in telecommunications engineering studies.

4.9 Precision Tests of Relativistic Gravity

The BepiColombo Mercury Orbiter Radio-science Experiment (MORE) team will carry out high precision tests of relativistic gravity in the most desirable "laboratory" in the Solar System, the gravitational field of the Sun. Being the innermost planet, Mercury is the ideal test mass for probing general relativity. Range and range-rate measurements from radio tracking a spacecraft in orbit around Mercury, with frequent superior solar

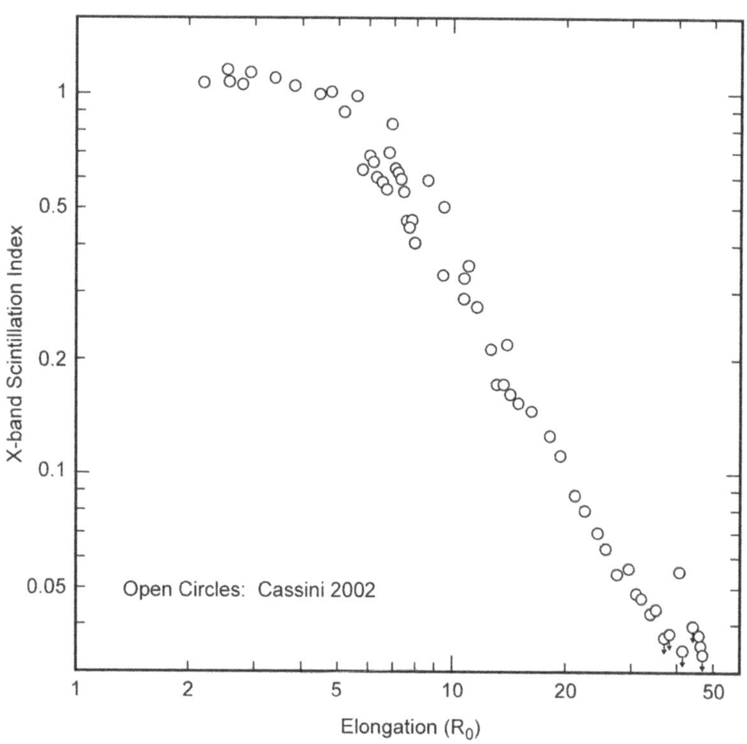

Figure 4.3 Scintillation index (rms intensity/mean intensity) for X-band data versus Sun-Earth-spacecraft angle (expressed as offset from the solar limb in solar radii). The dots are Cassini data taken in 2002. The scintillation index initially increases with increasing path-integrated turbulence (as solar offset becomes smaller). As the path-integrated turbulence becomes large, however, the scintillation index asymptotically saturates at unity ("strong" intensity scintillation (Marians 1975; Rumsey 1975)).

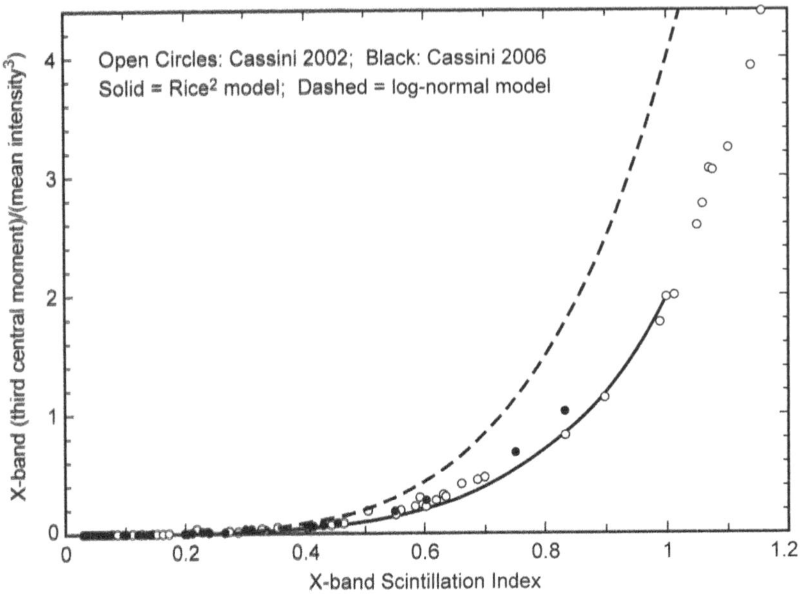

Figure 4.4 Relationship between normalized third central statistical moment of intensity fluctuations and the scintillation index (normalized rms intensity). This is useful for constraining the functional form of the scintillation probability distribution (in turn used, after correcting for different geometries, to predict telemetry link performance for tracking close to the Sun). The predictions of two idealized models ("Rice-squared" (solid line) and "log-normal" (dashed line); see, e.g., Armstrong 1993) are plotted. Either model fits reasonably well when the scattering is weak (scintillation index $\ll$1), but both fail when the scattering is strong: log-normal does not fit the data and the Rice-squared does not predict scintillation indices >1.

conjunctions, provides abundant occasions to explore the relativistic gravitational effects of the Sun and the structure of the solar corona. Figure 4.5 illustrates an example of the Radio Science investigations with BepiColombo. Figure 4.6 illustrates the MORE end-to-end instrumentation with two uplink radio signals transmitted simultaneously from a ground station and three coherent downlink signals that are coherently returned back by the spacecraft.

General relativity, Einstein's theory of gravity, has passed every test to date. The incompatibility of general relativity and quantum mechanics, however, has led scientists to question the ranges of their validity and to believe that either one or both will ultimately fail. Furthermore, cosmological observations that the universe undergoes phases of accelerated expansion provide compelling motivations to seek more accurate laws of gravity. The theory of general relativity will likely require a modification, such as the inclusion of a scalar term in the field equations. Deviation of the values of the parameterized post-Newtonian (PPN) parameters from those expected for general relativity at the level of 10^{-7}–10^{-5} is predicted, as discussed in Damour and Nordtvedt (1993). Experimental detection of violations of the theory would have significant implications in physics and cosmology. Such experiments require technological advances which have been slow and often very costly, with incremental improvements typically made on the time scale of a decade.

Figure 4.5 An illustration of the relativistic bending of a radio beam transmitted by the Mercury Planetary Orbiter and received on Earth.

A unique opportunity is available via the European Space Agency's (ESA) Mercury Planetary Orbiter (MPO), one of two spacecrafts comprising the multinational BepiColombo mission to Mercury; the second spacecraft is a Japanese Mercury Magnetospheric Orbiter. ESA selected the MORE team for the MPO to investigate relevant PPN parameters, solar oblateness, and possible time variations of the gravitational constant in addition to planetary geophysical objectives. MPO is specifically designed for Radio Science observations with flight instruments contributed by the Italian Space Agency (ASI). Compatible Radio Science ground instrumentation is proposed to be provided by NASA via the DSN.

4.10 Scientific Goals and Objectives

MORE's scientific goals are to carry out high-precision dynamic tests of relativistic gravity in an ideal laboratory and to characterize the structure of the solar wind in and out of the solar ecliptic. The conventional framework for discussing Solar System tests

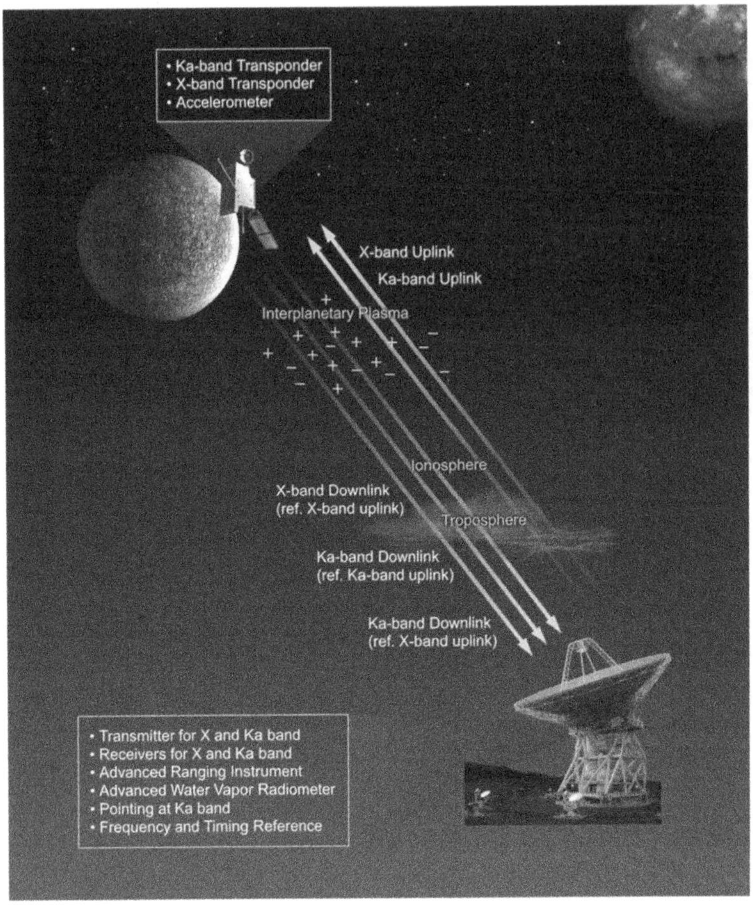

Figure 4.6 MORE end-to-end instrumentation: two uplink frequencies (X-band and Ka-band) are transmitted simultaneously from a ground station and three coherent downlink signals are coherently returned back by the spacecraft.

is the post-Newtonian parameterization. General relativity predicts definite values of the parameters but alternate theories of gravity predict deviations from these values. Nearly every metric theory of gravity can fit into the generalized 10-parameter PPN framework except for possible cosmological effects on the gravitational constant (Ashby et al. 2007). Of the 10 parameters, four are considered for improvement by MORE techniques, namely, the PPN γ, β, η, and α_1. In addition, the solar oblateness will be determined with improved accuracy, useful information will be obtained on the possible rate of change of the gravitational constant, and properties of the solar corona will be monitored accurately. Thus, the objectives of the MORE investigation are as follows.

4.10.1 Determine γ to an Accuracy of 2×10^{-6}

In the PPN formalism according to Shapiro (1964), Will (1971, 1993), and others, γ is a measure of how much space curvature is produced by a unit rest mass. The theory of general relativity, where $\gamma = 1$, predicts that a ray of light grazing the Sun is deflected by 1.75 arcsec and delayed in time by roughly 200 μs. Deflection and delay experiments place constraints on γ. The best accuracy measured to date for this parameter is 2.3×10^{-5} from the Cassini experiment (Bertotti et al. 2003). MORE will achieve an accuracy for γ of 2×10^{-6}, an improvement of one order of magnitude.

4.10.2 Determine β to an Accuracy of ~3×10^{-5}

In the same formalism, $\beta = 1$ for general relativity and is a measure of the nonlinearity in the superposition law for gravity. MORE will achieve an accuracy for β of ~3×10^{-5} or better. This is a significant improvement over the best current accuracy of 1.2×10^{-4} (Williams et al. 2006), which was derived by determining η and using the Cassini value for γ in the $\eta = 4\beta - \gamma - 3$ parameter relationship.

4.10.3 Determine η to an Accuracy of at Least 4.4×10^{-4}

If gravity is described by a metric theory, the PPN parameter η is a linear combination of the parameters γ and β and equals $4\beta - \gamma - 3$ in the Nordtvedt relationship. It addresses the difference between radial and transverse stress of gravity and bears on possible strong equivalence principle violation. MORE will achieve an accuracy for η of 4.4×10^{-4} (Ashby et al. 2007) to possibly 2×10^{-5} (Milani et al. 2002), comparable with, or an improvement over, the current accuracy of 5×10^{-4} (Williams et al. 2006).

4.10.4 Determine α_1 to an Accuracy of 7.8×10^{-6}

According to Will (2006), while the parameters γ and β are used to describe "classical" tests of general relativity and are, in some sense, the most important, they are the only non-zero parameters in general relativity and scalar-tensor gravity. The parameters α_1, α_2, and α_3 measure whether or not the theory predicts post-Newtonian preferred-frame effects. The MORE simulations by Milani et al. (2002) achieve a determination for the PPN preferred-frame parameter α_1 to an accuracy of 7.8×10^{-6}, a significant improvement of the present accuracy of ~10^{-4} (Will 2006).

4.10.5 Determine the Solar Oblateness to an Accuracy of 4.8×10^{-9}

Measurement of the relativistic PPN parameter (such as β) is inextricably connected with the solar quadrupole moment J_2, which contributes, just as the relativistic corrections do, to the advance of Mercury's perihelion. However, for solar studies, it is reasonable to also consider the accuracy with which J_2 can be determined if general relativity is assumed to be correct. The expected accuracy for J_2 will provide information about the differential rotation of the solar core and will be relevant to a better understanding of the structure of the deep interior of the Sun. MORE will determine the solar oblateness in a dynamical measurement to an accuracy of 4.8×10^{-9}, much more accurate than the present estimate of 4×10^{-8} (Milani et al. 2002).

4.10.6 Test Any Time Variation of the Gravitational Constant, *G*, to an Accuracy of 3×10^{-13} Per Year

Alternate theories of gravity include cosmologically evolving scalar fields that lead to time variability of fundamental physical constants such as the gravitational constant,

G. Lunar laser ranging experiments have placed limits on variation of $=(4 \pm 9) \times 10^{-13}$ per year (Williams et al. 2006), making the uncertainty 9×10^{-13}.

Before this result was obtained, Will (1993) summarized recent classical tests of post-Newtonian gravity and the improved observational constraints on time variation of the gravitational constant. This was accomplished by ranging measurements to the Viking spacecraft at Mars, lunar laser ranging measurements, and pulsar timing data. Will quotes a suggestion by Bender et al. (1989) of the possibility of reaching an accuracy of the order of 10^{-14} per year for a Mercury orbiter with ranging accuracy of the order of 20 cm.

MORE will improve on the constraints for detecting a variation of this fundamental "constant" to an accuracy of 3×10^{-13} per year from one year of measurements and will provide an independent check of the lunar laser ranging result; after 2 years, the accuracy would improve to 1×10^{-13} per year.

4.10.7 Characterize the Solar Corona

During the relativity observations, the Earth–spacecraft LOS necessarily passes close to the Sun. As the radio beams propagate through the solar corona, information about the near-Sun plasma is imposed on the radio signals. These radio-wave propagation effects are noise for the MORE relativity observations and will be estimated and then largely removed using the sophisticated MORE radio system (spacecraft and DSN components). To implement the required calibration, however, the plasma effects will be known with excellent accuracy. This presents an opportunity to use the radio-wave scintillations for science; the radio measurements contain near-Sun plasma information on spatial and temporal scales that cannot be measured by other techniques.

4.11 Comparison with Other Experiments

Dedicated missions to study relativistic gravity have been proposed in the past (e.g., Bender et al. 1989), but recent simulations (Milani et al. 2002; Ashby et al. 2007) indicated that precision radio tracking of a spacecraft inserted in a circular orbit around Mercury could lead to an improvement in the determination of many PPN parameters. Testing relativistic gravity was recognized as a crucial scientific objective of BepiColombo at the inception of the project, although it is primarily a planetary mission. The Ka-band Transponder and accelerometer instruments will allow the MORE team to carry out many classical tests of relativity in the best dynamic Solar System conditions. Testing general relativity, however, should not be viewed as a race between missions. Each experimental result, with important implications in physics and cosmology, would require confirmation by other missions. The science community benefits from multiple investigations in this field.

4.11.1 Cassini

Launched in 1997, the Cassini mission to Saturn had a long cruise period during which a relativity experiment was carried out during the solar conjunction period of

2002. Radio science data at X- and Ka-bands were acquired at the DSN station (Goldstone, California) equipped for precision measurements at Ka- and X-bands for approximately one month centered on the solar conjunction where the minimum impact parameter was 1.6 solar radii, and no occultation by the Sun. The experiment was performed using Doppler (range-rate) observations, a method that had not been used until Cassini for such experiments because of the overwhelming noise contribution of the solar corona. Cassini overcame this limitation by augmenting the standard X-band link with a high frequency Ka-band link, providing a multi-frequency link system that calibrated and removed the effect of the solar corona. This capability was enabled on Cassini by a Ka-band Translator payload provided by ASI (different design from the MORE Ka-band Transponder also provided by ASI). The experiment determined, in agreement with general relativity (Bertotti et al. 2003), the parameter $\gamma = 1 + (2.1 \pm 2.3) \times 10^{-5}$.

4.11.2 Gravity Probe B

Gravity Probe B (GPB), a polar orbiting spacecraft containing four very precise gyroscopes, sought direct measurement of the geodetic and frame-dragging precession of a gyroscope's spin axis as the spacecraft orbits the spinning Earth. The GPB measurement can be cast in terms of PPN parameters. The final results are not yet available, but MORE is expected to determine γ with substantially better accuracy.

4.11.3 Messenger

The Mercury Surface, Space Environment, Geochemistry, and Ranging (MESSENGER) mission's published science objectives (Solomon et al. 2001) do not include determination of PPN parameters or J_2 of the Sun. MESSENGER is equipped with only one radio link (X-band) so precision measurements would be severely degraded by the un-calibrated noise on Doppler and range from the solar corona. The absence of an accelerometer further limits the necessary precision calibration of non-gravitational forces acting on the spacecraft.

4.11.4 Lunar Laser Ranging

Williams et al. (2006) reported analysis of laser ranges to the Moon that provides increasingly stringent limits on any violation of the Equivalence Principle and also enable very accurate tests of relativistic gravity. They report a value for the Strong Equivalence Principle violation parameter η of $(4.4 \pm 4.5) \times 10^{-4}$. Williams et al. (2006) also reported the search for a time variation in the gravitational constant results in $=(4 \pm 9) \times 10^{-13}$ per year. Further measurements with greater accuracy in the next 5–10 years appear likely to improve the accuracies by factors of roughly 3–5.

4.11.5 Gaia

Launched in 2013, Gaia has been charting a 3D map of our galaxy and has provided unprecedented positional and radial velocity measurements with the accuracies needed to produce a stereoscopic and kinematic census of about one billion stars.

Additional scientific products include a number of stringent new tests of general relativity and cosmology. Gaia is expected to provide a precision of γ of 1×10^{-6} (Mignard and Klioner 2009), an improvement approaching two orders-of-magnitude better than the current best estimate (Bertotti et al. 2003). Gaia, however, will not address the other MORE science objectives.

4.12 MORE Summary

The BepiColombo Mercury Orbiter Radio-science Experiment team will carry out high precision tests of relativistic gravity in the most desirable laboratory in the Solar System, the gravitational field of the Sun. General relativity predicts definite values of the PPN parameters, but alternate theories of gravity predict deviations from these values. Of the 10 parameters, 4 are considered for improvement by MORE techniques, namely, the PPN parameters γ, β, η, and α_1. In addition, the solar oblateness will be determined with much improved accuracy, useful information will be obtained on the possible rate of change of the gravitational constant, and properties of the solar corona will be monitored accurately. Testing relativistic gravity was recognized as a crucial scientific objective of BepiColombo; the Ka-band Transponder and accelerometer instruments enable many tests of relativity. Testing general relativity should not be viewed as a race between missions. Experimental results with important implications in physics and cosmology would require confirmation by other missions, and the science community benefits from multiple investigations in this field. With BepiColombo, comparable accuracies would be achieved with quite different types of measurements for γ, β, η, and α_1, and the accuracy for J_2 of the Sun would be improved by a substantial factor.

4.13 Anomalous Motion of Pioneers 10 and 11

Observations of the Pioneer 10 and 11 spacecraft showed a large, anomalous, steady, and apparently sunward-directed acceleration of about 8.7×10^{-8} cm/s^2 (Anderson et al. 1998, 2002). This acceleration was unmistakably detected above the measurement noise. The cause of the acceleration was initially unclear. Possible systematic effects were discussed by Anderson et al. (2002), and the observations provoked speculation about "new physics." It is now believed that the cause was a systematic effect due to anisotropic radiation from the radioisotope thermal generators powering the Pioneer spacecraft (Turyshev and Toth 2010; Will 2014).

Appendix 4A

Solar Corona Observation Methodology Illustrated by Mars Express

Contributed by Stefan Remus[1]

4A.1 Formulation

Solar corona observations (SCOs) are often carried out by Solar System missions when a spacecraft's orbit places it in a solar conjunction configuration, behind the Sun as seen from Earth. Radio communication signals from the ground station to the spacecraft and/or from the spacecraft to the ground station travel through the solar corona, enabling scientific probing of the coronal environment. An SCO experiment can be carried out in the one-way link mode (defined in Chapter 5) or in the two-way (or three-way) mode when ranging is required.

Prior to the development of uplink (forward link) at X-band (7.1–8.4 GHz), SCO experiments used S-band (2.1–2.3 GHz) uplink and coherent S- and X-band downlinks (see Table 1.3 in Chapter 1). S-band is more susceptible than X-band to electron content variations in the uplink path as well as multipath effects that can cause loss of signal lock at the spacecraft. The susceptibility to charged particles is inversely proportional to the square of the carrier frequency, so lower frequencies lead to a higher impact of the charged media on the signal. Recent missions have been moving away from using S-band telecom links for either uplink or downlink in order to improve telecommunications and navigation performance as well as reduce this effect. The exceptions are missions with specific radio science (RS) experiments at various wavelengths; for example, Cassini used a carrier-only S-band transmitter specifically for radio occultation experiments.

Mars Express (MEX), Venus Express (VEX), and Rosetta (ROS) are examples of missions that used downlink carrier frequencies at S- and X-bands during SCO. Figure 4A.1 illustrates the principle for the measurements carried out when the spacecraft's apparent offset from the Sun is within 40 solar radii, or solar elongation angle within ~10°.

An example view of the experimental geometry from a ground station is shown in Figure 4A.2. The faint background shows a Large Angle and Spectrometric Coronagraph (LASCO) image with a coronal mass ejection (CME) and streamer to demonstrate typical structures present in thc region of space investigated with coronal radio sounding. The Earth to Mars distance while Mars was in solar conjunction, in this example, was ~2.4 astronomical units (au).

---

1 Edited by S. W. Asmar based on material originally prepared for the 2014 ESA document "Mars Express Solar Corona Experiment, A Cookbook," by S. Remus, M. Hahn, M. Bird, M. Pätzold, B. Häulser, S. W. Asmar, O. Witasee, and G. L. Tyler; https://www.cosmos.esa.int/documents/772136/977578/Solar_Corona_Experiment___A_Cookbook_17Mar2014.pdf/6cbdca81-9ca3-4240-87a5-017e380cdb14.

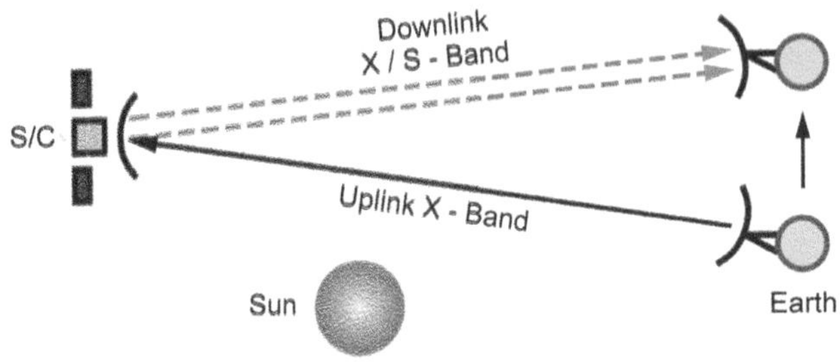

Figure 4A.1 SCO experiment: geometry and link configuration overview.

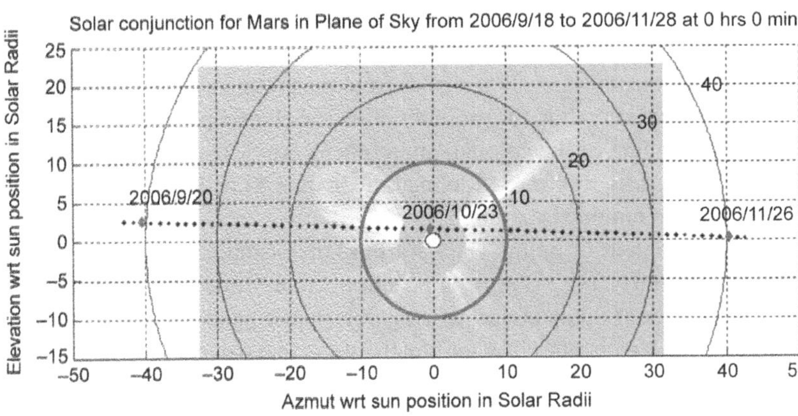

Figure 4A.2 Mars SCO of 2006 as seen from Earth in the plane of sky or solar ecliptic system.

The two radiometric tracking data types utilized for SCO are range (or ranging, in meters (m)) and Doppler (in Hertz (Hz), also known as range-rate); see additional details in Chapter 5 and in Moyer (2003). Ranging is a measurement of the distance along the line of sight (LOS) computed based on the difference between the time stamp of the transmitted and received ranging code modulated on the signal carrier. The round-trip light time (RTLT) of the ranging code, multiplied by the velocity of light, c, gives the range; ambiguities of the position of the spacecraft can be resolved in the data processing phase.

The Doppler data are based on the time measurement of the carrier signal phase, converted to a frequency. The transmitted frequency at time t_0 is compared with the returned signal frequency at t_0 + RTLT and the difference is the round-trip Doppler-shifted frequency. If measured at the spacecraft, it would be the one-way Doppler shifted frequency, approximately half the effect. In addition to the effects of the radial motion between the transmitter and the receiver, changes in ray path length due to the propagation through the media of interest are also mixed with this observable. The effect of the distance change between the transmitter and receiver is modeled accurately using navigational orbit determination tools, thereby isolating the important effect due to propagation through the charged medium for scientific analysis.

For Doppler measurements, the cycles of carrier phase along the ray path are counted as a function of time without resolving the cycle ambiguity (based on the wavelength). Therefore, only the changes from a given point in time, such as the beginning of signal acquisition, can be measured, not the total number of cycles. For range measurements, the group delay is computed to an accuracy dependent on the ranging code (as described in Chapter 5) and a unique solution can be obtained for the spacecraft radial (or LOS) distance within the ambiguity associated with the highest ranging signal component.

4A.2 Total Electron Content from Ranging Data

The number of electrons in a column along the LOS between the spacecraft and ground station, also known as the plasma, can be derived from the additional delay of the ranging signal compared to the delay in the absence of electrons (free space). The SCO ranging measurement is conducted at two frequencies in order to deduce the frequency-dependent (dispersive) plasma effects from the frequency-independent effects due to the relative motion. The plasma contributions to the ranging delay are higher at S-band than X-band, while the free space travel time is the same at the speed of light.

Figure 4A.3 illustrates the geometrical separation of uplink and downlink due to the movement of the ground station during the time span of the RTLT, typically tens of minutes. Among the limiting factors for the accuracy of the spacecraft topocentric range are the signal-to-noise ratio (SNR) and integration time of the highest tone. For example, for a 1-mega-Hertz (MHz) ranging tone, a phase uncertainty of 1 degree results in a position LOS error of about 1 m.

The total electron content (TEC) is the integrated number of electrons along the ray path, with dimensions of per inverse meters squared. The S- and X-band frequencies (*fs*, *fx*) are known precisely and a linear relation between the path length change and the TEC along the downlink is

$$TEC_{down} = K_{mission}\Delta S_{down\,plasma} = K_{mission}c\Delta\tau, \tag{4A.1}$$

where the constant K is unique for every mission due to assigned frequency channels to the mission by spectrum managers. $\Delta\tau$ is the group delay difference between the S- and X-band ranging observables. For MEX, the nominal S- and X-band frequencies are 2296.432 and 8420.432 MHz, respectively, leading to

$$TEC_{down\_MEX} = 4.240\,\Delta\tau_{down\,plasma}, \tag{4A.2}$$

where TEC is in hexems (1 hexem = 10^{16} electron per m^2) and the differential delay time, $\Delta\tau$, is in nanoseconds (ns). The change in travel time due to the number of

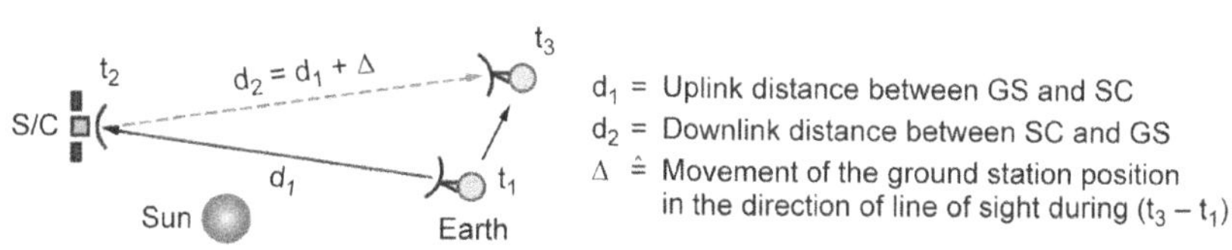

Figure 4A.3 Ranging geometry between the spacecraft (SC) and ground station (GS). The quantity Δ can be positive or negative.

electrons in the propagation path (N_e is the electron density in a cubic-meter) can be expressed as a change in the path length, S_{plasma} divided by c, where S_{plasma} is

$$S_{plasma} = \frac{40.31}{f^2}\int_0^S N_e ds = \frac{40.31}{f^2} TEC. \tag{4A.3}$$

Bird (1982) posit that both group and phase delays of a signal traveling through a plasma environment are given by

$$T_{gr/ph} = \int \frac{ds}{v_{gr/ph}} = \frac{S}{c} \pm \frac{40.31}{cf^2}\int_0^S N_e ds = \frac{S}{c} \pm \frac{40.31}{cf^2} TEC. \tag{4A.4}$$

The travel distance of the uplink and downlink path is (refer to Figure 4A.3)

$$S_{measured\ phase} = d_1 + s_{up\ plasma} + d_2 + s_{down\ plasma}. \tag{4A.5}$$

Note that d_1 is the distance in meters that the transmitted signal travels before being received at the receiver; the transmitter continues to move after transmission and the distance changes. For the return path with Earth and the spacecraft in constant motion, this measurement approach applies equally for the S- and X-band signals, canceling each other out. However, the time rate of distance change is of interest.

The plasma contribution can also be expressed by the number of electrons in the path per square meter. The total number of electrons is usually counted in units of hexem (hexem = 10^{16} el/m^2). The plasma path delay difference between S- and X-band signals, assuming the paths are equal, which is a good approximation for a tenuous plasma (see Bertotti and Giampieri 1998), is

$$\begin{aligned}\Delta S_{S-X} &= \left|d_1 + s_{up\ plasma} + d_2 + s_{down\ plasma}\right|_{S-band\ ranging} \\ &\quad - \left|d_1 + s_{up\ plasma} + d_2 + s_{down\ plasma}\right|_{X-band\ ranging}. \\ \Delta S_{S-X} &= S_{S\ up\ plasma} - S_{X\ up\ plasma} + S_{S\ down\ plasma} - S_{X\ down\ plasma}\end{aligned} \tag{4A.6}$$

MEX can receive uplink at either S- or X-band and can coherently transmit at both bands simultaneously. For this configuration, by analyzing the dual frequency downlink only, the path difference observed between the S- or X-band ranging signals, as explained earlier, can be transformed to the number of electrons in the path (TEC) according to

$$\Delta s_{down\ plasma} = 40.31\left(\frac{1}{f_s^2} - \frac{1}{f_x^2}\right) TEC_{down} \tag{4A.7}$$

and

$$\Delta \tau_{down\ plasma} = \frac{40.31}{c}\left(\frac{1}{f_s^2} - \frac{1}{f_x^2}\right) TEC_{down}. \tag{4A.8}$$

TEC_{down} is defined as $\int_{SC}^{GS} N_e ds$ (where SC is the spacecraft (transmitter) and GS is the ground station (receiver) and both are a function of time).

As noted earlier, MEX can receive uplink at either S- or X-band and can coherently transmit at both bands simultaneously. So, the MEX two-way dual-frequency links do not allow a determination of TEC on the uplink. Independent values of the uplink and downlink TEC would require five different radio links, such as two signals for the uplink and three signals for the downlink; as in the case of the system on the Cassini, Juno, and BepiColombo missions (see Chapters 4 and 5).

4A.3 Change in Total Electron Content from Doppler Data

As discussed in the previous section, coherent dual-band downlink can be used to study the plasma content along the LOS by isolating and removing the effect of the dispersive media. Because of the ambiguity in the measurement of the total phase of the received carrier Doppler profile, it is only possible to determine the change in TEC relative to the beginning of the recording, which is set to zero.

Advantages of the carrier signal Doppler measurements over coded ranging measurements are the higher SNR and the short wavelength yielding a much smaller error bar for the TEC values. The observed differential Doppler frequency shift follows the change in electron content as a function of time and the received signal frequency at the ground station is (note the designation of the time derivative of TEC):

$$
\begin{aligned}
f_{rec} &= f_0 - \frac{v_{rel}}{c} f_0 + \frac{40.31}{c}\frac{1}{f_0}\frac{d}{dt}\int_{SC}^{Earth} N_e dy (+other) \\
&\cong f_0 - \frac{v_{rel}}{c} f_0 + \frac{40.31}{c}\frac{1}{f_0}\dot{TEC}
\end{aligned}
\tag{4A.9}
$$

and

$$
f_{rec} = k_{X,S}\left(f_{up} + f_{up\_Dop} + \Delta f_{up\_plasma}\right) + f_{down\_Dop} + \Delta f_{down\_plasma}. \tag{4A.10}
$$

$\dot{TEC}$ corresponds to the time derivative of the TEC in line of sight and the constant $k_{X,S}$ is the coherent frequency multiplier or transponder ratio at the spacecraft:

$$
k_{X_{down}/X_{up}} = \frac{f_{X_{down}}}{f_{X_{up}}} = \frac{880}{749} k_{S_{down}/X_{up}} = \frac{f_{S_{down}}}{f_{X_{up}}} = \frac{240}{749}. \tag{4A.11}
$$

The frequency ratio between S- and X-band is

$$
k_{S/X} = \frac{f_S}{f_X} = \frac{3}{11}. \tag{4A.12}
$$

Using these constant relations, the downlink plasma residual calculated from the differential Doppler of the downlink signals is

$$
\begin{aligned}
\Delta f_{Plasma} &= f_{rec\_S} - \frac{3}{11} f_{rec\_X} \\
\Delta f_{Plasma} &= \frac{\dot{s}}{c} f_S + \frac{40.31}{c}\frac{1}{f_S}\dot{TEC} - \left(\frac{3}{11}\frac{\dot{s}}{c} f_X + \frac{40.31}{c}\frac{f_S}{f_X^2}\dot{TEC}\right)
\end{aligned}
\tag{4A.13}
$$

$$\Delta f_{Plasma} = \frac{40.31}{c} f_S \left(\frac{1}{f_S^2} - \frac{1}{f_X^2} \right) T\dot{E}C \tag{4A.14}$$

$$T\dot{E}C \cong \frac{\Delta TEC}{\Delta t} = \frac{c}{40.31} f_S^{-1} \left(\frac{1}{f_S^2} - \frac{1}{f_X^2} \right)^{-1} \Delta f_{Plasma}. \tag{4A.15}$$

These quantities are known and the integration of this equation with time yields $TEC(t)$ from the beginning of the measurement:

$$TEC = \left[\frac{40.31}{c} f_S \left(\frac{1}{f_S^2} - \frac{1}{f_X^2} \right) \right]^{-1} \int_{t_0}^{t} \Delta f_{Plasma} dt + TEC_0. \tag{4A.16}$$

The value TEC_0 can only be achieved by a two-frequency band ranging measurement or is reset to zero for every Doppler data arc.

4A.4 Electron Density

In an early study of the electron density distribution of the solar corona, Baumbach (1937) used coronal brightness data, assuming spherical symmetry, which were transformed via an Abel integral to the electron density distribution with the following radial dependence:

$$N(r) = 10^8 \left(\frac{0.036}{r^{1.5}} + \frac{1.55}{r^6} + \frac{2.99}{r^{16}} \right). \tag{4A.17}$$

Allen (1947) improved the empirically based model, now known as the Allen-Baumbach model, by separating the source of the coronal brightness into two parts: sunlight scattered on electrons (Thomson scattering) and Fraunhofer emission lines from solar corona constituents. Latitudinal effects were also included by Tyler et al. (1977) and Pätzold et al. (1995) using

$$F(\phi) = \sqrt{A^2 \cos^2 \phi + \sin^2 \phi}. \tag{4A.18}$$

Muhleman and Anderson (1981) used a Gaussian latitudinal variation and obtained the following parametric fit to the Viking coronal sounding data:

$$N_e = \frac{1.32 \cdot 10^6}{\left(r / R_\odot \right)^{2.7}} e^{-\left(\frac{\phi}{\phi_S} \right)^2} + \frac{2.3 \cdot 10^5}{\left(r / R_\odot \right)^{2.04 \pm 0.02}}. \tag{4A.19}$$

The solar latitude is represented by ϕ and the best fit was received at $\phi s = 8° \pm 3°$. r is the distance of the LOS to the Sun center and R is the solar radius. Bird et al. (1994) adapted the following equation to the special geometry applicable to the Ulysses solar conjunction in 1991:

$$N(r,\phi) = \left[N_A \left(\frac{R_\odot}{r} \right)^6 + N_B \left(\frac{R_\odot}{r} \right)^\alpha \right] F(\phi). \tag{4A.20}$$

It was found that $F(\phi)$ and r^{-6} were negligible for the ranging data set recorded during the 1991 Ulysses conjunction. This simplified the equation and the integration along the downlink ray path from Ulysses to Earth yield:

$$TEC = K(\alpha) N_B R_\odot \left(\frac{R_\odot}{r} \right)^{\alpha-1} \tag{4A.21}$$

with

$$K(\alpha) = \int_{-\pi/2}^{\pi/2} \cos^{\alpha-2} \phi \, d\phi = \sqrt{\pi} \left[\Gamma\left(\frac{\alpha-1}{2} \right) / \Gamma\left(\frac{\alpha}{2} \right) \right]. \tag{4A.22}$$

Bird (1982) presented a summary of SCO measurement models up to the time of that publication. Figure 4A.4 shows a contour plot of the TEC in the ecliptic and polar planes based on the density model derived from Tyler et al. (1977).

4A.5 Coronal Mass Ejections

CMEs are massive eruptions of plasma that separate from the Sun and travel through the Solar System. In case the CME travels through the LOS between Earth and spacecraft during a two-way tracking pass (see Figure 4A.5), both the uplink and downlink signals are affected by the change in the TEC. Each part of the moving CME that crosses the radio link is sounded twice, once in the uplink and once in the downlink. The CME signature modulated near time t_0 on the signal will be seen in the received data at the station on the downlink at t_1 and a second time at t_3 after the signature on the uplink travels to the spacecraft and back to the station.

In general, it is possible to observe downlink signals with two ground stations at the same time, as shown in Figure 4A.6. The setup allows calculating the CME velocity by using the time difference of the CME features on the two different downlink paths on a one-way dual-frequency link. This would be a highly efficient method to calculate the CME velocities since the distance between the ground stations and in turn, between the signal paths, is very well known, and the bending of the signal is equal with the exception of a small change of TEC versus time that occurs between the locations or areas of the modulations. The drawback is the limited observation time due to visibility from two different Deep Space Network (DSN) stations at different longitudes in combination with the earth rotation.

The solar conjunction geometry shown in Figure 4A.7 is special as the transmitting and receiving units have equal parallel velocities, unlikely in reality although useful for a theoretical study. The figure illustrates that all signals traveling in time along the blue and black lines, which will be received in the future at the ground station or at the spacecraft, are for a snapshot, at time t_0, lined up without a geometrical distance between the uplink and downlink signal positions. The example uplink and downlink positions are represented by yellow and blue markers, respectively; the yellow markers are on top of the blue markers, thus the blue markers are not visible in the figure. This geometry configuration does not allow determining the CME velocity.

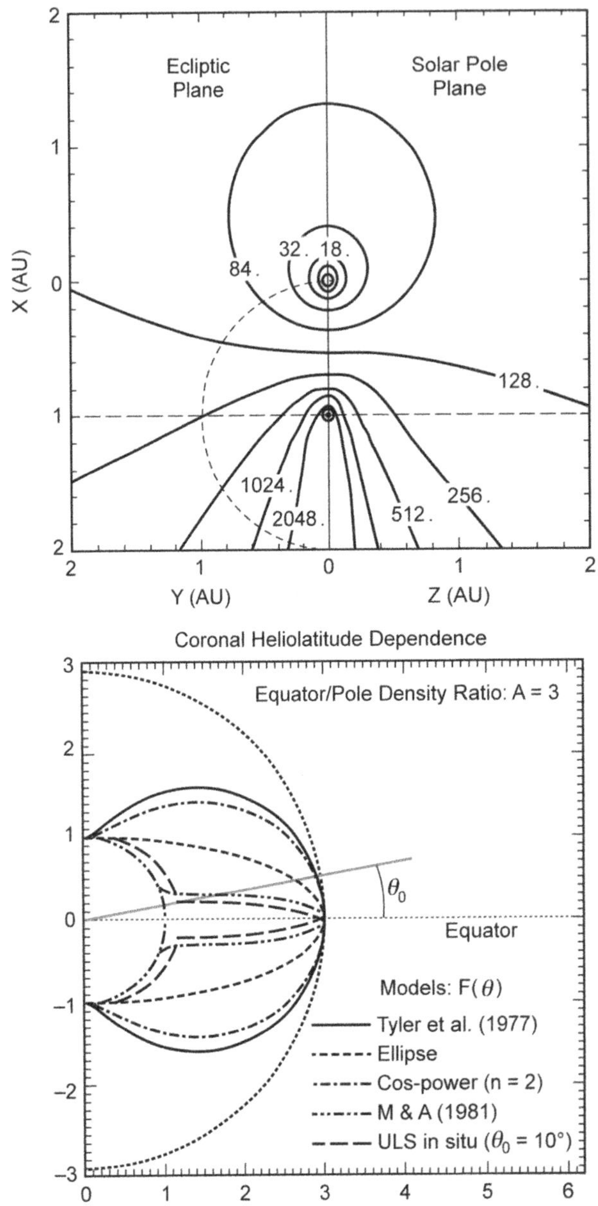

Figure 4A.4 Upper panel: TEC in the x–y solar ecliptic plane and the x–z solar polar plane from Bird (1982) in hexem, based on a model from Tyler et al. (1977). Lower panel: Various models for the heliospheric latitude dependency as modified from Bird et al. (1996b).

Figure 4A.8a illustrates another model where the relative velocity of the station is zero with respect to the Sun. The locations of the signals that were sent in the past but have not reached the receiver are traveling along the black lines (downlink) and along the blue lines (uplink). The positions of the shown signal locations are separated by $\Delta t = LT/6$ on the example paths and are marked blue or yellow. A theoretical

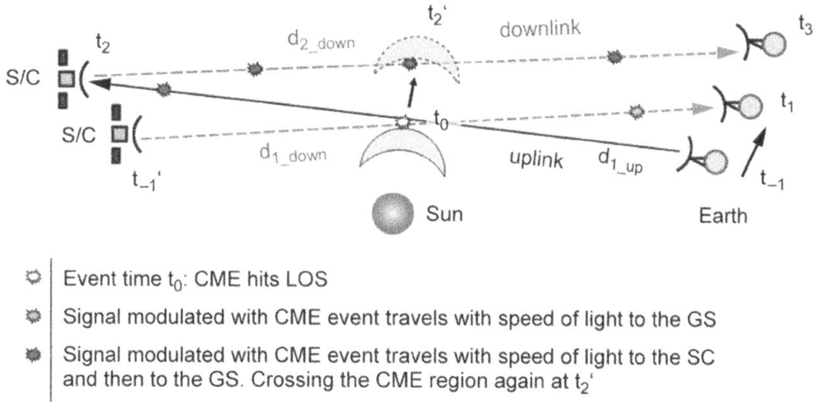

Figure 4A.5 Propagation of the CME and the modulated radio signal versus time. Two scenarios are described: (1) the modulation event at t_0 indicated by the yellow mark travels along the downlink path indicated by the green mark toward the ground station, and (2) blue markings indicate upward effects from the ground station to the spacecraft that are "transponded" back to the ground station. Since time has elapsed, on that downlink path, the CME is sounded again at time t_2'. In time symbols, (1) $t_{-1}' \rightarrow t_0 \rightarrow t_1$ and (2) $t_{-1} \rightarrow t_0 \rightarrow t_2 \rightarrow t_2' \rightarrow t_3$.

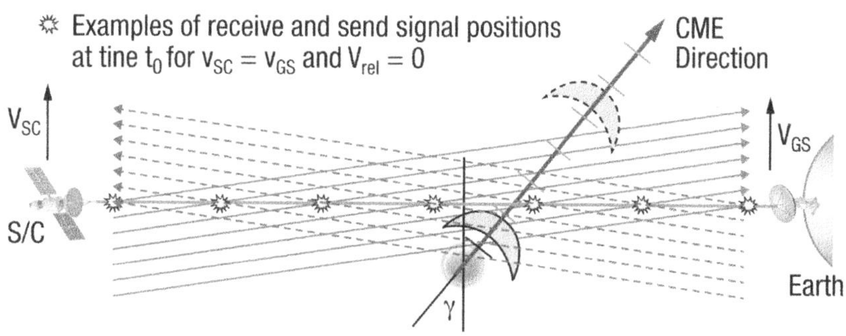

Figure 4A.6 Geometry for the velocity determination of a CME with two ground stations.

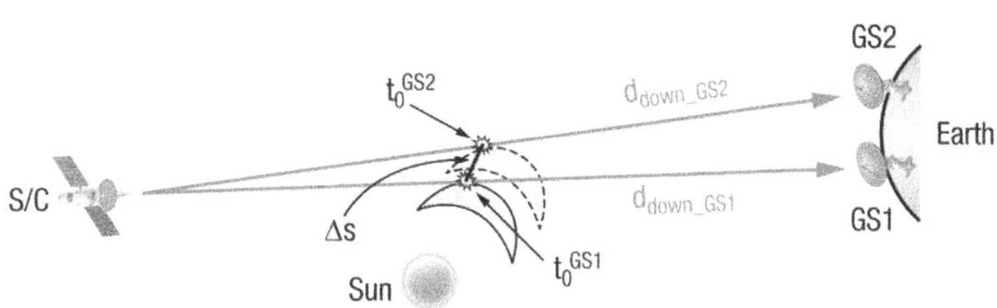

Figure 4A.7 Schematic for the position of the signals at a defined time. The velocities of the ground station and spacecraft are equal and parallel. The red line indicates all microwave signal positions that were sent in the past from the spacecraft and the station and will arrive in the future at the station and the spacecraft. The yellow markings indicate the positions of the signals that will be received in the future at the ground station and at the satellite. Along the black lines, the position is given for the downlink and along the blue lines, the positions for the uplink. This special example does not allow CME velocities to be estimated because there is no location and time difference between the modulation of the uplink and downlink signal.

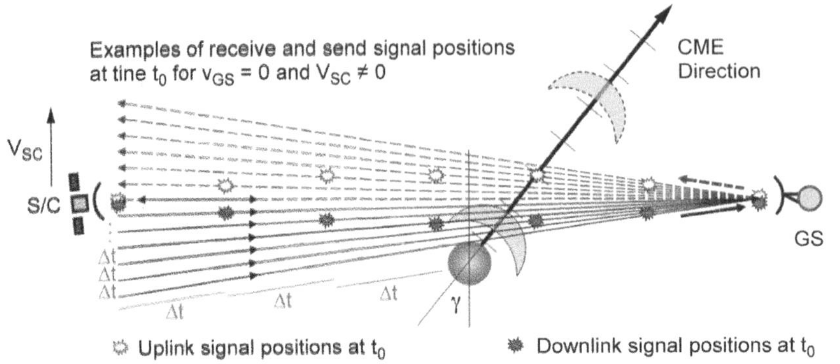

Figure 4A.8a Another example for the location of the uplink and downlink signals at time t_0. This example reflects the normal case with nonlinear velocity changes between GS and SC. The example radio signal positions (yellow and blue markers) will only be a bit more twisted in space. The uplink and downlink modulation do not take place at the same position and thus, not at the same time.

connection of the yellow uplink markers would reflect the location of the continuous signal that would be received at the spacecraft in the future and, for the blue markers, would show the continuous signal positions that would be received in the ground station in the future for this time t_0. The figure clearly shows that the modulation of the signals due to the crossing CME on the uplink and downlink takes place at different times and locations. Thus, the uplink and downlink signals have a separation in space, and features in the electron content of the CME are therefore separated in time on the received uplink and downlink signals. The structure of the whole electron content is seen (received) in the microwave phase shift twice. The two modulated plasma shifts are from different times and locations on the uplink compared to the downlink.

Figure 4A.8b has been adapted to an ingress and egress case for an SCO campaign at Mars at a constant time t_0, while Figure 4A.9 reflects the signal links and positions that are involved in an event that occurs at t_0.

It is assumed that the CME velocity vector penetrates in a radial direction from the Sun and that the direction is determined through additional information. If the direction is not known, an approximated velocity can be calculated for the signal path tangential point in order to determine points P($t_{0\,up}$) and P($t_{0\,down}$). This information allows reconstruction of the CME crossing event on the uplink and downlink (Figure 4A.9). All times and locations are known and the velocity of the CME can be calculated using

$$v_{CME} = \frac{\Delta s_{CME}}{\Delta \tau_{CME}} = \frac{P_r\left(t_{0_{up}}\right) - P_r\left(t_{0_{down}}\right)}{t_{0_{up}} - t_{0_{down}}}. \tag{4A.23}$$

$P_r(t)$ is the radial distance from the Sun at time t.

The fast TEC changes reveal finer structures in the corona, such as waves or boundaries, but the locations of the changes are local and the interpretation may not be unique; for example, the TEC shows some variation after the front of a CME passes the radio link. These small changes are possibly due to a change of electron density in the tail itself or a diameter change of the tail with no change in the electron number density.

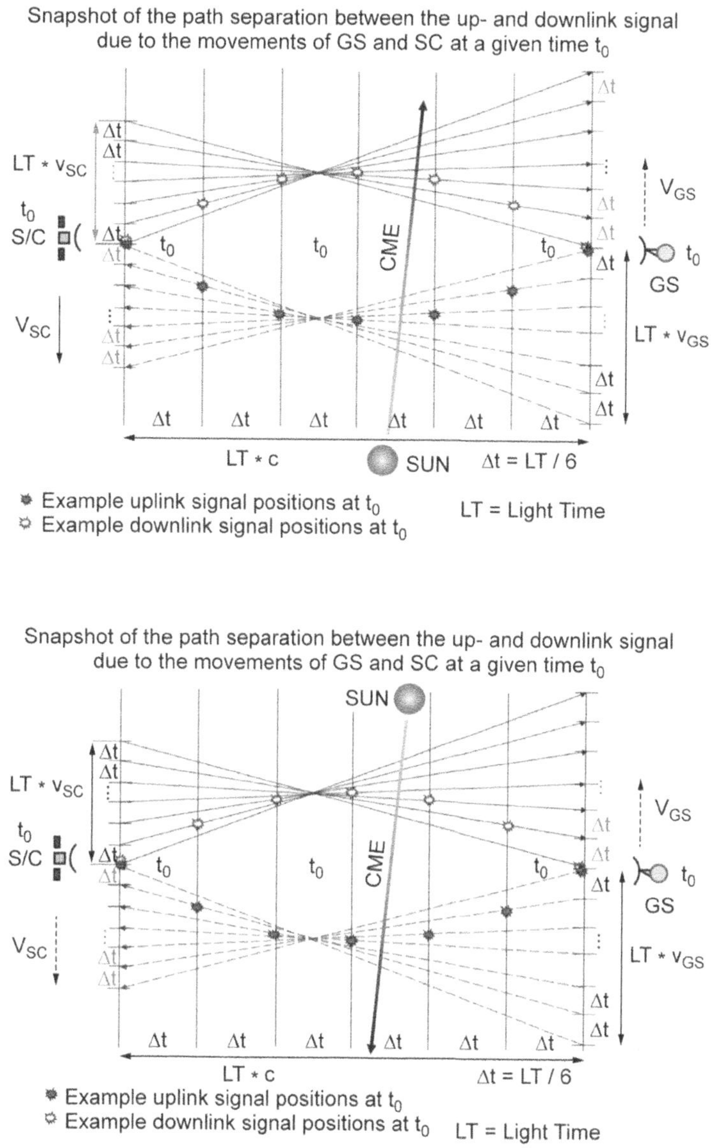

Figure 4A.8b SCO drawing for a spacecraft at Mars distance, which accounts for the orbit velocities of Mars and Earth. The example shows the location of the uplink and downlink signals at time t_0 in the ecliptic plane (no *z*-direction movement). The left part of the figure represents the egress SCO with the Sun at the bottom. Due to the higher velocity of the GS, the signal tangential point distance to the Sun is increasing while the figure slowly rotates around the Sun. The right part shows the SCO ingress case. The difference between the two cases is, which link is crossed by the plasma flow first. In the egress case, the CME modulates the downlink earlier than the uplink signal. The modulation of the ingress links is reversed.

Dividing the received data set into segments containing features may reveal a velocity series over time. Note that the uplink and downlink paths cause different bending that has to be carefully analyzed. Furthermore, the difference of the bending at S- and X-bands contributes to the offset between uplink and downlink signal phase path

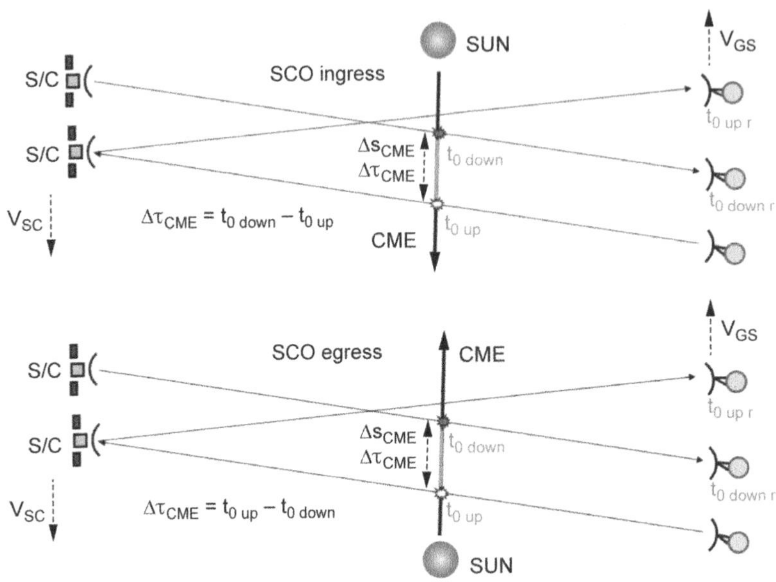

Figure 4A.9 Location of the CME direction in space and time for the general case of different velocity vectors for the spacecraft and ground station. The drawing shows an observation for the SCO ingress and egress geometry. For the egress case, the downlink is modulated first and the uplink a bit later by the CME plasma-free electron content. See also the right-hand image in Figure 4A.8b.

lengths, Δs_{CME}. The typical configuration for SCOs enables a check on the derived CME velocity, which should be equal for the $X_{up} - X_{down}$ and $X_{up} - S_{down}$ analysis.

The TEC on the downlink path derived by the differential ranging enables approximation of the bending difference and the ray path distance at the downlink between S- and X-band TEC features of the less noisy Doppler data. Figure 4A.10 shows the bending of a radio signal for the S, X, and K frequency bands as a function of the ray parameter (tangential distance of the signal path from the Sun) in solar radii. For example, S- and X-band signals at 5 radii will have a bending difference of ~10^{-6} degrees, resulting in a distance of ~ 150 km between the rays (for a station position at 1 au and spacecraft at infinity). When applying a speed of 400 km/s for the CME, the time difference on the downlink is ~0.25 s, which may be detectable in the downlink bandpass-filtered TEC residuals by a correlation of S- and X-band data.

4A.6 Separation of Uplink and Downlink Effects from Plasma

As noted in the previous section, the methodology assumes that the S- and X-band signals propagate along the same downlink path and that modulation occurs at the same time. However, this is not absolutely true because the different bending paths at closest approach are separated and the moving plasma creates the events on the signal at slightly different times. Though not explored here, it may be possible to find an iterative process that accounts for detected differences in the modulation of the downlink bands.

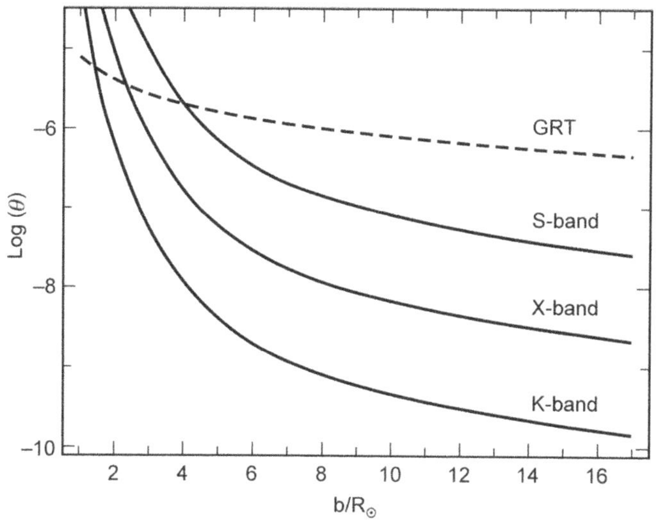

Figure 4A.10 Radio signal deflection due to the solar corona on the S-, X-, and K-band signals as shown in Bertotti and Giampieri (1998). GRT is the general relativity theoretical bending.

With the constant ratios between the uplink and downlink frequencies, where $f_S/f_X = 3/11$ from Equation (4A.12):

$$f_S^2\left(\frac{1}{f_S^2}-\frac{1}{f_X^2}\right)=\frac{112}{121}, \tag{4A.24}$$

we get for the S-band plasma downlink contribution:

$$\begin{aligned}\Delta f_{Plasma,down,X} &= \frac{40.31}{c}\frac{1}{f_s}\dot{I} = \frac{40.31}{c}\frac{1}{f_s}\frac{\Delta I}{\Delta t} \\ &= \frac{40.31}{c}\frac{1}{f_s}\frac{c}{40.31}\frac{1}{f_S}\left(\frac{1}{f_S^2}-\frac{1}{f_X^2}\right)^{-1}\Delta f_{Plasma}. \\ &= \frac{121}{112}\Delta f_{Plasma} \\ &= \frac{121}{112}\left(f_{rec\,S}-\frac{3}{11}f_{rec\,X}\right)\end{aligned} \tag{4A.25}$$

And similar for the X-band downlink contribution:

$$\begin{aligned}\Delta f_{Plasma,down,X} &= \frac{40.31}{c}\frac{1}{f_X}\dot{I} = \frac{40.31}{c}\frac{1}{f_X}\frac{\Delta I}{\Delta t} \\ &= \frac{40.31}{c}\frac{f_S}{f_X}\frac{c}{40.31}\frac{1}{f_S^2}\left(\frac{1}{f_S^2}-\frac{1}{f_X^2}\right)^{-1}\Delta f_{Plasma}. \\ &= \frac{3}{11}\frac{121}{112}\Delta f_{Plasma} \\ &= \frac{3}{11}\frac{121}{112}\left(f_{rec\,S}-\frac{3}{11}f_{rec\,X}\right)\end{aligned} \tag{4A.26}$$

Finally, the X-band uplink plasma contribution on the signal can be calculated from Equation (4A.9) after a correction for the movements of the spacecraft and station and further calibrations ($fx_{down}/fx_{up} = 880/749$ from Equation 4A.11):

$$\Delta f_{Plasma,up,X} = \frac{749}{880}\left(\Delta f_{Plasma,X} - \Delta f_{Plasma,down,X}\right) = \frac{40.31}{c}\frac{1}{f_{up,X}}\dot{I}_{up} \tag{4A.27}$$

The uplink and downlink data sets still have a strong correlation on the solar corona electron content because the separation in time and location of the paths is small with respect to the change of the features in time and with location. The data are used for the feature velocity calculation.

4A.7 Earth Atmospheric Correction

The Earth's atmosphere is part of the radio propagation path and the path length through the atmosphere changes over time due to the change of the antenna elevation as well as the temperature, pressure, and humidity of the air. Since the frequency dependence of the neutral atmosphere refractive index is negligible, a correction for the differential range or frequency is not needed. The atmospheric contribution to the delay for the one-way single-frequency (uplink or downlink) path is calculated by using standard Earth atmosphere models. Four functions are typically created for the characterization of the neutral Earth's atmosphere:

- Zenith dry model
- Zenith wet model
- Mapping function for the dry model
- Mapping function for the wet model

Hoffman-Wellenhof et al. (1997) describe a model for the atmospheric path increase that combines the elevation angle and the vertical profile in one equation for the dry part and another equation for the wet part. The dry part is described by

$$\Delta s_{dry}(t) = \frac{10^{-6}}{5}\frac{77.64\frac{p}{T}}{\sin\left(\sqrt{E^2+6.25}\right)}\left[40136+148.72(T-273.16)\right], \tag{4A.28}$$

and the wet part by

$$\Delta s_{wet}(t) = \frac{10^{-6}}{5}\frac{-12.96T+3.718\cdot 10^5}{\sin\left(\sqrt{E^2+2.25}\right)}\frac{e}{T^2}11000, \tag{4A.29}$$

where the elevation angle E is in degrees, the temperature T is in Kelvin, the pressure p is in mbar, and the water vapor partial pressure e is in hectopascal.

The relation between the relative humidity h (a percentage), as given in the meteorological data files, and the temperature T provides the water vapor partial pressure e as

$$e = 6.108\cdot 10^{-2}\cdot h\cdot e^{\frac{17.393(T-273.15)}{T-33.95}}. \tag{4A.30}$$

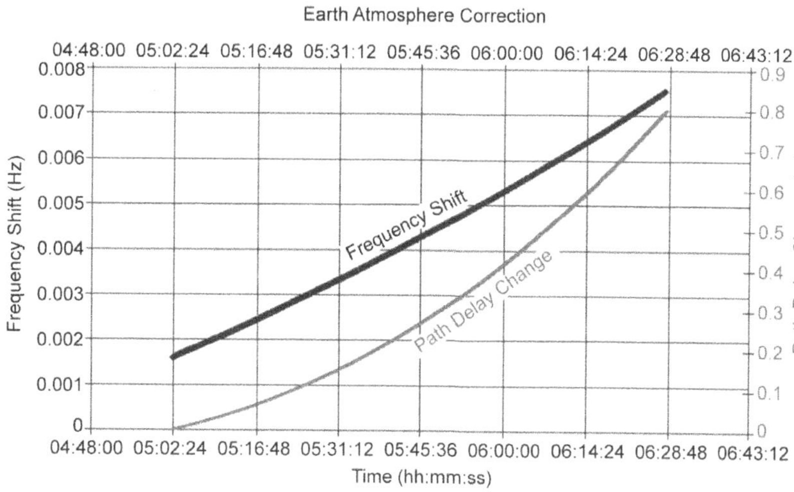

Figure 4A.11 A representative Earth atmospheric correction for the frequency residuals from a Mars Express SCO pass. The blue values represent the frequency shift and the red values the path delay change over 1.5 h.

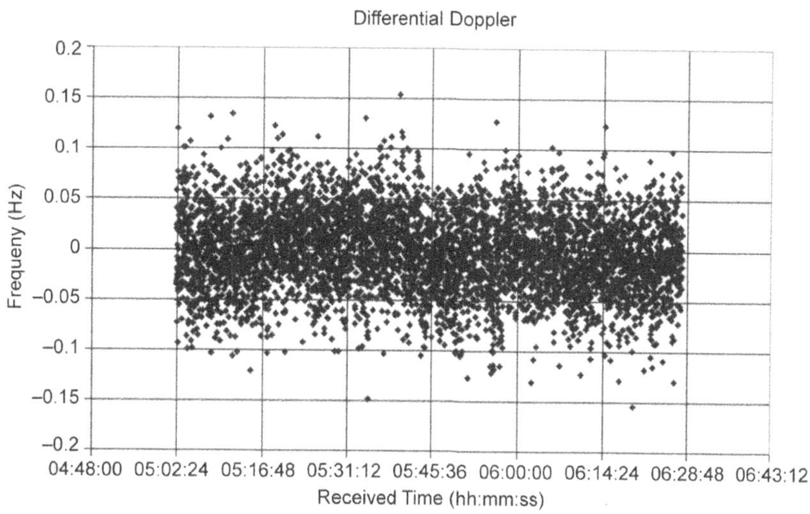

Figure 4A.12 Mars Express SCO example of differential Doppler residuals.

The total time delay for the uplink and downlink is

$$\tau_{Troposphere} = \frac{1}{c}\left[\left(\Delta s_{dry} + \Delta s_{wet}\right)_{uplink} + \left(\Delta s_{dry} + \Delta s_{wet}\right)_{downlink}\right]. \tag{4A.31}$$

A plot of the frequency shift and the path delay change is shown in Figure 4A.11.

4A.8 Example Data

Example plots generated from three level 2 data sets are shown in the next three figures. The differential Doppler (Figure 4A.12) contains the change in TEC information along the downlink and can be used for further analysis. The residual calibrated

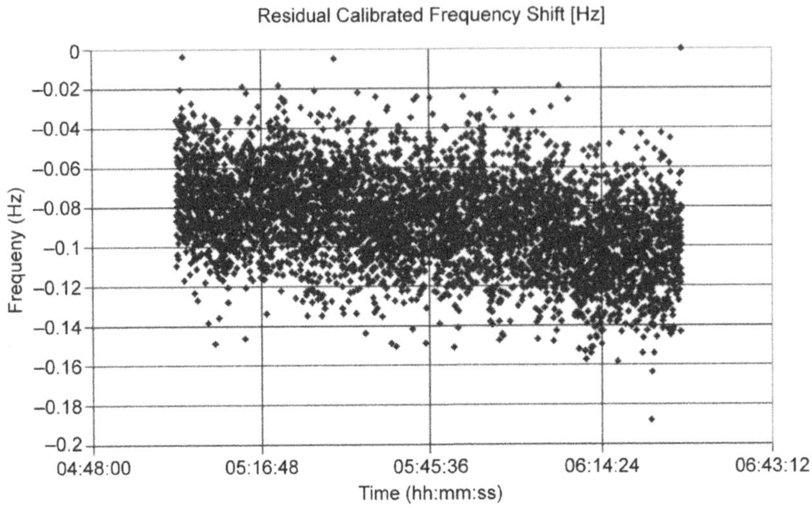

Figure 4A.13 Mars Express SCO example of residual calibrated frequency shift.

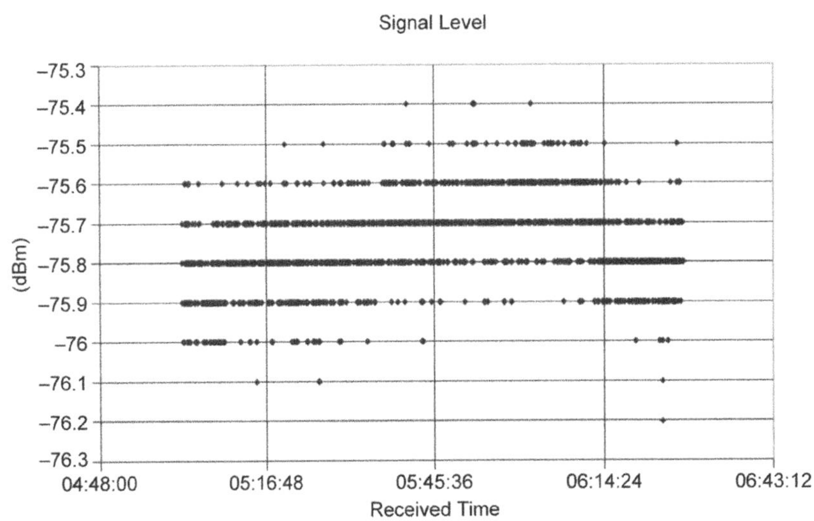

Figure 4A.14 Mars Express SCO example of signal level data.

frequency shift (Figure 4A.13) still has a nonzero offset and drift, in contrast to the differential Doppler. Because of this offset and drift, the residuals do not reflect the true change in TEC at this stage of the processing. The signal-level plot (Figure 4A.14) reflects the resolution of the automatic gain control (AGC) data. Due to the almost constant power level and the small noise level (±0.2 dB), the quantization into 0.1 decibel (dB) steps is obvious in the high-resolution plot. The signal level shown here excludes the internal amplification of the receiving system and is thus not the absolute received power level at the antenna.

The TEC described in Equation 4A.16 is shown in Figure 4A.15 from differential Doppler data. TEC_0 was not available for that pass from differential ranging data. The

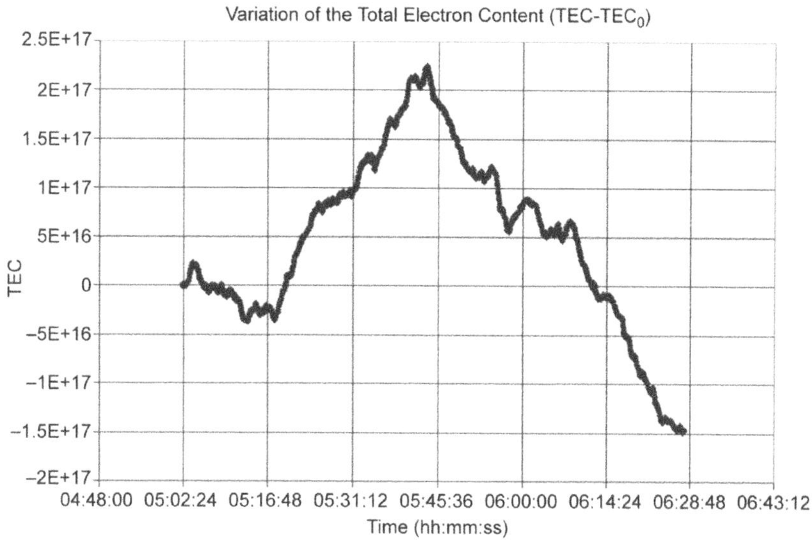

Figure 4A.15 Variation of the TEC versus time. The integrated differential Doppler shows the variation of the total amount of electrons in the path. The absolute value of the column density can only be calculated with differential ranging values.

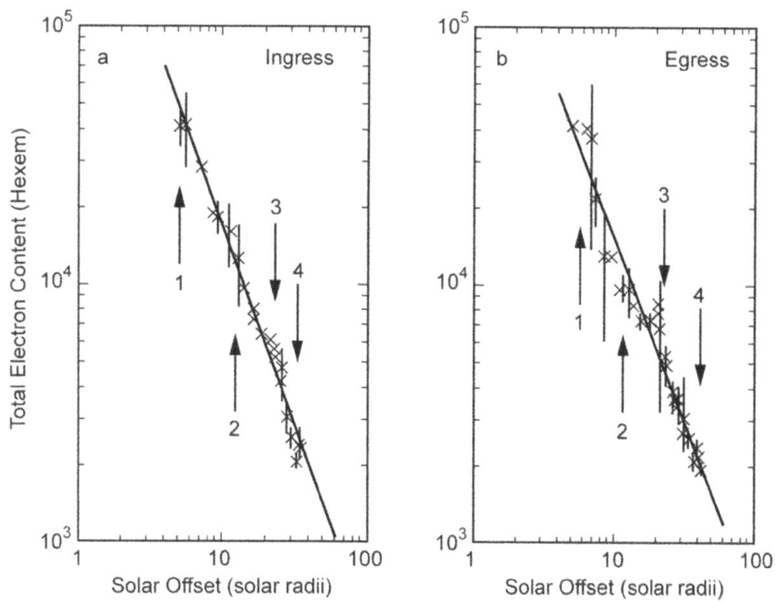

Figure 4A.16 TEC as function of the ray path proximate point distance to the Sun.

differential ranging values during a conjunction period show how the TEC is changing with respect to the distance of the ray path from the Sun.

Figure 4A.16 shows the results of the dual-frequency ranging measurements recorded during the Ulysses 1991 conjunction. A mean value of the TEC is computed for each tracking pass and the data are fitted separately for both egress and ingress phases.

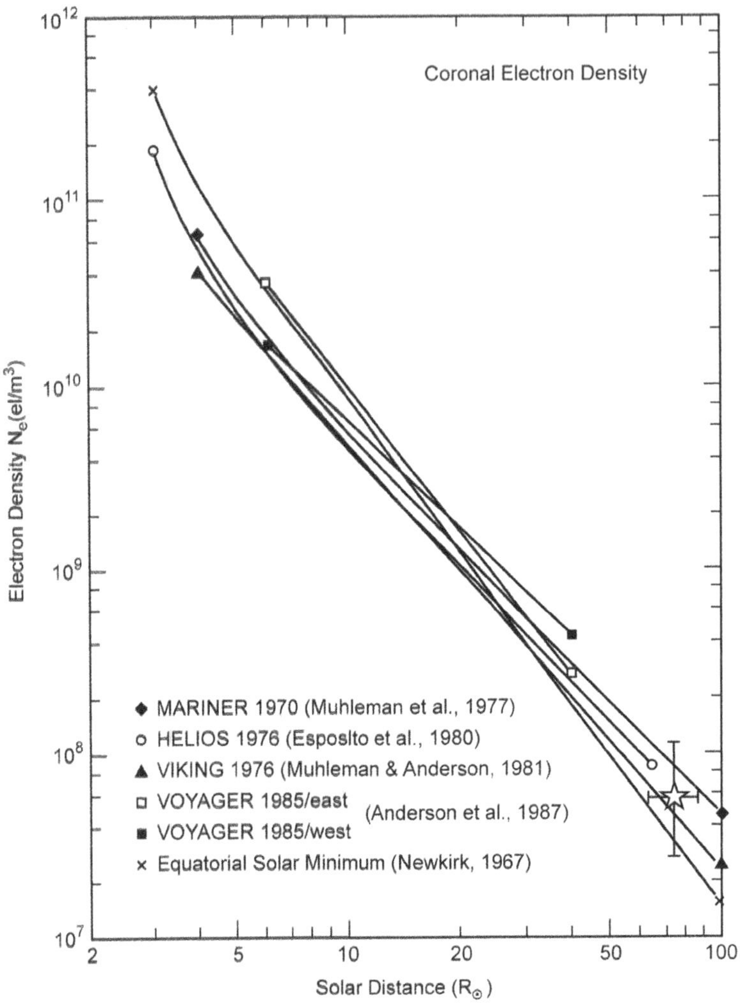

Figure 4A.17 An example for the electron density as a function of solar distance in Sun radii.

The total number of electrons available in the differential ranging data is also needed to calculate the approximate electron density of the tangent point and therefore the density change with respect to the distance to the Sun. It is possible to do this via a regression fit method used on the total electron content or the Abel inversion can be used on the integral. Both methods are based on a spherical electron density distribution. Figure 4A.17 shows some results of conjunctions done on various missions. More details regarding the image are given in Bird and Edenhofer (1990). Use of the Abel inversion, for example, in the field of tomography and atmosphere sounding, is discussed in detail in literature.

Appendix 4B

Faraday Rotation Methodology Illustrated by Magellan Observations

4B.1 Formulation

The physics of the solar corona is critically dependent on measurements of the magnetic field. Since there has not been in-situ measurements of the heliospheric magnetic field inside 0.3 astronomical unit (au), one method of measuring the coronal magnetic field is to observe the Faraday rotation (FR) effect, named after Michael Faraday, that occurs when an electromagnetic wave travels through a magnetized medium and its plane of polarization is rotated by an amount proportional to the strength of the magnetic field. The solar corona is magnetized and, when occulted by it, a spacecraft transmitting polarized radio waves to Earth provides an excellent tool to investigate the FR effect.

The Magellan spacecraft, a project of the National Aeronautics and Space Administration (NASA), orbited Venus for 5 years starting in 1990 and conducted radar imaging of the surface of the cloud-shrouded planet. The images and other information were returned to Earth via radio communications at two frequencies and received by the antennas of the Deep Space Network (DSN). The radio signals were linearly polarized at 2.3 GHz carrier frequency (S-band) and right-hand circularly polarized (RCP) at 8.4 GHz carrier frequency (X-band) with a clearly detectable left-hand circularly polarized (LCP) component as well. The DSN used open-loop receivers (see Chapter 5) to record the radio signals at the 70-m diameter antennas, where four channels of simultaneous reception of the right- and left-hand polarizations of the S- and X-band signals are possible. The four signals are designated S-band right-hand circularly polarized (SRCP), S-band left-hand circularly polarized (SLCP), X-band right-hand circularly polarized (XRCP), and X-band left-hand circularly polarized XLCP.

In 1992, Magellan entered a superior conjunction period, centered on June 13, and came as close as 0.6 solar radii from the Sun's center as viewed from Earth. The spacecraft was occulted by the photosphere for almost 2 days and the elongation angle of the signal ray path to the Sun was less than 3 deg for ~23 days. The combined motion of Venus and Earth around the Sun caused the apparent position of Magellan to move relatively slowly behind the Sun from ingress on the west solar limb to egress on the east solar limb. The signals propagating from the spacecraft to Earth traveled through the solar corona and were received and recorded at the ground stations as separate data sets, two wavelengths of two polarizations. The high sampling rate open-loop data were later processed by a software digital phase-locked loop. The phase difference between the right-hand and left-hand components was computed and used to obtain information about the magnetic field in the solar corona.

Magellan marked the first use of two-wavelength polarized links enabling simultaneous acquisition of both FR and electron content data, the latter being a dispersive phenomenon that is isolated using dual frequency links. Chapter 4 notes the previous FR experiments on planetary missions (see Levy et al. 1969; Stelzreid 1970; Volland et al. 1977; Pätzold et al. 1987). Figure 4B.1 shows a graphical representative of the Magellan FR experiment showing the rotation angle before and after propagation of an electromagnetic wave in a magnetized medium.

4B.2 Coronal Radio Sounding

Goldstein (1969) obtained experimental data about the radial dependence of spectral broadening and its relationship to the integrated coronal electron density. As shown in Chapter 4, numerous experiments were conducted during the solar conjunctions of several spacecraft. In addition to spectral studies, range measurements as a function of solar offset distance provided an excellent method of mapping the electron density in the solar corona by examining the delays in the group velocity of the signal. In 1970, Muhleman et al. (1977) used this mapping method during the solar conjunctions of Mariners 6 and 7. This tool becomes more powerful when two carriers at two frequencies (S- and X-bands) are transmitted from a spacecraft since the difference between measurements provides a direct measurement of total columnar electron density, or electron content.

The Viking spacecraft were equipped with this dual-frequency capability, and the 1976–1977 and 1978–1979 Viking solar conjunctions provided excellent electron density measurements (Tyler et al. 1977). Dual-frequency ranging observations acquired during solar occultations of the Voyager spacecraft in 1985 and 1988 were reported by Anderson et al. (1987c) and Krisher et al. (1991), respectively. Ulysses dual-frequency ranging data were obtained by Bird et al. (1994) and Pätzold et al. (1995).

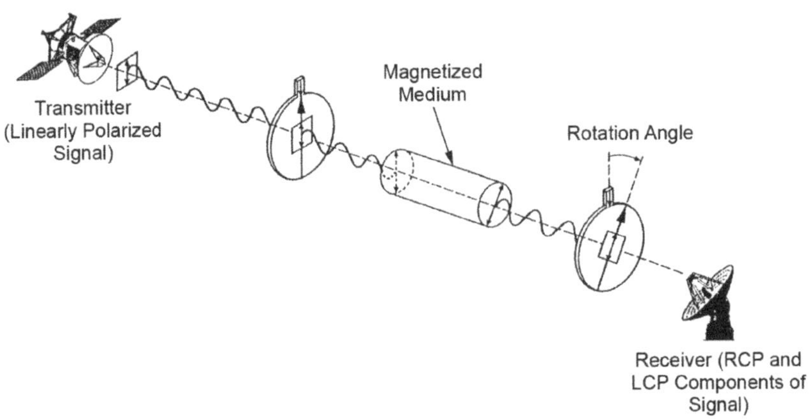

Figure 4B.1 A graphical representative of the Magellan FR experiment showing the rotation angle before and after propagation of an electromagnetic wave in a magnetized medium.

While range measurements are associated with the signal's group velocity, Doppler measurements are associated with the signal's phase velocity. Doppler data are also radially dependent and the highly dynamic solar wind causes temporal variations; statistical studies show that the occurrence of Doppler noise transients strongly depends on the phase of the solar cycle (Woo 1988). The two most utilized measurement techniques take advantage of the fact that electromagnetic waves' group (v_g) and phase velocities (v_p) are slowed and sped up, respectively, by electron concentration. The group and phase velocities equal the speed of light in a vacuum. In a plasma medium, however, the propagation relations have the following approximation (Bird 1982; Kraus 1966) for a carrier frequency much larger than the plasma frequency:

$$v_g \cong c\left(1-\frac{1}{2}\frac{Ne^2}{4\pi^2\varepsilon_o mf^2}\right) \text{ and } v_p \cong c\left(1+\frac{1}{2}\frac{Ne^2}{4\pi^2\varepsilon_o mf^2}\right), \tag{4B.1}$$

where e is the electron charge in Coulomb, ε_o is the free space permittivity in F/m, m is the electron mass in kg, N is the number of electrons per unit volume, and f is the signal frequency ($f_S = 2.29$ GHz; $f_X = 8.4$ GHz).

The group and phase time delays are associated with these quantities and defined by the following integrals:

$$\tau_g = \int \frac{ds}{v_g} \text{ and } \tau_p = \int \frac{ds}{v_p}. \tag{4B.2}$$

The limits of integration are the position of the transmitter and receiver; the integration is along the radio ray path. The total electron content in the medium is defined as the number of electrons per unit volume integrated along the signal ray path from the transmitter to the receiver. It is designated as I:

$$I = \int N ds. \tag{4B.3}$$

Using the values of the constants above, and examining only the change in the phase velocity and time delay, gives (see also Bird and Edenhofer 1990)

$$\Delta v_p = \frac{40.3cN}{f^2} \text{ and } \Delta\tau_p = \frac{40.3I}{cf^2}. \tag{4B.4}$$

Since the observable is the received signal phase with respect to a signal in free space, or an undisturbed signal starting at some arbitrarily specified time, this can be converted to phase by multiplying both sides by the carrier frequency to express it in terms of cycles $\Delta\phi$:

$$\Delta\phi = \frac{40.3I}{cf}. \tag{4B.5}$$

Due to the dispersive nature of plasma, a comparison of delay measurements at two different signal carrier frequencies yields information about the columnar electron content along the ray path. Specifically, a comparison of the range measurements at two frequencies yields absolute measurements of the electron content, while a comparison of the phase measurements at two frequencies yields changes in the measurements of the electron content. This is because phase is multivalued and one cannot count the number of cycles that need to be added to obtain the total electron content.

4B.3 The Faraday Rotation Effect

A linearly polarized wave can be decomposed into two circularly polarized components of opposite sense, and if the wave propagates through magnetized plasma in interstellar space, the two circularly polarized components would have different phase velocities in the plasma. The change in phase velocity is due to the change in the index of refraction of the plasma material compared to free space. Along the line of sight, all the electrons gyrate either clockwise or counterclockwise, depending on the direction of the magnetic field, effectively rotating the plane of polarization of the linearly polarized wave by an amount proportional to the magnetic field intensity, as well as the electron density in the plasma. For a given electron density, the observed rotation would be at its maximum when the direction of the magnetic field is parallel to the line of sight and vanish when the field is perpendicular to the line of sight. This is seen in the integrand of the FR expression as a dot product between the magnetic field and line of sight vectors. Transverse rotation would be dominated by the longitudinal rotation (see also Bird and Edenhofer 1990).

The FR technique could be applied to a configuration where a spacecraft in solar occultation transmits a radio signal that is linear, or at least elliptical in polarization, to a receiving station on Earth. The FR measurement consists of a determination of the received polarization angle and subsequent comparison with the known transmitted polarization angle. The difference between the two angles, or the amount of rotation, is the FR experienced by the signal along the propagation path from transmitter to receiver in a magnetized medium.

The FR is given to a very good approximation by (Asmar and Herrera 1995)

$$FR = \frac{K}{f^2} \int N\vec{B} \cdot d\vec{s} \; [\text{radians}], \tag{4B.6}$$

where $K = \dfrac{e^3}{8\pi^2 m^2 c \varepsilon_o} = 2.36 \times 10^4$ radians Hz^2 m^2/T (T = Tesla), $\vec{B}$ is the magnetic field, and $d\vec{s}$ is the integration path element along the signal ray path.

The solar corona can be assumed to be tenuous plasma with free electrons and ions. Only the electrons, however, need to be considered in the FR since the motion of ions, due to their much larger mass, is at least three orders of magnitude less than that of electrons (Rohlfs 1986). From this equation, FR is inversely proportional to frequency squared, or directly proportional to the wavelength squared, $FR \propto \lambda^2$. Since the FR is a true angle, it is different in each radio band, that is, the phase difference changes with the wavelength, and since it is considered to be the difference between the plane's polarization angle at transmission and reception, then:

$$FR_S = \alpha_S^r - \alpha_S^t \;\; and \;\; FR_X = \alpha_X^r - \alpha_X^t, \tag{4B.7}$$

where α_S^r and α_S^t are the S-band polarization angles at reception and transmission, respectively, and analogously for α_X^r and α_X^t. In order to obtain the rotation information independent of the signal's wavelength, we examine a rotation measure (RM, Kraus 1966) such that

$$FR_S = RM \times \lambda_S^2 \;\; and \;\; FR_X = RM \times \lambda_X^2, \tag{4B.8}$$

where the wavelengths are

$$\lambda_S = \frac{c}{f_S} = 0.131\text{m and } \lambda_X = \frac{c}{f_X} = 0.036\text{m}. \tag{4B.9}$$

The RM is derived from the FR but not expressed as a function of the signal frequency. Thus,

$$RM = K' \int NBds, \tag{4B.10}$$

where $K' = 1.49 \times 10^{-11}$ deg/T.

RM can be expressed as

$$RM = \frac{FR_S - FR_X}{\lambda_S^2 - \lambda_X^2}. \tag{4B.11}$$

As the ray path gets closer to the solar disk (e.g., solar offsets around five solar radii), the quantity $FR_S - FR_X$ can exceed 180 , the notorious $\mp\pi$ ambiguity. The magnitude of the rotation at S-band, however, will always be greater than at X-band due to the dependence of α on λ. Care must be taken in accounting for the multiple-cycle rotation.

Experimentally, the fundamental observables are the four phase quantities, namely, the phases of the XRCP, XLCP, SRCP, and SLCP signals, called ϕ_X^R, ϕ_X^L, ϕ_S^R, and ϕ_S^L. They are related to the polarization angles as follows:

$$\alpha_S^r = \frac{\phi_S^R - \phi_S^L}{2} \text{ and } \alpha_X^r = \frac{\phi_X^R - \phi_X^L}{2}. \tag{4B.12}$$

This leads to

$$FR_S = \frac{\phi_S^R - \phi_S^L}{2} - \alpha_S^t \text{ and } FR_X = \frac{\phi_X^R - \phi_X^L}{2} - \alpha_X^t. \tag{4B.13}$$

By design of the spacecraft transmitting antenna, the difference between α_S^t and α_X^t is constant, which is negligible for this spacecraft, so the last two terms of the above equation drop out; non-negligible values would lead to an offset in the results. Thus, the expression of the desired parameter in terms of the observables and known constants is

$$RM = \frac{\frac{\phi_S^R - \phi_S^L}{2} - \frac{\phi_X^R - \phi_X^L}{2}}{\lambda_S^2 - \lambda_X^2}. \tag{4B.14}$$

4B.4 Measurement of the Total Electron Content

The magnetic field cannot be extracted directly as shown in Section 4B.3 because the contribution of the electron density is yet to be isolated. This could be done by measuring the total electron content simultaneously with the FR measurement. From Section 4B.2, the integrated electron content and corresponding phase of the carrier frequency, in cycles, are

$$I = \int Nds \text{ and } \Delta\phi = \frac{40.3\Delta I}{cf}. \tag{4B.15}$$

In the presence of two coherently related radio signals of different frequencies, the dispersive total electron content can be measured by differential ranging data. Ranging data, however, were not available for this experiment while Doppler data were; so, changes in the total electron content were available from the differential Doppler observables.

By design of the spacecraft transponder, coherency between the S-band and X-band downlink signals is obtained by the relationships, since $f = \frac{d\phi}{dt}$:

$$\frac{\phi_S}{\phi_X} = \frac{3}{11} \; and \; \frac{f_S}{f_X} = \frac{3}{11}. \tag{4B.16}$$

The differential phase is

$$\phi_{dif} = \phi_S - \frac{3}{11}\phi_X, \tag{4B.17}$$

or,

$$\phi_{dif} = \frac{37.3\Delta I}{cf_S}. \tag{4B.18}$$

Solving for ΔI,

$$\Delta I = \frac{cf_S}{37.3}\phi_{dif}. \tag{4B.19}$$

Converting from units of cycles to degrees:

$$\Delta I = 9.65cf_S\phi_{dif}. \tag{4B.20}$$

This expresses a physical and measurable parameter of interest from the observed data. It can be modified to follow common literature showing the electron content in units of electrons per cubic meter (*N*) times meter (*ds*) and expressing *ds* in solar radii ($Rs = 6.9598 \times 10^8$ m):

$$\Delta I = 1.39 \times 10^{-8} cf_S\phi_{dif} (\text{elec. Rs m}^{-3}) \tag{4B.21}$$

4B.5 Combining the Faraday Rotation and Total Electron Content

The expressions for the RM and the total electron content have been derived and can be combined. The mean value of the parallel component of the magnetic field can be isolated as follows:

$$FR = \frac{K}{f^2}\langle B\rangle \int N ds, \tag{4B.22}$$

and the RM expressed as:

$$RM = K'\langle B\rangle I. \tag{4B.23}$$

Solving for the mean value magnetic field:

$$\langle B\rangle = \frac{RM}{I}\frac{1}{K'}. \tag{4B.24}$$

This is the field line of sight field and can be positive or negative. The total electron content can be treated as

$$I = I_o + \Delta I, \tag{4B.25}$$

where I_o is a constant value at the beginning of a tracking pass, and ΔI represents the (typically small) changes in I_o with time. So,

$$\langle B \rangle = \frac{RM}{I_o + \Delta I} \frac{1}{K'}. \tag{4B.26}$$

Employing the expressions for RM and ΔI, we get:

$$\langle B \rangle = \frac{\frac{\phi_S^R - \phi_S^L}{2} - \frac{\phi_X^R - \phi_X^L}{2}}{\lambda_S^2 - \lambda_X^2} \frac{1}{I_o + 1.39 \times 10^{-8} c f_S \phi_{dif}} \frac{1}{K'}, \tag{4B.27}$$

which reduces to

$$\langle B \rangle = 220.13 \frac{\left(\phi_S^R - \phi_S^L\right)}{-\left(\phi_X^R - \phi_X^L\right) 1.05 \times 10^{-10} I_o + \left(\phi_S^R - \frac{3}{11}\phi_X^R\right)}. \tag{4B.28}$$

This equation represents a new data type in solar research that contains the four phase data sets and eliminates the need to process, model, and remove the complex spacecraft effects whose signature can be difficult to separate from the true Doppler or media propagation effects. The motion of the spacecraft due to orbital dynamics, trajectory-correcting maneuvers, or the effect of moving parts such as the fluid in the hydrazine tank or leaking thruster, etc., would move the electrical phase center of the spacecraft antenna relative to the spacecraft's center of gravity. These effects drop out by using the difference in phases between the two wavelengths, leaving the scientific observable without the need for additional corrections.

Bird and Edenhofer (1990) give a widely used empirical representation called the Allen–Baumbach formula (in units of electrons per cubic meter), for a spherically symmetric distribution:

$$N = \left(\frac{2.99}{r^{16}} + \frac{1.55}{r^6}\right) \times 10^{14}, \tag{4B.29}$$

where r is the distance from the solar center in units of solar radii. They further describe various modifications and simplifications of that formula, such as a valid estimate of the total electron content (integral of electron density) for a spherically symmetric coronal density and constant outflow velocity to be

$$I(R) = N \cdot R \cdot \theta, \tag{4B.30}$$

where R is the distance in solar radii and θ is the angle in radians formed by the Earth–Sun–spacecraft lines, and π is near superior conjunction. For this purpose, the results of the coronal sounding experiment performed by the Ulysses spacecraft in 1991 are used to obtain the total electron content. Bird et al. (1994) used dual-frequency ranging data to obtain information on both solar limbs.

Earth's ionosphere is another plasma medium present between the spacecraft transmitter and the ground station receiver. The contribution of the ionosphere to the FR, however, is usually negligible for this experiment for solar offsets below 10 solar radii. The reason is that the ionosphere changes at a slower rate than the observed coronal effect and follows the diurnal variations.

4B.6 Instrument Overview: The Magellan Spacecraft

The Magellan Venus orbiter carried a radar system operating at a 12-cm wavelength capable of imaging most of the surface of the cloud-shrouded planet at a resolution better than 300 m. The spacecraft was three-axis stabilized with an attitude control system for orientation via gyroscopes, and maneuvering was performed with computer-controlled reaction wheels and thrusters. Tracking of the orbiter's coherent radio signal permitted observations of small accelerations related to inhomogeneities in the planet's gravitational field (Konopliv et al. 1999).

The telecommunications system was centered around two redundant transponders capable of receiving and transmitting at S-band (13 cm wavelength) and/or X-band (3.6 cm wavelength) links. The transponder includes a receiver, command detector, exciter, and low-power amplifier. Two high-power traveling wave tube assemblies enabled transmission at a peak telemetry rate of 268.8 kb/s. The high-gain antenna (HGA) was used for both communications with Earth and radar mapping. A medium-gain and a low-gain antenna were also available. Table 4B.1 shows selected parameters of the Magellan radio system.

The S-band signal transmitted by Magellan's high-gain antenna was linearly polarized. The X-band signal is RCP but had a strong LCP component. Figure 4B.2 shows the signal strength for all four channels: XRCP, XLCP, SRCP, and SLCP, recorded by the radio science instrumentation of the DSN. The Magellan FR observations were possible at both S- and X-bands, which made it a unique instrument for such an

Table 4B.1 Selected parameters of the radio system.

	S-band	X-band
Magellan spacecraft		
Nominal transmit frequency (MHz)	2298	8426
Nominal wavelength (cm)	12.97	3.54
RF output power (W)	5	20
HGA beamwidth at S-band (deg)	2.2	0.6
HGA gain–transmit, decibel isotropic (dBi)	36	48
Effective isotropic-radiated power (EIRP) (dBm)	71.4	90.5
Transmit polarization	Linear	RCP and LCP
HGA diameter (m)	3.7	3.7
70-m Deep space station		
Antenna gain–receive (dBi)	63.3	74.2
Half power beamwidth (deg)	0.108	0.031
System noise temperature (K)	~20	~20
Polarization	RCP and LCP	RCP and LCP

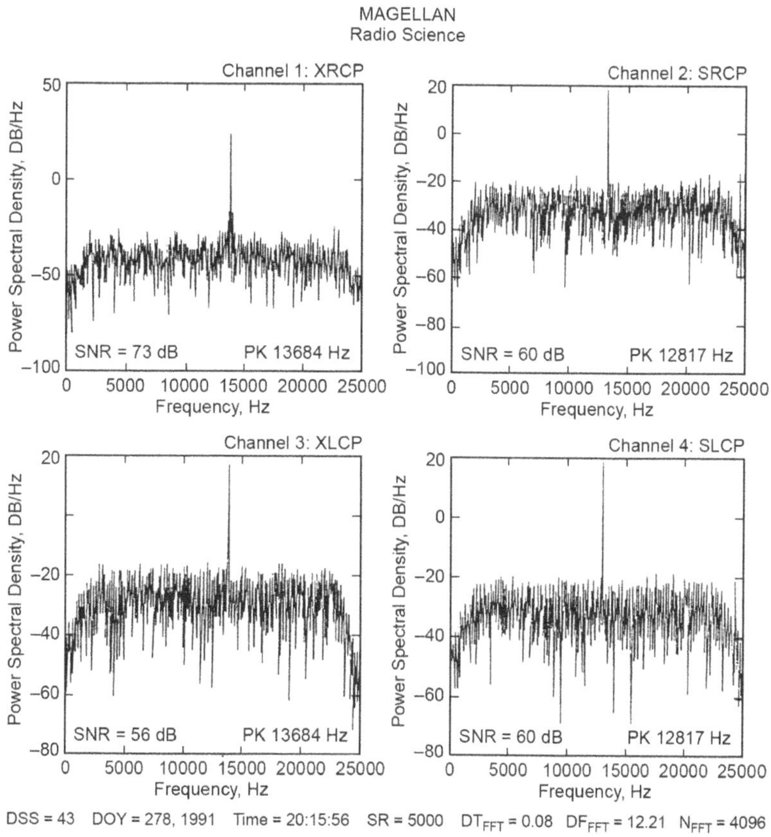

Figure 4B.2 Representative relative strength of the peak signal carrier to the noise floor for XRCP, XLCP, SRCP, and SLCP channels recorded by the radio science instrumentation of the DSN during the Magellan FR observations on day of year (DOY) 278 in 1991. Each panel shows the SNR in dB.

experiment. Although Magellan did not incorporate a ranging system, dual-frequency Doppler observations can always provide the all-important changes in electron content. In addition, the presence of both S- and X-band signals was used to completely eliminate any effect of spacecraft spin or nongravitational motion.

4B.7 Instrument Overview: The Deep Space Network

The DSN's 70-m deep space stations supported the Magellan FR experiment. Table 4B.1 shows selected parameters of the 70-m stations relevant to this experiment. The large parabolic surface of the antenna focuses the incoming RF energy onto a subreflector, which is adjusted in position to optimize the transfer of energy to the system housed inside the structure. During transmission, RF energy is focused on the subreflector that is adjusted to optimize the transfer of energy to the primary reflector and radiated from there to the target spacecraft. There are several ways to keep the massive antennas properly pointed at the spacecraft, as described in Chapter 5. In one method, the acquired signal is conically scanned (CONSCAN) by the antenna, and the feedback from a closed-loop receiver is applied to an algorithm which compares the scan pattern to the received signal level. The antenna then compensates to point the scan center at the apparent direction of the spacecraft signal. In cases of excessive signal dynamics or low received signal levels, as is the case in many occultation experiments, CONSCAN cannot be used. Instead, the antenna is blind-pointed where pointing angle predictions are computed using an ephemeris provided by the spacecraft navigators.

Once the signal has been focused by the front-end of the antenna, it is channeled to a low-noise amplifying maser and then to a receiver. There are two forms of reception: closed-loop reception that provides the capability for rapid acquisition of a spacecraft signal and telemetry lockup, and open-loop, which does not have a feedback loop to track the received signal but relies on frequency predictions to remain tuned to the incoming signal. The signal is down-converted from RF to IF as explained in Chapter 5 and transmitted to the signal-processing center where it is further down-converted to video band.

The RF-IF downconverters in the 70-m stations have four IF channels that can be recorded simultaneously. They are the right- and left-hand circular polarizations of the S- and X-band signals (i.e., XRCP, SRCP, XLCP, and SLCP). A processor digitizes and records the output from the IF-VF downconverters in a preselected bandwidth and tunes the receiver based on predictions generated based on the latest trajectory information available for the spacecraft. Models of the planetary atmospheres are also included in the predict-generation software for most of the planets in the solar system.

4B.8 Data Processing and Results

The raw data were digital samples of the electromagnetic field received at the ground stations and delivered from the DSN on nine-track magnetic tapes. The processing software was composed of two programs. The, FR phase-locked loop (FRPLL) is a digital phase-locked loop routine whose input is the raw open-loop voltages sampled at a rate of 1000 samples per second, and output are the time-tagged frequency, phase, and amplitude of the detected signal, at a rate of five points every second. No decimation of the input data was performed. Due to signal spectral broadening effect of the solar corona on the radio signal, the loop bandwidth was set to 10 Hz to maintain lock onto the signal. The loop's automatic gain control bandwidth was set to a one-second time constant. For each data set, the FRPLL program was run four times (once for each channel) producing four output files. The second program, FARaday ROTation (FAROT), read those four files simultaneously, used the values of the signal's frequency, and converted them to rotation angles, labeled ϕ_X^R, ϕ_X^L, ϕ_S^R, and ϕ_S^L, to results.

The following parameters are plotted for each data set: the frequency for each channel (S-band, 2298 MHz; X-band, 8425 MHz), the XRCP-XLCP frequency and polarization angle, and SRCP-SLCP frequency and polarization angle, normalized difference of the frequency and phase $\left(S-\frac{3}{11}X\right)$, the line-of-sight component of the magnetic field at S- and X-bands, the RM at S- and X-bands, the two-wavelength RM (in degrees per square meter, although labeled degrees for lack of space), and the two-wavelength mean-value magnetic field (in units of micro-Tesla), all as a function of time (in seconds past midnight).

Table 4B.2 Summary of the results of the Magellan FR experiment showing mean magnetic fields in the solar corona in micro-Tesla along with a comparison with results from the Helios experiment.

Radii	Asmar	Data from Pätzold et al. (1987)
1	5.8	–
2	5.0	–
3	6.5	40
4	15.0	20
5	–	10
6	–	7
7	7.5	4
8	3.8	3
9	2.1	2
10	–	1

Table 4B.2 lists a summary of the mean-value magnetic field averaged for each data set. The results are shown as a function of solar offset, a distance in units of solar radii. The table includes a model for the total electron content taken from an experiment conducted in August 1991 using the Ulysses spacecraft and extrapolated for the missing points.

4B.9 Conclusion

The physics of the solar corona is critically dependent on the magnetic field, but there are no in-situ measurements of the heliospheric magnetic field inside 0.3 au. Bird and Edenhofer (1990) state that the magnetic field is primarily generated within the solar convection zone by differentially rotating plasma. There are two distinctly different magnetic field configurations: high-brightness loops with both magnetic footprints on the solar surface (closed-type) and low-brightness regions where the magnetic field has one footprint on the Sun and the other dangling out into interplanetary space (open-type).

They further state that closed-field line configurations rarely extend beyond two solar radii from the Sun. It is commonly mentioned in publications (e.g., Volland et al. 1977) that the electron density as well as the radial component of the interplanetary magnetic field decreases proportionally to the square of the distance from the Sun, and that at distances greater than about two solar radii, spiral effects dominate and that the electron structure in this region of the corona may be complicated. An experimentally determined form for the magnetic field in this region during quiet conditions was inferred from the magnetic configuration observed on the photosphere.

FR is an integrated product of the component of the magnetic field along the signal ray path and the electron density. With knowledge of the FR and the electron content, the mean longitudinal magnetic field is directly obtained. This assumed that the radio signal's frequency is larger than both the plasma frequency and the electron gyrofrequency. A new formulation was shown to produce the mean magnetic field based on knowledge of the four polarization angle data sets (S- and X-band RCP and LCP) and a model of the total electron content.

The most commonly quoted results of mean-value magnetic field measurements in the solar corona are those by Pätzold et al. (1987) from the Helios mission. Figure 4B.3 compares the results of the Magellan and Helios FR experiments. Very good agreement is seen in the range four–nine radii and, at three radii, differences are apparent and attributed to the assumptions made for the total electron content, where the Helios team used a power-law approximation of an electron content model. Extrapolated Ulysses measurements were used for the Magellan results. At one and two radii, there is no available Helios data. It should be noted that the Helios experiment was conducted near solar minimum and the Magellan experiment was conducted near solar maximum, yet the data do not reveal a noticeable statistical difference, which should be examined further.

The results are divided into two categories: the three–nine radius region showing relatively good agreement with the Helios results, and the one–two radius region where no comparison can be made. The second category is in the closed-type field and it is possible that a different type of physical phenomena governs the behavior of the magnetic field near the solar surface, as opposed to the better understood physics of the open field addressed by the Helios experiment in the 3–10 radii region.

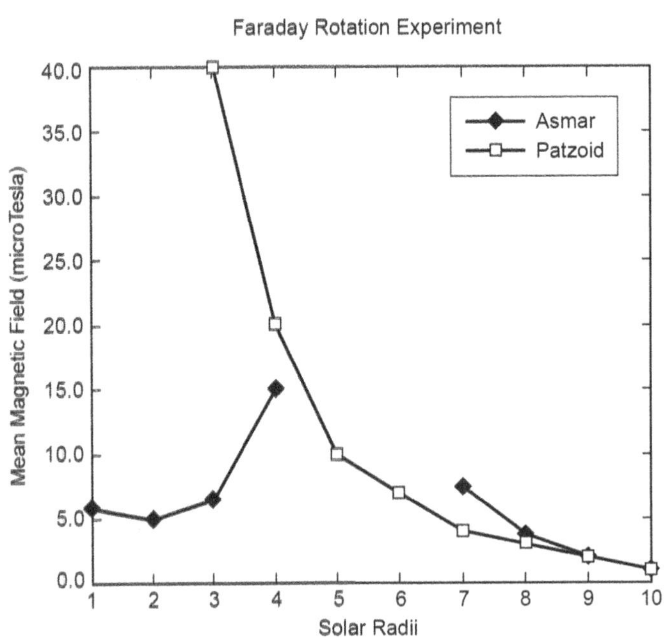

Figure 4B.3 Graphical representation of Table 4B.2, the summary of the results of the Magellan FR experiment showing mean magnetic field in the solar corona as a function of solar radii in units of micro-Tesla along with a comparison with results from the Helios experiment.

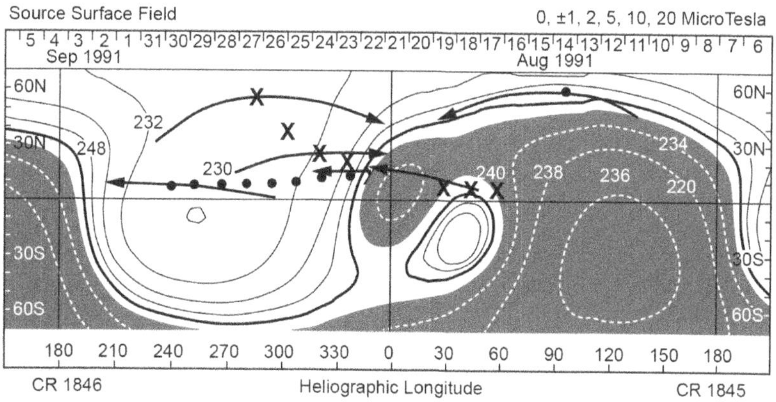

Figure 4B.4 Figure from Bird et al. (1994) shows a contour map of the coronal magnetic field at 2.5 solar radii during the time of the Ulysses experiment indicating that the magnetic field is not uniform.

Figure 4B.4 (from Bird et al. 1994) shows a contour map of the coronal magnetic field at 2.5 solar radii during the time of the Ulysses experiment. It indicates that the magnetic field is not uniform and heliographic longitudes or latitudes can explain the differences in magnetic field strength as the spacecraft moves behind the Sun, sounding different longitudinal coronal regions. This implies that at close proximity to the solar surface, the location of the sounding region is another variable to be studied in future work.

Appendix 4C

Precision Doppler Tracking of Deep Space Probes and the Search for Low-frequency Gravitational Radiation

Contributed by J. W. Armstrong[1]

4C.1 Background

Gravitational waves (GWs) are propagating, polarized, gravitational fields—"ripples" in the curvature of space time (Thorne 1987). They are expected to be produced over a broad band of Fourier frequencies. They were first directly observed on September 14, 2015, by the Laser Interferometer Gravitational-Wave Observatory (LIGO) in the high-frequency band (Abbot et al. 2016).[2] At the time of this writing (early 2016), there is considerable interest in this new observational window into the Universe.

GWs are predicted by all relativistic theories of gravity, although different theories of gravity predict different properties of the radiation (e.g., polarization states). The waves are propagating space–time strain, so their amplitudes are dimensionless strain amplitudes. GWs cause fractional frequency shifts of electromagnetic waves exchanged between separated test masses and cause differences in the rates at which separated clocks keep time.

In general relativity, GWs are propagating solutions of the Einstein field equations (Misner et al. 1973). These waves carry energy and momentum. Like electromagnetic waves, they are transverse to the propagation direction, have two independent polarization states, and propagate at the speed of light. However, unlike electromagnetic waves, GWs couple very weakly to matter. GWs are generated at potentially detectable levels only by very massive sources undergoing extreme dynamics, such as astrophysical sources.

Because of their weak interaction, GWs propagate unchanged from their sources, essentially unaffected by scattering or absorption by intervening matter. Hence, GWs preserve information—not obtainable by any other means—about the deep interiors of high gravity, high velocity objects. As emphasized by Thorne (1987), direct detections and detailed study of gravitational waveforms are interesting for two reasons. First, GWs provide a direct strong-field test of fundamental physics, and second, they provide a new approach for observational astronomy with information intrinsically different from that available with electromagnetic observations.

---

1 An expanded version appears in Armstrong (2006).

2 Although GWs were directly detected in 2015 by LIGO, they were indirectly inferred via changes in the orbital period of the binary system containing PSR 1913 + 16 (Hulse and Taylor 1975).

This section describes a radio science application using a distant spacecraft electromagnetically tracked with a precision Doppler system on Earth as a detector of low-frequency[3] GWs.

4C.2 Response of Spacecraft Doppler Tracking to Gravitational Waves

Precision spacecraft Doppler tracking is done in the "two-way" mode: an Earth station transmits; the spacecraft phase coherently transponds the signal it receives; and the Earth station receives the returned signal. The Doppler shift is estimated by comparing the ground-received frequency with the frequency of the transmitted signal. Although the measurement is two way, the theory of a Doppler GW detector is most easily built up by generalizing the GW response of one-way Doppler measurements. Kaufmann (1970) calculated the redshift fluctuation (fractional frequency fluctuation) caused by GWs on a one-way Doppler link in the context of a proposed Earth-based GW detector using the Mossbauer effect. At about the same time A. J. Anderson (1971) commented on ≈100-s fluctuations in Mariner 6's Doppler time series. He suggested that those fluctuations might be related to Weber-bar events. R. W. Davies (1974) surveyed the prospects for GW detection using deep space probes. He noted the sensitivity of Doppler measurements, identified several important noise sources, and derived the response in two-way Doppler for GWs incident normal to the Earth–spacecraft line. F. B. Estabrook and H. D. Wahlquist (1975) derived the general GW response for a spacecraft Doppler detector, signal and noise transfer functions, and the spectral distribution of Doppler fluctuations that would be produced by an isotropic GW background. They also calculated the GW sensitivity then-achievable by Doppler tracking (and the prospects for improving it) and pointed to the utility of simultaneous tracking of several spacecraft (Wahlquist et al. 1977; Estabrook 1978; Estabrook et al. 1979). The first systematic GW observations with deep-space tracking were made in the 1980s (Hellings et al. 1981). Subsequent technical improvements between the early 1980s and 2001–2003 led to a ≈1000-fold sensitivity improvement.

Unlike ground-based detectors, the Earth–spacecraft distance can be large compared with the relevant GW wavelengths. Thus, times-of-flight of both GWs and radio waves within the detector are important in the analysis of detector response. Consider the Earth and a spacecraft at rest with respect to one another and separated by a distance $L = cT^{2}/2$ where T^{2} is the two-way light time. The deep space network's (DSN) precision Doppler tracking system continuously transmits an almost monochromatic microwave signal having center frequency ν_0 up to the spacecraft. The spacecraft coherently transponds the signal it receives, retransmitting it back to the Earth. The ground station computes the two-way Doppler time series by comparing the frequency

3 Gravitational wave bands conventionally divide along different detector technologies (Thorne 1987). Future extremely low-frequency (~10^{-18}–10^{-15} Hz) searches are based on mapping the intensity and polarization of the cosmic microwave background; very low-frequency observations (~10^{-9}–10^{-6} Hz) use pulsar timing observations; low-frequency (~10^{-6}–10^{-1} Hz) observations currently use spacecraft Doppler tracking (in the future, a laser interferometer in space); high-frequency (~1 to ~1000 Hz) observations involve ground-based laser interferometers.

of the signal it is transmitting with the frequency of the signal it is receiving. The two-way fractional frequency fluctuation, $y_2(t) = [\nu(t - T_2) - \nu(t)]/\nu_0$, measures the relative dimensionless velocity of the Earth and the spacecraft: $2\Delta v/c = \Delta\nu/\nu_0$. In an idealized system (no noise, no systematic errors, and no GWs), this time series would be zero.

A transverse, traceless GW incident on this system causes perturbations in the two-way Doppler time series (Estabrook and Wahlquist 1975):

$$y_2^{gw} = \frac{(\mu-1)}{2}\Psi(t) - \mu\Psi\left(t - \frac{(\mu+1)}{2}T_2\right) + \frac{(\mu+1)}{2}\Psi(t - T_2) \quad (4C.1)$$

where $\mu = \mathbf{k} \cdot \mathbf{n}$, $\mathbf{n}$ is a unit vector from the Earth to the spacecraft, k is the GW wave vector, and $\Psi(t) = (\mathbf{n} \cdot \mathbf{h}(t) \cdot \mathbf{n})/(1 - \mu^2)$, and $\mathbf{h}$ is the metric perturbation.

The GW is assumed plane-parallel, transverse, and traceless (Misner et al. 1973). (The tensor wave $\mathbf{h}(t)$ is a linear combination of the "plus" and "cross" polarization states; the mathematical formulation in this notation is given in detail by Wahlquist (1987).) The Doppler responds to a projection of the time-dependent wave metric, in general producing a three-pulse response to the incident gravitational radiation (Figure 4C.1). That is, an impulsive GW has copied three times in the Doppler time series. The amplitudes and locations of the Doppler pulses depend on the arrival direction of the GW with respect to the Earth–spacecraft line, the two-way light time, and the GW's polarization state. From Equation (4C.1) the sum of the three pulses is zero. Thus, for waves comparable to or larger than the Earth–spacecraft distance, there

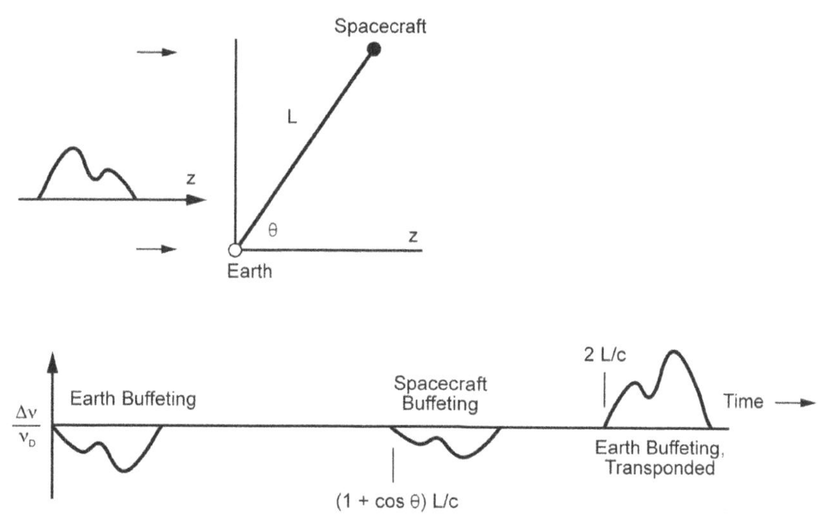

Figure 4C.1 Schematic response of two-way Doppler tracking to a burst GW. The Doppler responds with three pulses having amplitudes and relative locations depending on GW arrival direction, the two-way light time, and the wave's strain amplitude and polarization state. The sum of the three pulses is zero, so the pulses overlap and partially cancel when the characteristic time of the GW pulse is comparable to or larger than the two-way light time. This results in loss of signature and loss of sensitivity, so that the low-frequency band edge of the detector response is roughly $c/(2l)$.

is loss of signature and partial cancellation of the response in the Doppler time series. This partial cancellation sets an approximate low-frequency band edge: GWs with Fourier frequencies smaller than about $1/T^2$ will be suppressed in the detector output.

In practical GW observations (spanning 20–40 days), T_2 and the orientation of the Earth–spacecraft vector on the celestial sphere change with time. This modifies the idealized GW response (it is no longer strictly time-shift invariant) with consequences for search strategies for long-lived signals.

4C.3 Noise in Doppler GW Observations and Their Transfer Functions

Signals in a Doppler time series compete with noise (Asmar et al. 2005b). Both the level of noise sources and their transfer functions to the Doppler are relevant to signal detection: the Doppler system has a three pulse response to GWs while the principal noises couple to the Doppler with one- or two-pulse time-domain transfer functions (Estabrook and Wahlquist 1975; Wahlquist et al. 1977; Estabrook 1978). The important noises at current levels of sensitivity are frequency and timing system (FTS) noise, radio wave propagation noise, and antenna mechanical noise. In principle, stochastic motion of the spacecraft is also important (but not at current levels of sensitivity). Figure 4C.2 schematically shows the transfer functions of these noises to two-way Doppler observations.

The magnitudes of the noises for Cassini-era observations are discussed in Armstrong (2006). The aggregate spectrum of Doppler residuals ($y_2(t)$ after removal of orbital Doppler and all calibrations have been applied) is shown in Figure 4C.3.

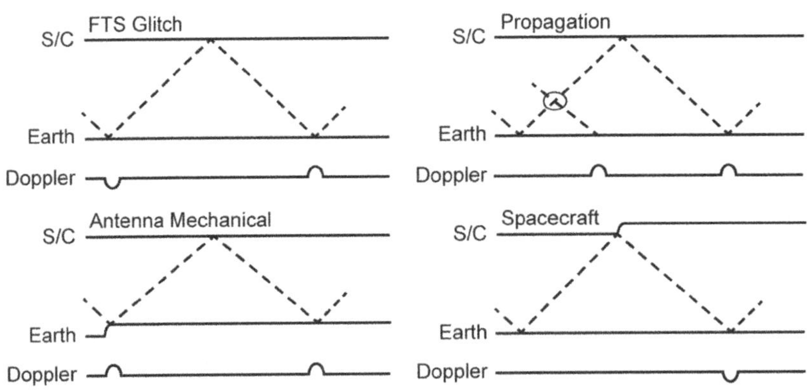

Figure 4C.2 Schematic transfer functions of noises to the two-way Doppler link, adapted from Estabrook (1978). For each type of disturbance, a separate space–time diagram (space vertically, time horizontally) is shown. Radio waves propagate continuously up to and down from the spacecraft. Some of these are represented as the dashed lines to illustrate the indicated Doppler frequency perturbations, e.g., a momentary glitch in the FTS affects the frequency reference for both the receiver and the transmitter. This shows up as an immediate effect in the received Doppler (difference between the transmitted and received frequency). Because the glitch also affects the transmitted frequency, it shows up again, but with the opposite sense, in the Doppler after T^2. These various noise responses contrast with the three-pulse GW response in Figure 4C.1.

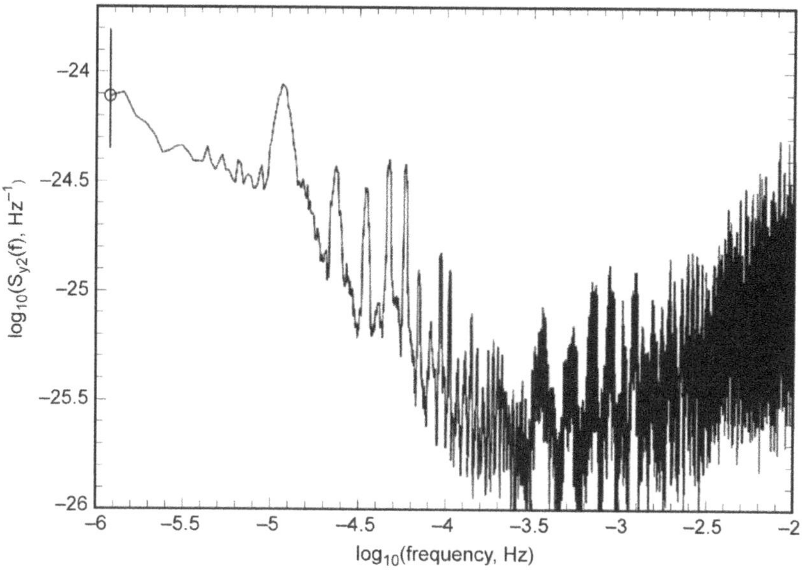

Figure 4C.3 Power spectrum of the two-way fractional frequency fluctuations (calibrated for plasma and troposphere) for 2001–2002 Cassini solar opposition observations (adapted from Armstrong et al. 2003). The spectrum has been smoothed from the intrinsic resolution of the observations, approximately 1/40 days, to a bandwidth ≈3 μHz to reduce estimation error.

The low-frequency part of spectrum in Figure 4C.3 (the region between about 10^{-6}–10^{-4} Hz) is roughly an inverse power law continuum plus spectral lines. The lines in the unsmoothed spectrum are at or near the formal Rayleigh resolution limit of the (40-day) observation.

Their apparent width in Figure 4C.3 is due to the spectral smoothing used to reduce estimation error (Jenkins and Watts 1969; Percival and Walden 1993). The lines are all very near one cycle/day and its harmonics. The multi-link plasma correction removes, essentially perfectly, the plasma noise (Bertotti et al. 1993, 2003; Tortora et al. 2004) so that all random processes entering this spectrum are nondispersive. At frequencies greater than about $1/T_2$ there is clear cosinusoidal modulation, a characteristic of positive correlation in the time series at lag $\tau = T_2$. That is, based on the transfer function, the process causing this modulation is either antenna mechanical noise or residual (after correction, by the water–vapor radiometer-based Advanced Media Calibration (AMC) system Resch et al. 2002; Keihm et al. 2004) tropospheric noise. The level is too large to be residual tropospheric scintillation[4] but is consistent with independent measurements of antenna mechanical noise (Otoshi and Franco 1992). Note that several minima of the mechanical noise transfer function minima at odd multiples of $1/T_2$ are visible in Figure 4C.3. The spectrum continues to be dominated by a process having mechanical noise transfer function up to at least 0.01 Hz.

4 Also, the modulation is imposed on a process having a general spectrum that rising with increasing Fourier frequency. The spectrum of tropospheric noise falls with increasing Fourier frequency (see, e.g., Armstrong and Sramek 1982).

4C.4 Detector Performance

Doppler GW data taken between ~1980 and the present have become increasingly sensitive principally due to engineering improvements in spacecraft and in the DSN. Voyager observations in the early 1980s were limited by a combination of plasma scintillation noise in the S-band uplink and spacecraft buffeting noise from its thrusters. Although the data volume was small, Voyager observations were used in the first formal search for low-frequency burst waves (Hellings et al. 1981). The Pioneer spacecraft were spin-stabilized, resulting in lower spacecraft buffeting noise, but were again sensitivity-limited by plasma propagation noise in the S-band radio links. Despite this, Pioneer data were able to exclude—observationally—the putative sinusoidal GW emission from Geminga (Anderson et al. 1984) and to place the then-best limit on a GW background (Anderson and Mashoon 1985). Ulysses observations in 1992 had the longest data arc to that date, which motivated innovations in signal processing, and resulted in the then-best sensitivity to periodic and chirp waveforms (Bertotti et al. 1992b). Mars Observer was the first spacecraft to have X-band on both uplinks and downlinks, resulting in plasma noise reduction by a factor of $(3/11)^2$. In 1993, the Mars Observer, Galileo, and Ulysses projects performed the first (and so far only) coincidence experiment in the low-frequency band (Iess and Armstrong 1996). This coincidence was used successfully to disqualify an apparently statistically significant event in the one time series and was the first observation where polarization-state coupling was used explicitly in the analysis. Galileo observations were done in 1994–1995. Unfortunately, Galileo's high-gain antenna failed to open, requiring low-power and S-band-only observations. Mars Global Surveyor observations in 1997 were done with X-band links on both the downlink and the uplink (but away from solar opposition, so the benefit of the relative immunity to plasma noise was only partially realized), and they were the only observations where spacecraft engineering telemetry was used to correct the Doppler data for slow systematic spacecraft motion. Those data also showed strong correlation at the two-way light time, pointing to the importance of tropospheric calibration (Asmar et al. 2005b).

The sensitivity examples discussed below, however, use the very high-quality Cassini data. The Cassini data set is distinguished in two ways. Cassini had a very sophisticated multi-link radio system, shown schematically in Figure 4C.4. This allowed essentially perfect plasma scintillation correction. Additionally, the AMC system (Resch et al. 2002; Keihm et al. 2004) allowed excellent tropospheric scintillation corrections.

It is conventional to discuss sensitivity separately for three different waveforms:

1. Periodic and quasi-periodic waves.
2. Burst waves.
3. Stochastic waves.

4C.4.1 Periodic and Quasi-periodic Waves

Sinusoidal wave sensitivity is traditionally stated as the amplitude, h, of a GW required to achieve a specified signal-to-noise ratio (SNR), as a function of Fourier frequency (e.g., Wahlquist 1987). Conventionally, for statements about sensitivity, the signal is

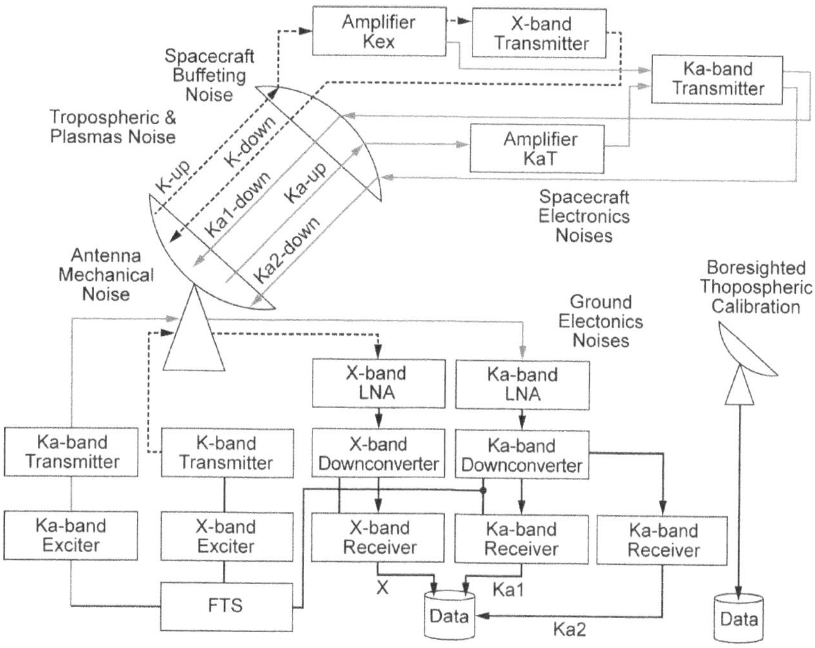

Figure 4C.4 Schematic Cassini signal flow to emphasize subsystem groupings that affect specific radio links. (KaT = Ka-band translator, KEX = X-to-Ka translator, LNA = low-noise amplifier).

averaged over the celestial sphere and over polarization state (Estabrook and Wahlquist 1975; Wahlquist 1987). Figure 4C.5 shows all-sky sensitivity using a smoothed version of the actual Cassini noise spectrum (black curve, for the Cassini 2001–2002 observations) and for reference the model noise spectrum based on pre-observation estimates of the system performance (Tinto 2002). Cassini achieved $\approx 10^{-16}$ all-sky sensitivity over a fairly broad Fourier band (Figure 4C.5).

4C.4.2 Burst Waves

GW burst waves (i.e., waves that are sensibly "on" only over a finite time interval) could be produced for example in the final coalescence of a binary black-hole system. Analysis for bursts has usually employed matched filtering with assumed waveforms and targeted sky-directions (Iess and Armstrong 1996). An all-sky measure of burst sensitivity, however, is the characteristic strain sensitivity as a function of Fourier frequency (Thorne 1987): $h_c = (2fS_y/R_2(f))^{1/2}$. Here S_y is the spectrum of the two-way Doppler residuals and R_2 is the sky-averaged GW response function (Estabrook and Wahlquist 1975). Figure 4C.6, adapted from Armstrong et al. (2003), shows the all-sky characteristic strain sensitivity for Cassini 2001–2002 observations.

Burst sensitivity is not uniform over the sky, however. (For example, the sensitivity for waves having wave vector parallel or antiparallel to the Earth–spacecraft vector is zero.) The sensitivity depends on waveform shape, the two-way light time, and the positions of the source and spacecraft on the celestial sphere. As an example

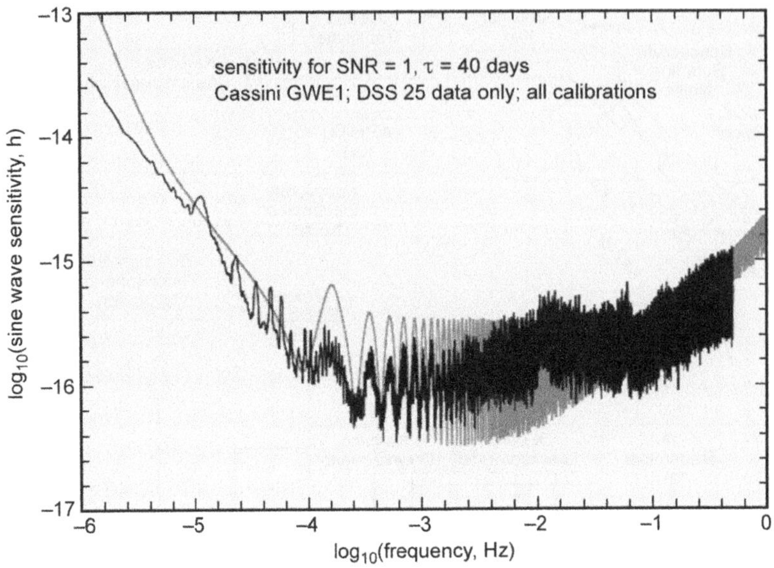

Figure 4C.5 All-sky sensitivity to sinusoidal GWs, Cassini 2001–2002 observations.

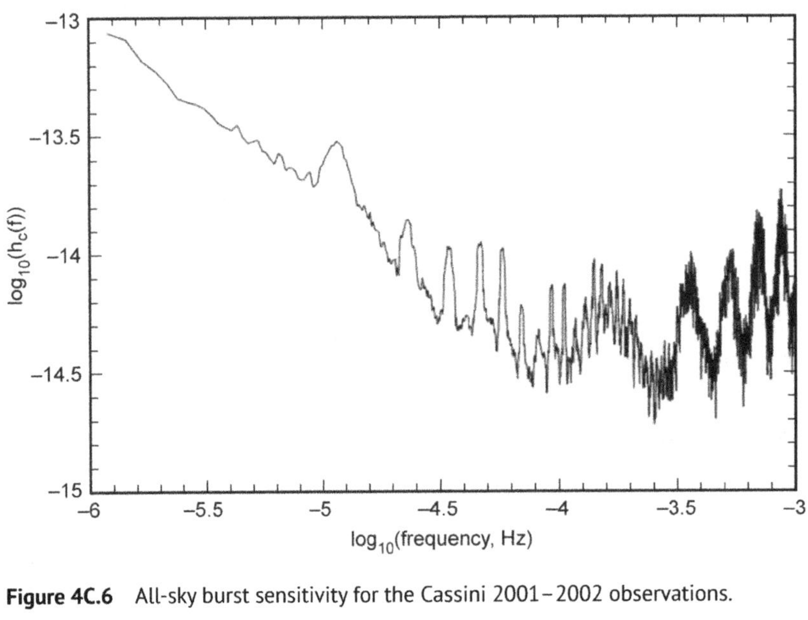

Figure 4C.6 All-sky burst sensitivity for the Cassini 2001–2002 observations.

of the angular sensitivity, Figure 4C.7 shows contours of constant SNR for matched filtering for a mid-band burst wave.

4C.4.3 Stochastic Waves

Stochastic GWs (random fluctuations that persist for times comparable to, or longer than, the observing interval) could be produced by an ensemble of incoherently

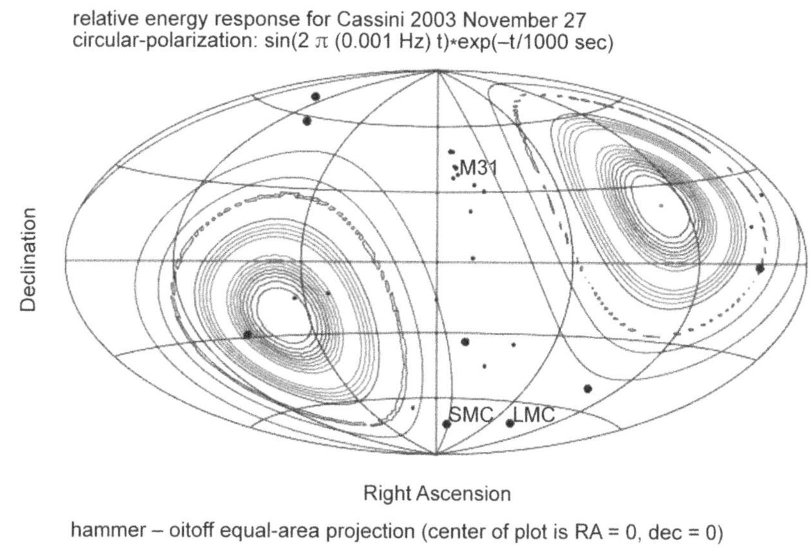

Figure 4C.7 Contours of constant SNR in a matched filter for a burst wave of assumed waveform $h+(t) = sin(2\pi\,0.001Hzt)exp(-t\,/\,1000s)H(t)$, with the cross-polarization given by the Hilbert transform of $h+$ (figure adapted from Iess et al. 1997). Cassini November 2003 geometry is assumed (Cassini's position on the celestial sphere is marked with the red dot; positions of the Local Group of galaxies are marked with black dots). Contour levels are at 0.1 of the maximum. Red contours are between 0.9 and 0.5 of the maximum SNR; blue contours are for 0.4–0.1 of the maximum. This illustrates that angular response to a midband burst varies strongly over the sky.

radiating binary black holes (Jaffe and Backer 2003; Wyithe and Lobe 2003). The level of a stochastic GW background is conventionally expressed in two ways. One is the power spectrum of strain required to give the observed spectrum of two-way Doppler fluctuations. The other is in terms of the energy density of the GWs normalized by closure density (Thorne 1987). Figure 4C.8 shows the upper limit to the GW background, expressed in terms of energy density, from Cassini 2001–2002 data; these are about 1000 times more sensitive than previous limits from Doppler tracking.

4C.5 Sensitivity Improvements in Future Doppler GW Observations

Improving the sensitivity of Doppler observations by a factor of ten (e.g., to $\sim 10^{-16}$ for bursts and $\sim 10^{-17}$ for periodic waves) would require upgrades to several spacecraft and ground subsystems. As discussed in more detail by Armstrong (2006), FTS noise is not currently a limiting factor, but it would be in 10-fold better observations. It is thought, however, that significantly better practical FTS systems are technologically feasible. One published report, Takamoto et al. (2005), suggests that Allan deviations better than 10^{-18} are possible. Ground electronics noise (Abbate et al. 2003) would have to be improved by about a factor of ≈ 3, and tropospheric scintillation corrections would have to be improved by about a factor of ≈ 10. The former may be achievable by even more careful design. The latter would require either substantial improvement in

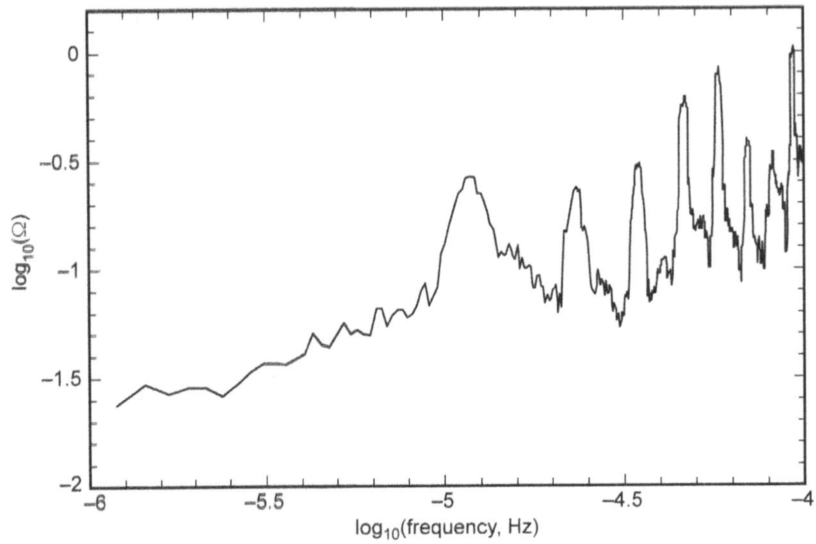

Figure 4C.8 Upper limits to the energy density of GWs, in bandwidth equal to center frequency, relative to closure density (adapted from Armstrong et al. 2003). These data are from Cassini 2001–2002 and assume an isotropic GW background.

water–vapor radiometer technology, a DSN antenna situated at high altitude (above most of the troposphere), or perhaps multi-station observations (discussed below). A Cassini-class multi-link radio system already cancels the plasma noise to very low levels, and a Cassini-like system would probably be adequate as-is for 10× better Doppler stability. Unmodeled spacecraft motion—at least Cassini's unmodeled motion—would only have to be modestly improved (≈2×; Won and Lee 2001) for 10^{-16} Doppler.

The largest required improvement would be in reducing ground antenna mechanical noise by a factor of ~20. It is impractical to build large antennas that are significantly more stable than the current antenna mechanical noise limit of $\approx 2 \times 10^{-15}$ at 1000 s. Thus, a calibration/removal scheme, perhaps by simultaneous three-way tracking with a stiffer antenna (Armstrong et al. 2008), may be more feasible. If the three-way tracking station were also at high altitude, or otherwise benign site for tropospheric scintillation, the lower noise of the three-way site would appear in the spacecraft Doppler. Such a scheme is being investigated by the DSN (in 2016) for possible improvement of the BepiColombo mission's Doppler quality.

There are currently no spacecraft missions requiring, for any other reasons, $\sim 10^{-16}$ tracking stability, so these ideas have not received engineering attention to the point of including them operationally. Although clearly difficult, an order-of-magnitude sensitivity improvement in spacecraft Doppler tracking is possible with reasonable extensions of existing technology. However, to achieve sensitivity improvement several orders of magnitude better than Cassini will probably require different technologies, as well as moving all the test masses into space. The laser interferometer space antenna (LISA) mission, currently in the planning and technology development stage, is such a mission (Gair et al. 2013).

5

Technologies, Instrumentation, and Operations

5.1 Overview

End-to-end instrumentation for a radio science (RS) experiment comprises distant spacecraft communicating via radio links with specialized stations on Earth or, in some cases, with each other (see Figures 1.6 and 1.7 in Chapter 1). As illustrated in Figure 5.1, for most RS experiments, the science observables are acquired directly at the ground stations, unlike typical scientific payloads in which stand-alone units on board the spacecraft acquire data that are subsequently transmitted to ground stations as data packets ready to deliver to science investigations. For RS experiments, the ground station plays a critical role as the bigger part of a science instrument distributed across the Solar System.

The performance of each component of the RS end-to-end instrument affects the quality of the scientific outcome and requires proper design, development, implementation, testing, calibration, and operational planning. Furthermore, specialized techniques are used to calibrate the effects of the media in the radio link's propagation path, namely, Earth's neutral atmosphere, ionosphere, and the interplanetary plasma as shown in Figure 1.6. Additional techniques have been developed to constrain and calibrate nongravitational forces acting on the spacecraft in order to isolate the planetary gravitational effects being investigated.

All Solar System missions are equipped with radios for data communications, which is the basic instrumentation needed for RS observations. However, aside from opportunistic observations, acquiring high-quality RS data can require supplementary flight and/or ground elements such as ultrastable oscillators (USOs), Ka-band translators (KaTs), spacecraft open-loop receivers (SOLRs), or ground station water vapor radiometers (WVRs), among other instrument options.

At a minimum, sufficient testing, planning, and systemic operational procedures are required for ensuring experimental success and safeguarding data quality. These procedures vary in complexity depending on the science objectives, as described in the following examples. Procedures can involve both spacecraft and ground system configurations and sequences, as well as teams of science, engineering, and operations specialists.

Since the Mariner program six decades ago, most Solar System missions have included RS observations as part of their prime science objectives. Opportunities for all investigations on missions are provided through a NASA solicitation process

Radio Science Techniques for Deep Space Exploration, First Edition. Sami W. Asmar.

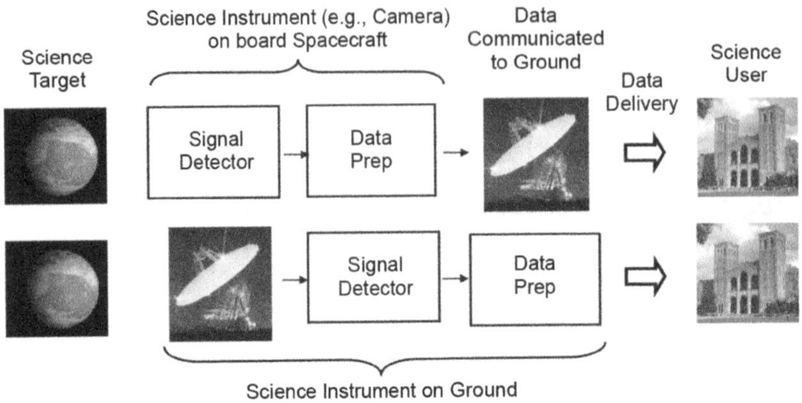

Figure 5.1 The role of the Deep Space Network station for data communications for science instruments on board spacecraft (upper panel) and as a fundamental element of the instrument for RS experiments (lower panel).

that invites proposals that can achieve the agency's science goals and objectives. Once a RS proposal is selected and a project office is formed, a team of scientists, engineers, and mission designers collaborate to optimize the trajectory of the spacecraft as well as the overall performance of the flight and ground systems to enable and optimize RS observations. For missions that do not start out with RS objectives in the design phase, RS observations are often added in an opportunistic manner or as part of an extended mission operations phase.

Because there is wide range of approaches to proposing and carrying out RS observations, as detailed in the following sections, the total cost of experiments varies depending on the functional and performance requirements to achieve the science objectives. In some cases, the cost of flight instrumentation to one space agency is reduced if partner agencies contribute modular elements. Additionally, the cost of the ground instrumentation to one mission is reduced if elements are inherited from previous missions and integrated into the infrastructure.

5.1.1 End-to-end Instrumentation Overview

5.1.1.1 Flight Segment

Appendix 5A presents the functional RS instrumentation block diagrams for representative planetary missions over the past six decades: Mariner 10, Galileo, Mars Global Surveyor (MGS), Cassini, Juno, BepiColombo, and JUpiter ICy moons Explorer (JUICE). These examples show the fundamental deep space telecommunications equipment common to all, as well as system evolution with the cases supplemented by RS-specific elements, such as USOs and KaTs.

At the heart of the flight segment is a transponder, which is redundant due to its criticality to the mission, that carries out the functions of reception and transmission of communications signals, modulation and demodulation, and tracking, telemetry, and command (TT&C). The transponder also supports RS experiments and navigation (flight dynamics) functions via Doppler and ranging radiometric observables.

To prevent interference between Solar System missions, through international agreements among space exploration agencies, transponders use frequencies carefully selected and channels assigned from the deep space bands. The three portions of the radio-frequency (RF) spectrum supported by NASA's Deep Space Network (DSN) are S-, X-, and Ka-bands, listed in Table 1.3. Near-Earth bands for lunar missions are also supported by the DSN. The ultrahigh-frequency (UHF) band (~401 MHz) is used only for spacecraft-to-spacecraft proximity links such as those between Mars landers and Mars orbiters. The UHF radios such as Electra are typically separate from the transponders but can be functionally combined as demonstrated by the Iris radio on the Mars Cube One (MarCO) mission.

Contemporary missions have used the Deep Space Transponder (DST) and the Small Deep Space Transponder (SDST); see Edwards et al. (1998) and Chen et al. (2000). Some spacecraft (e.g., New Horizons) use a transceiver instead of a transponder. Haskins and DeBoy (2007) describe the development of transceivers and contrast their functions with those of the transponders. For additional details on deep space communications systems, see Taylor (2016) and Yuen (1983).

In a transponder, the downlink signal carrier is generated based on the uplink signal it receives. The downlink frequency changes in proportion to the uplink frequency, thereby permitting direct two-way tracking of the Doppler effect. Note that when the arriving carrier is modulated with data, then the outgoing carrier can also be modulated; this is often called bent pipe operation. In a transceiver, on the other hand, the receiver and exciter functions are separate and self-contained to enable a more modular manufacturing approach. Furthermore, the transceiver requires an onboard local oscillator (LO) to provide reference frequencies for both the receiver and transmitter modules; this LO can be a USO (preferred for RS), as in the case of New Horizons.

Software-defined radios (SDRs) carry out the same functions as transponders via software that can be updated even in flight, providing tremendous flexibility to the missions; see Duncan et al. (2014) and Satorius et al. (2006). The Iris SDR, for example, has flight heritage on the MarCO mission (see Chapters 1 and 6) as a deep space technology demonstration. Electra and the universal space transponder (UST) are other SDRs; Electra already has extensive flight heritage for proximity link communications on many Mars missions.

The UST, a next-generation transponder with telecommunication, navigation, and RS functions for deep space missions, combines modular hardware architecture and in-flight reprogrammability. The UST has the potential to become the single future SDR to support direct-to-Earth and relay communication requirements (Pugh et al. 2017). Another available SDR with flight heritage is the Frontier Radio (Haskins et al. 2016), currently on the Parker Solar Probe and the Emirates Mars Mission, Hope. The Frontier Radio is also planned for use on other missions, including Europa Clipper. The European Space Agency (ESA) has a transponder product line for their deep space missions that has also evolved to include X- and Ka-band functions and is currently proposed for future Solar System missions in the formulation phase.

In addition to transponders/transceivers and their power amplifiers, communication systems include antennas and other microwave components. A typical Solar System mission carries one high-gain antenna (HGA) and several low-gain antennas (LGAs) distributed around the spacecraft structure. Some missions also use a medium-gain antenna (MGA). For details on use of a HGA (4-m diameter) for RS observations on Cassini, see Kliore et al. (2004) and Asmar et al. (2014b), and for the use of an HGA (2.5 m diameter) on Juno, see Vacchione et al. (2012) and Asmar et al. (2017b).

Another spacecraft subsystem that is key to the success of RS experiments on Solar System missions is the attitude and articulation control subsystem (AACS). This subsystem controls the critical functions of the dynamical stability of spacecraft structures as well as the pointing of antennas. For example, Cassini was a very large spacecraft with an AACS that met stringent requirements for accurate pointing and pointing stability for communications, navigation, RS, and optical remote sensing instruments. Lee and Hanover (2005) and Pilinski and Lee (2009) describe the common two control modes, control via reaction-wheel assembly (RWA) and control via a system of thrusters.

RS experiments require the use of RWA because they provide higher dynamical stability. Firing thrusters or desaturating the RWA cannot be carried out during precision RS observations as they cause significant Doppler noise events. Nonoptimum pointing or attitude jitter can degrade the science experiments, so investigators place strict performance specifications on the AACS. This subsystem is also required for limb-track maneuvers (see Chapter 2) that optimize radio occultation (RO) experiment performance by dynamically pointing the spacecraft HGA toward the limb of the target planet to address the signal refraction in the atmosphere (Roth et al. 2005).

5.1.1.2 Ground Segment

The ground segment is composed by the worldwide network of DSN stations, as well as similar networks run by other space agencies (e.g., European, Japanese, Indian, and others). These ground stations include large antennas, transmitters, receivers, and various additional equipment that are custom-made or commercially available. The DSN receivers are developed at the Jet Propulsion Laboratory (JPL) according to the international standards of the Consultative Committee for Space Data Systems (CCSDS) to enable interoperability among space agencies and are customized to meet stringent performance requirements.

The contemporary DSN includes one 70-m diameter station at each of its three complexes and several 34-m diameter stations (see Figure 5.2). The stations are capable of transmission and reception at various frequency bands allocated by international agreements.

The frequencies of the signals transmitted by the DSN antennas (called uplink or forward link) are related to the frequencies of the received signals (called downlink or return link) by specific mathematical relationship referred to as the turnaround ratio or transponder ratio (see Table 1.3). The most common deep space band for TT&C on contemporary deep space missions is X-band. Some missions add Ka-band downlink as an additional link for high-rate telemetry. S-band TT&C is available for legacy planetary missions and many lunar missions. For additional details on DSN systems, see the DSN Telecommunications Link Design Handbook (810-005) and DSN Mission Interface Design Handbook (810-007), as well as Taylor (2016).

DSN systems are referenced to highly stable atomic standards and use accurate antenna pointing systems. In addition, they have complex supporting subsystems in the RF and digital data handling required for station operations, as well as calibration systems required for RS data enhancement.

In the case of the Cassini ground segment, in order to make the ground elements compatible with the flight system, a Ka-band transmitter and receiver were installed at

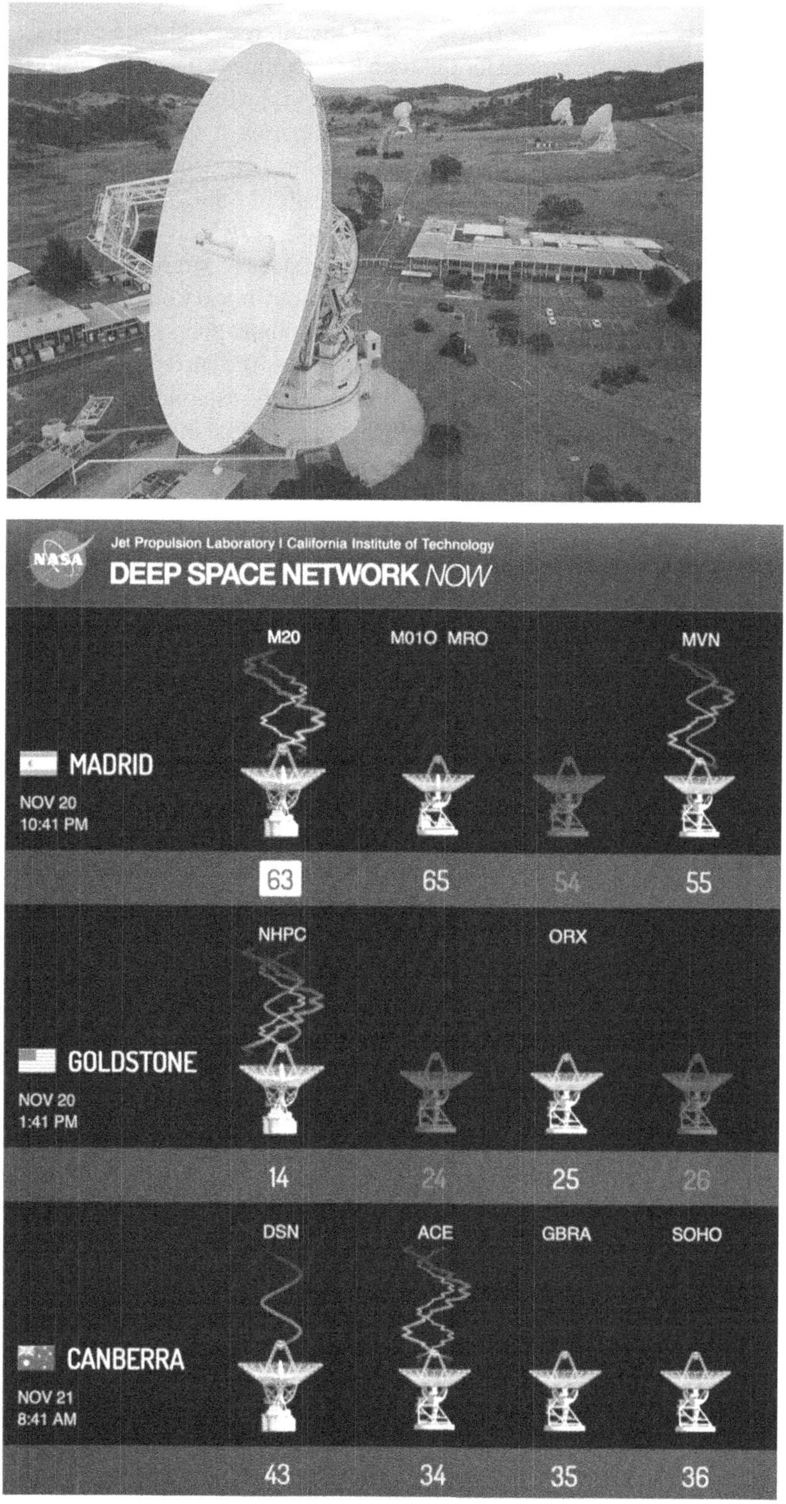

Figure 5.2 The upper panel shows the 70-m diameter Deep Space Network antenna in the Canberra, Australia, complex in the foreground with several 34-m diameter beam-wave guide stations in the background. The lower panel shows a graphical representation of the currently active DSN stations.

one 34-m diameter beam-wave guide (BWG) DSN station at the Goldstone complex, designated DSS-25, and later upgraded for the Juno RS experiment. There are plans to install Ka-band transmitters at the other two complexes in the future. The reception capability at Ka-band is already available at all DSN complexes. A special media calibration system (MCS), at the core of which is a custom-made advanced water vapor radiometer (AWVR), was developed and installed at the site of DSS-25 (see Figure 5.3 and Section 5.5.6) for precision Ka-band RS experiments.

Communicating via optical links with Solar System missions is part of the planning for future ground stations (Hemmati 2006). The Deep Space Optical Communications (DSOC) package, developed at JPL, is a laser space communication system that can increase the data rates one or more orders of magnitude over the current RF systems. A flight demonstration of DSOC is planned for NASA's Psyche asteroid mission circa 2023 (Biswas et al. 2018). Ground-based optical telescopes will be initially used with

Figure 5.3 Upper panel is DSS-25, a 34-m diameter station at NASA's Deep Space Network, Goldstone, California, capable of Ka-band uplink and downlink. In the drawn circle is the advanced water vapor radiometer (AWVR) showing relative location and size to the DSN station. Lower panel is a close-up of the AWVR showing the relative size to two radio scientists.

the possibility of adding optical surfaces and systems to the DSN in the future. In the optical communications era, the RS community will carry out equivalent optical link science as outlined in Chapter 6 and in Asmar et al. (2019).

5.1.2 Experiment Error Budgets

The success of RS experiments relies on a thorough understanding and modeling of error sources and on applying calibration techniques whenever possible. Key concepts to understanding RS observables are phase and group velocity as well as the frequency dependence of dispersive propagation media. These concepts apply to the proper calibration of observables and to the design of an budget, such as the motivation for use of Ka-links that are less susceptible to charged particle effects in the interplanetary plasma and planetary ionospheres.

Cassini, Huygens, Juno, Mars Reconnaissance Orbiter (MRO), Venus Express (VEX), Gravity Recovery and Interior Laboratory (GRAIL), BepiColombo, JUICE, and Europa Clipper are examples of missions that underwent detailed studies of error budgets prior to flight system development as well as in-flight experimental configuration to optimize RS observations. ESA's BepiColombo and JUICE missions, for example, build on the experiences with Ka-band links and advanced calibration techniques used by NASA's Cassini and Juno missions, which had significant contribution from European investigators. The Europa Clipper studies have led to design choices of X-band uplink with coherent X- and Ka-band downlink signals.

Significant improvement using Ka-band in precision radiometric investigations is achieved when there is Ka-band uplink and coherent downlink. A Ka-band downlink signal that is referenced to an X-band uplink signal is limited by the noise on that X-band uplink and provides a very marginal advantage for RS experiments compared to having Ka-band uplink.

5.1.2.1 Cassini

Thorough prelaunch studies were carried out in response to science requirements from the Cassini science team. The final design included the TT&C uplink and downlink functions at X-band, supplemented by a KaT, a coherent carrier-only (no modulation) Ka-band translator/transponder, an S-band transmitter, and a state-of-the-art USO. Independent of the KaT, another Ka-band power amplifier was included for a separate Ka-band downlink referenced to either the X-band uplink or the USO, as detailed in Kliore et al. (2004) and shown in Figure 1.7. The Cassini KaT and HGA were among equipment contributed by the Italian Space Agency (ASI).

The end-to-end error budget developed in the design phase (see Riley et al. 1990) was validated in flight prior to arrival at Saturn during the Cassini Gravitational Wave Experiment (GWE), which imposed the most stringent Doppler requirements to date on Solar System missions. The measured results are detailed in Asmar et al. (2005b) along with a detailed error budget; results showed that the best achieved Doppler data quality (or Doppler noise) had an Allan deviation measure of frequency stability of $\sim 3 \times 10^{-15}$ at 1000-s integration time. This value corresponds to better than an outstandingly small 1 μm/s velocity error.

Propagation noise in the signal's frequency (or phase) profile is the result of a random index of refraction fluctuations along the signal ray path due to phase scintillations as the radio wave passes through the troposphere, ionosphere, and the interplanetary plasma (Armstrong and Woo 1981; Armstrong et al. 1995; Abbate et al. 2003). The charged electrons in Earth's ionosphere give rise to both amplitude and phase scintillations.

Variations in the water vapor content of Earth's troposphere cause a path delay in addition to amplitude variations. Some variations can be calibrated using available technologies and models, and the unmodeled components can limit the precision of RS measurements. To calibrate these effects, two technologies, multilink systems and WVRs, have been developed and successfully used on the Cassini and Juno missions and are planned for use on future missions.

Calibration of a medium comprising charged particles exploits the dispersive properties of the medium, that is, the dependence of the plasma phase scintillation on the carrier wavelength. The effect can be largely isolated with two or more radio links at different wavelengths via the multilink system described in Bertotti et al. (2003) and Tortora et al. (2004). Higher frequency links are less susceptible to plasma noise; so Ka-band, which is ~4 times higher in frequency than X-band, is theoretically less susceptible by a factor of 16 (in practice, better to approximate as an order of magnitude). Combining coherent X- and Ka-band links in the system is very effective at calibrating this effect.

After calibrating the plasma phase scintillation effect, the remaining phase variation can be attributed to nondispersive effects on the propagating signals observed as a path delay such as Earth's troposphere. A DSN-based AWVR would determine the total water vapor content along a line of sight (LOS) by measuring the strength of microwave emissions from water molecules. By copointing the radiometer with the antenna tracking a spacecraft, the signal delay is measured along the same path, or column of air. Investigations that require high-precision radiometric observables, especially using Ka-band links, benefit from this advanced WVR because Ka-band links are less susceptible to a charged-particles effect making the tropospheric effect more pronounced.

5.1.2.2 Juno

After the experience gained from the Cassini mission, similar studies were carried out for the Juno mission to Jupiter with one its primary objectives being an investigation of the interior structure via precision Doppler measurements of the gravity field. Juno also carried a KaT to supplement the X-band TT&C links. Appendix 5A outlines the end-to-end Juno RS instrument system. After Asmar et al. (2017b), the DSN tracking station records the transmitted X- and Ka-band signal frequencies, as well as the frequencies received a round-trip-light time (RTLT) later as a function of time. The Doppler shift is then computed using light-time iteration algorithms described in Moyer (2003). The Doppler shift of the X-band carrier frequency is used for spacecraft navigation (Thompson et al. 2014) as well as for RS while the Ka-band equipment is used to provide improved Doppler accuracy specifically for the Jupiter gravitational field investigation.

The Juno trajectory was designed so that close flybys of Jupiter, or perijoves, take place within view of DSS-25 at the Goldstone complex in California, the only DSN station worldwide capable of transmitting the Ka-band signal to the spacecraft. At perijoves designated for RS observations, the spacecraft HGA, which is located along the spacecraft spin axis, is pointed toward Earth. DSS-25 transmits signals at both X-band and Ka-band; the X-band signal is fed to the SDST that detects the carrier signal and generates a coherent frequency for transmission back to Earth via the HGA amplified by a traveling-wave-tube amplifier. The Ka-band signal is fed to the KaT that detects the carrier frequency and generates a Ka-band transmission back to Earth via the HGA, amplified by a solid-state power amplifier (SSPA).

For both Cassini and Juno, when no Ka-band signal uplink is available from the ground station or the spacecraft is not in view of DSS-25, but X-band uplink is available, the spacecraft's X-band transponder can generate a Ka-band downlink in addition to the primary X-band downlink signal. This Ka-band signal (sometimes referred to as X up-/Ka down-signal) is useful for providing partial calibration of the effect of charged particles on the Doppler measurement.

5.1.2.3 MRO

The MRO gravity science team placed performance specifications on the X-band two-way coherent tracking link during the mission's development phase. An in-depth study of the Doppler error budget was carried out (Lee 2002) and demonstrated that the large and dynamically complex spacecraft performance was capable of meeting the Doppler noise specification placed by the RS investigators without introducing additional flight instrumentation or modifying the design of the spacecraft. The in-flight radiometric observables were indeed of high scientific quality (Zuber et al. 2007a, 2007b; Konopliv et al. 2011b). Note that MRO also carried a Ka-band downlink instrumentation (no Ka-band uplink) as a technology demonstration that did not work (Shambayati 2007).

As done by the Cassini team, Lee (2002) divided the Doppler measurement errors into six categories:

- Thermal noise
- Solar plasma
- Ionosphere
- Troposphere
- Spacecraft delay variation
- Ground station delay uncertainty

Lee (2002) concluded that the one-sigma two-way X-band Doppler error at 10 s integration time can meet the requirements, assuming Sun-Earth-probe (SEP) angles more than 45 deg and elevation angles more than 30 deg, for a root mean square (RMS) of ~0.15 mm/s:

- ~0.1 mm/s for the spacecraft contribution
- ~0.09 mm/s for the media contributions
- ~0.05 mm/s for the ground contribution

5.1.2.4 GRAIL

The design of the GRAIL system is detailed in Park et al. (2012) and Asmar et al. (2013b) along with quantification of errors for precision Doppler in spacecraft-to-spacecraft links or crosslinks. Measurement errors in the integrated system are grouped into either dynamic or kinematic effects. Dynamic effects are forces acting on the spacecraft that will perturb the spacecraft orbit and are treated as nongravitational accelerations due to:

- Spacecraft thermal radiation (the largest dynamic error).
- Solar radiation pressure.
- Lunar albedo.
- Lunar thermal emission.
- Accelerations due to other unmodeled nongravitational forces.

Kinematic effects are parameters directly affecting the inter-spacecraft range-rate measurement and spacecraft-to-DSN Doppler observables. They include:

- Timing offset.
- Payload bias.
- USO drift.
- Spacecraft attitude jitter.
- Offset of the phase center from the line connecting the center of mass from each spacecraft, minimized by actively pointing the spacecraft to align the phase centers.
- Thermal effects that result from the spacecraft going between sunlight and shade during every revolution around the Moon. These effects depend on the incident Sun angle, called "amplitude cosine and sine once per revolution" in some literature.
- Changes in the temperature of the antennas and other exposed spacecraft elements that will slightly shift the antenna phase center.
- Fuel sloshing in the spacecraft tank, found to be negligible for this particular design.

5.1.2.5 BepiColombo and JUICE

ESA's BepiColombo mission to Mercury has an instrument design that builds on the Cassini and Juno advances and experience with the multifrequency link system. The mission also adds an additional capability, precision ranging at Ka-band, as shown in Appendix 5A, the first in deep space exploration; see Iess and Boscagli (2001) and Section 5.3.6 for the advanced ranging breakthroughs with BepiColombo. The RS error budget for the BepiColombo Mercury Orbiter Radioscience Experiment (MORE) is based on the following estimates (see Chapter 4 for the driving scientific objectives):

- Range measurements: 10 centimeters (cm) (one-sigma one-way)
- Troposphere line-of-sight range calibration: 2 cm
- DSN station location: 3 cm
- Earth orientation: 1.5 cm
- Range-rate measurements: 1×10^{-14} at 1000 s
- Troposphere line-of-sight range-rate calibration: 1×10^{-14} at 1000 s
- Spacecraft accelerometer: 3×10^{-8} m/s^2

The JUICE mission also continues in the tradition of Cassini, Juno, and BepiColombo with a multilink system. The Gravity and Geophysics of Jupiter and the Galilean Moons (3GM) investigation addresses the gravity of Ganymede, Callisto, and Europa and atmosphere of Jupiter with two payload elements, a KaT and a USO. The KaT's range rate is expected to be accurate to ~3 μm/s at 1000 s integration time, two-way, and range accurate to 20 cm, two-way.

5.2 Key Concepts and Terminology

Key principles commonly found in RS literature include the usage of stable clocks, how signal operational modes are named, how the two primary reception modes impact the usability of RS data, and the carrier modulation schemes. Figure 5.4 illustrates these principles and the branching options for spacecraft, ground stations, and combinations of links between them.

5.2.1 The Allan Deviation for Frequency and Timing Standards

Oscillators and clocks used in communications, navigation, and RS experiments are referred to as frequency and timing standards (FTS). They are present at the spacecraft and ground stations since they are required on both ends of any radio link. They provide the primary timekeeping system called the Coordinated Universal Time (UTC) (Wertz and Sarzi-Amade 2011), which is within 1 s of the mean solar time at zero

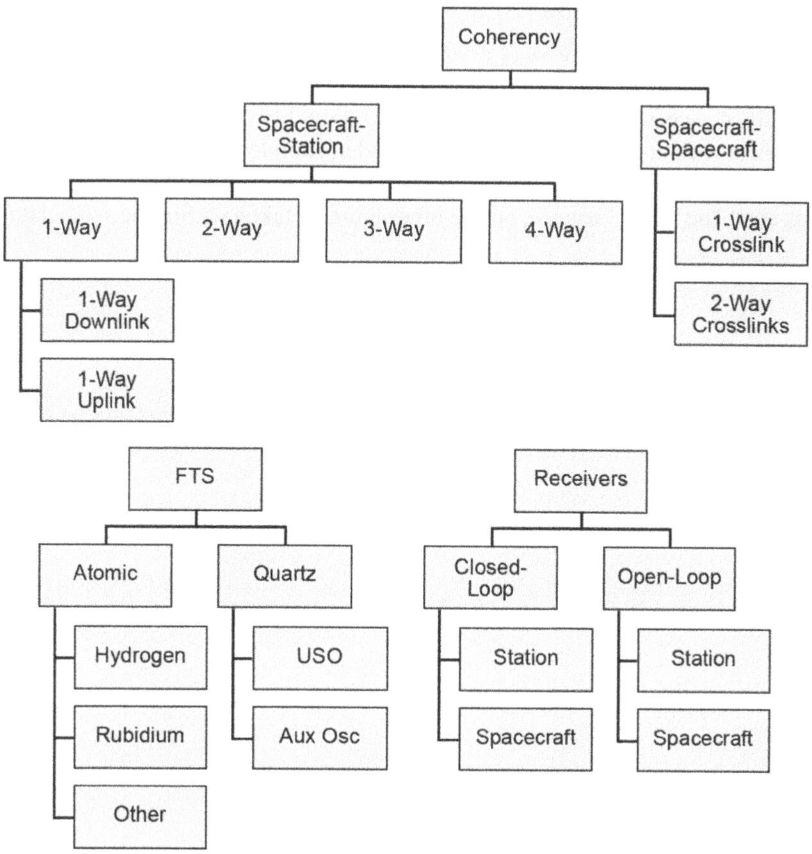

Figure 5.4 Concepts of link coherency. Radio links are either between spacecraft and ground stations (more common case) or spacecraft to spacecraft (less common case) and can be in the various "way" modes. The frequency and timing system (FTS) reference as well as the mode of reception are additional concepts for configuration instrumentation to carry our RS experiments.

longitude or Greenwich Mean Time (GMT). Mission operation teams apply occasional corrections to the spacecraft running clock time and typically report events using the Earth Received Time (ERT) in UTC, which accounts for the one-way light time between the spacecraft and ground stations (Doody 2011).

In RS systems, the operational coherency mode (see Section 5.2.2) determines which FTS, atomic at the ground station or quartz at the spacecraft, is the source of the transmitted and received signals. Figure 5.5 compares the quality of different FTSs in use. This quality is expressed in terms of the Allan deviation frequency stability, which has become the common metric for FTS quality.

The Allan deviation (Allan 1966; Allan et al. 1997), the square-root of the Allan variance, is a modified statistical deviation. It is named after its developer David Allan who reflected, decades after introducing his formulation, that "nature gives us many non-stationary and chaotic processes. If we can properly characterize these processes, then we can use optimal procedures for estimation, smoothing, and prediction ... the Allan variance was developed to this end for the timing and the telecommunication communities" (Allan 2016).

Allan came to the realization that white phase noise, white frequency noise, flicker frequency noise, and other types of noise do not converge when using traditional statistical tools. This made it necessary to consider an alternative characterization that can also account for drift, bias, initial phase/frequency error, time errors, systematic behavior of the device, test conditions, and the operational environment.

The statistical variance is a measure of the spread of samples in a data set, which is needed to measure the spread between multiple data sets by comparing the variances of the data sets. The Allan variance, on the other hand, is taken within the same data

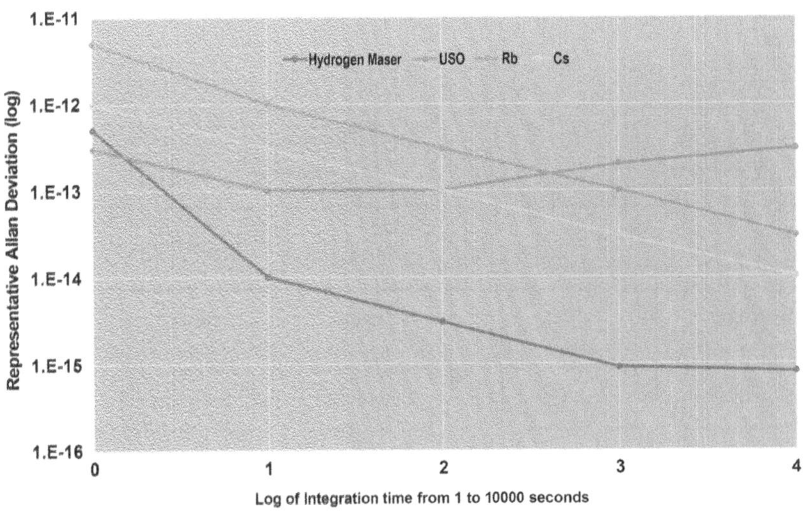

Figure 5.5 Comparison of typical performance of common frequency and timing references relevant to RS experiments. Hydrogen masers are used in the deep space network stations and ultrastable oscillators (USO) represent the top-of-the-line models flown on recent planetary missions. Rubidium (Rb) and Cesium (Cs) standards are used in other space applications.

set as the number of data points increases with function. So, it is taken such that each measurement has been averaged over a given time interval and taken in an adjacent series of data points.

The Allan variance at longer measurements can better isolate the true behavior of the device under test; random or white noise decreases due to statistical averaging and this helps reveal other effects. Integration times (τ) of the measurements are typically reported at 1 s, then, for the same data, at 10 s, then 100 s, and 1000 s. This is why four measurements or specifications are quoted when characterizing the stability of an FTS.

Since any known long-term drift can be accounted for, a key implication of the Allan variance is the attention to the uncertainty or jitter as an intrinsic measure of an FTS quality. A device can have a significant error in the absolute frequency but still be very stable or could drift at a significant but constant and known rate and be considered of good quality.

According to Allan et al. (1997), for x_n, x_{n+1}, and x_{n+2} being three sequential measurements of a device (imperfect) output frequency spaced by a measurement interval τ, the difference between x_{n+1} and x_n is defined to be the fractional frequency change, y_n:

$$\begin{aligned} y_n &= (x_{n+1} - x_n)/\tau \\ y_{n+1} &= (x_{n+2} - x_{n+1})/\tau^{\cdot} \end{aligned} \tag{5.1}$$

If $\Delta x_n = x_{n+1} - x_n$, called the first finite difference, then y_n can be written as

$$\begin{aligned} y_n &= \Delta x_n / \tau \\ y_{n+1} &= \Delta x_{n+1} / \tau^{\cdot} \end{aligned} \tag{5.2}$$

If the raw data points are one point per second and $\tau = 10$ s, for example, x_n, x_{n+1}, and x_{n+2} are each an average of the points in seconds 1–10, 11–20, and 21–30 s, respectively. The Allan variance is sometime called the two-sample variance because the normalized frequency departure is averaged over intervals of two consecutive samples; so, note that it takes 20 s (2τ) to measure y_n.

From the one interval to the next, the change in y_n is

$$y_{n+1} - y_n = \Delta y_n. \tag{5.3}$$

The difference of the difference, or second difference, is

$$\Delta y_n = \Delta^2 x_n / \tau. \tag{5.4}$$

The sum of the squares of these second differences can be computed up to N_{-2}, where N is the number of measurements in the series, and divided by $2 \times (N_{-2})$, the number of entries in the sum. The equation for the two-sample Allan variance may be written as (where the bar indicates an average value)

$$\sigma_y^2(\tau) = \frac{1}{2}\langle(\bar{y}_{n+1} - \bar{y}_n)^2\rangle = \frac{1}{2\tau^2}\langle(x_{n+2} - 2x_{n+1} + x_n)^2\rangle. \tag{5.5}$$

When plotting Allan deviation versus τ on a log–log scale, different types of noises can be distinguished as a function of τ dependence, or power law noise, as illustrated in Figure 5.6. Furthermore, the spectral density, $S(f)$, is a measure of the distribution of power present at different Fourier frequencies that compose the nominal carrier frequency.

While the Allan deviation addresses stability at 1 s and longer in the time domain, phase noise is typically expressed in the frequency domain and characterizes sub-second quality performance. The equation giving the Allan variance in terms of the spectral density is

$$\sigma_y^2(\tau) = 2\int_0^{f_h} S_y(f) \frac{\sin^4(\pi f \tau)}{(\pi f \tau)^2} df, \tag{5.6}$$

where f_h is the high-frequency cutoff for the applicable measurement system (Allan et al. 1997).

One commonly found definition of the phase noise is the frequency domain representation of rapid random fluctuations in the phase of a waveform; fluctuations are caused by time domain instabilities called jitter. Phase noise is also the single sideband power spectral density or the signal's power intensity in the frequency domain.

As noted, phase noise describes the stability in the frequency domain while jitter describes the stability in the time domain. Many sources note that RF engineers

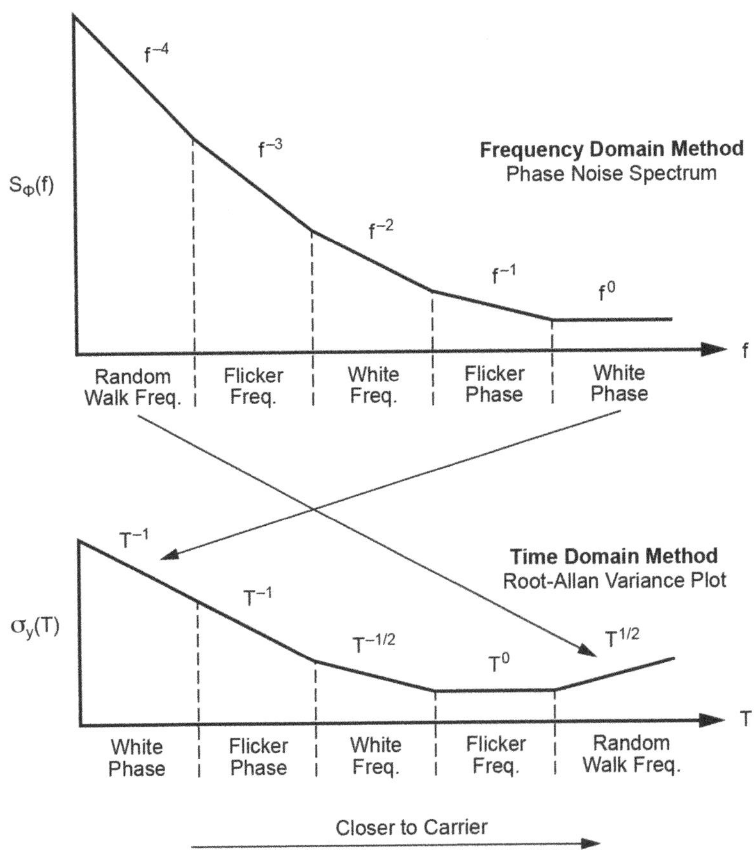

Figure 5.6 Plot of noise spectra in the frequency and time domains showing the various types of noise (Vig 2001).

generally speak of the phase noise of an oscillator while digital system engineers speak of the jitter of a clock. In the time domain, jitter combines noise sources with effects manifested as a function of time such as

- Johnson noise: random variation of voltage due to thermal noise.
- Shot noise: noise that can be mathematically modeled by a Poisson process.
- Flicker (pink noise): electronic noise with a 1/f power spectral density.
- Crystal defects and aging.

5.2.1.1 Atomic FTS

The most accurate FTS is the atomic standard, which is based on the oscillatory quantum transition properties of atoms, such as hydrogen, cesium, or rubidium. When an electron in the atom transitions from a higher to lower energy level, it emits electromagnetic energy as a photon that can be harnessed by a device such as a maser. A “maser” is coined from the acronym microwave amplification by stimulated emission of radiation, similar to the word laser. A maser locks to the frequency of the electromagnetic energy from the electron transition, which is highly reproducible under controlled temperature and other environmental conditions. The modern definition of the second is linked to this reproducibility and to the transition frequency of cesium-133 atomic standard.

Haibo et al. (2009) examined the different advantages in timekeeping between hydrogen and cesium standards and reported that the hydrogen standard has better Allan deviation in the short term than the cesium standard. The DSN stations use hydrogen masers; see Figure 5.7 for stability performance. The hydrogen maser has a typical Allan deviation of ~3×10^{-15} at an integration time of 1000 s, accurate to about 1 s in 10 million years. Newer technologies include Mercury ion and optical ion traps that can be smaller in size for possible flight on spacecraft.

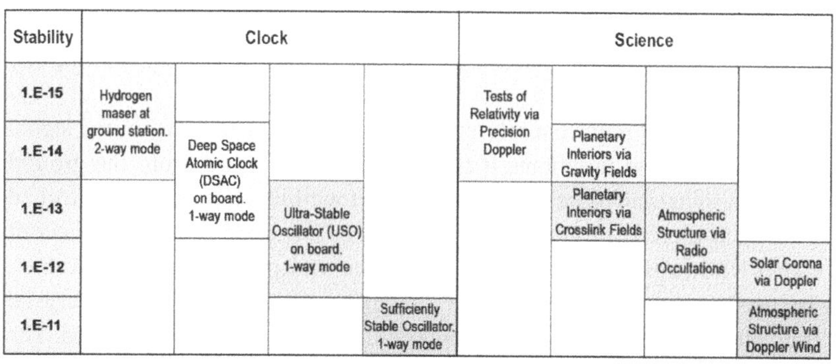

Figure 5.7 Types of RS experiments and frequency stability requirements, at 100-s integration time with scientific application and instruments (onboard USO vs. ground station's hydrogen maser). Note that the 1E-13 USO can be used for radio occultation investigations as well as planetary interior studies from precision gravity field measurements in crosslink experiments (e.g., GRACE and GRAIL). Non-relativistic studies of the solar coronal can be in one-way mode using a USO in the 1E-12 to 1E-13 for the Doppler observable; however, the ranging observable requires two-way mode.

Cesium and rubidium standards have been used in space around Earth, but not on Solar System missions, with the exception of the rubidium USO on the Huygens Titan probe, described in Section 5.3.1.1. Ely et al. (2014) describe NASA's development of a flight atomic FTS. The Deep Space Atomic Clock (DSAC) is a mercury ion atomic FTS with a built-in quartz USO, comparable in performance to the hydrogen maser at the ground station, providing a stability level that enables one-way radiometric tracking data with accuracy equivalent to the current two-way tracking mode. This would allow a shift to a more efficient and flexible one-way deep space navigation and RS architecture. Use of DSAC on board future Solar System missions can potentially increase the quantity of radiometric data (and quality referenced to currently available one-way data) and reduce mission cost.

Prestage and Weaver (2007) describe how all small space atomic standards are passive devices that do not themselves oscillate but monitor and correct the slow frequency changes of a quartz oscillator and describe how the new DSAC is a combination of quartz and atomic technologies. Since the advantage of atomic over quartz FTS is higher stability at long integration times, the combination of the two uses the advantage of the better quartz USO performance at short integration times and the atomic standards at the long integration times.

5.2.1.2 Quartz FTS

A quartz oscillator uses the material property of piezo-electricity where a resonating crystal generates an electric signal of a stable frequency. This property has been harnessed and fine-tuned to develop highly stable quartz crystal oscillators and clocks for numerous applications including communications and space operations. Lu and Czanderna (2012) point out that the piezoelectric property is accounted for by crystal symmetry. Early experimentation established that the orientation of the quartz crystal plate with respect to the crystallographic axes had an important effect on the resonance frequency; this orientation is called the crystal cut.

For the high-quality FTS, the design process optimizes the crystal growth and crystal cut processes, along with carefully designed electronic components and thermal controls. Quartz standards are typically smaller in size and mass and lower in cost than atomic standards and are easier to accommodate on board spacecraft exploring the Solar System.

Spacecraft transponders require a reference signal generator to perform the signal reception and transmission functions. If there is an uplink signal from the ground station, the uplink signal can be used to generate the downlink via a voltage-controlled oscillator (VCO). If there is no uplink, an onboard FTS is needed. If the spacecraft carries a quartz USO (external to but connected to the transponder via coaxial cable), it would be used to generate the one-way downlink. Otherwise, a quartz crystal auxiliary oscillator (often shortened to aux osc) built into the transponder is used. The auxiliary oscillator is not the preferred source for RS observations because among the three options (VCO, USO, or aux osc), it has the worst frequency stability, varying between 10^{-9} and 10^{-10} Allan deviation at 100 s (Asmar and Kursinski 1991).

Quartz crystal oscillators offer excellent short-term frequency stability, as well as low-power consumption, while atomic FTSs offer excellent long-term stability, but high-power consumption. So, there is a need for a technology that combines the advantages of both types. The chip-scale atomic clock (CSAC), which is compact and requires low power, offers the approximate performance of an atomic FTS and compactness of a quartz-based FTS (DeNatale et al. 2008).

5.2.2 Signal Operational Modes

For communications between spacecraft and ground stations, the operational mode, also called the coherency mode, can be in one-, two-, three-, or four-way modes, defined in the following subsections and illustrated in Figure 5.8. Spacecraft-to-spacecraft crosslinks also have their own set of modes. For RS experiments, a mode is chosen to maximize the quantity and quality of the returned data. Note that variations of these modes can be developed for new scenarios; for example, one can conceive of a five-way mode if the four-way mode is modified such that the receiving station is different from the transmitting station. Crosslink modes can have variations if multiple spacecraft in a constellation feed signals to each other, such as the Laser Interferometer Space Antenna (LISA) mission.

5.2.2.1 One-way

When the radio signal is generated on board the spacecraft, referenced to either the auxiliary oscillator built into every transponder or a USO, and transmitted to the ground station, this configuration is a one-way downlink. Radio occultation experiments are configured in the one-way mode only if there is a USO on board the spacecraft since the stability of the typical auxiliary oscillator does not meet RS specifications. If there is no USO on the spacecraft, RS investigators resort to the two-way mode, as discussed in Section 5.2.2.2.

A variation of this configuration is the one-way uplink mode, a path reversal. A signal is generated at the ground station and transmitted to a spacecraft where it is recorded. The advantage is the significant increase in the signal-to-noise ratio (SNR) since the DSN station can transmit thousands of watts (~20 kW) whereas the signal transmitted by a spacecraft is typical ~20 W; the difference in the SNR is ~3 orders of magnitude.

For distant planetary targets, and due to the one-over-distance-squared power loss, one-way uplink is preferred and has been used by the New Horizons mission during the Pluto encounter. The reason other very distant missions have not used the uplink mode is the need for an instrument on board the spacecraft that functions as an open-loop (OL) receiver, recorder, and processor. This was developed for the New Horizons mission by Stanford University (Tyler et al. 2009) and emulates the functions of the open-loop receiver (OLR) in the DSN.

Due to historical practice, it is common to see the terminology "two-way coherent" and "one-way noncoherent" in some literature because the two-way link is coherent with the highly stable atomic reference of the ground station while the one-way downlink is not. The one-way downlink signal is technically coherent but its coherency is from an onboard oscillator.

In the one-way downlink mode, it is assumed that there is no uplink signal from the station. However, the Voyager and Galileo spacecraft used a configuration that is not commonly used by other missions. The spacecraft can be in the one-way downlink mode even if there is an uplink from the ground station. This was exercised when the one-way downlink referenced to the spacecraft USO was required for RO experiments

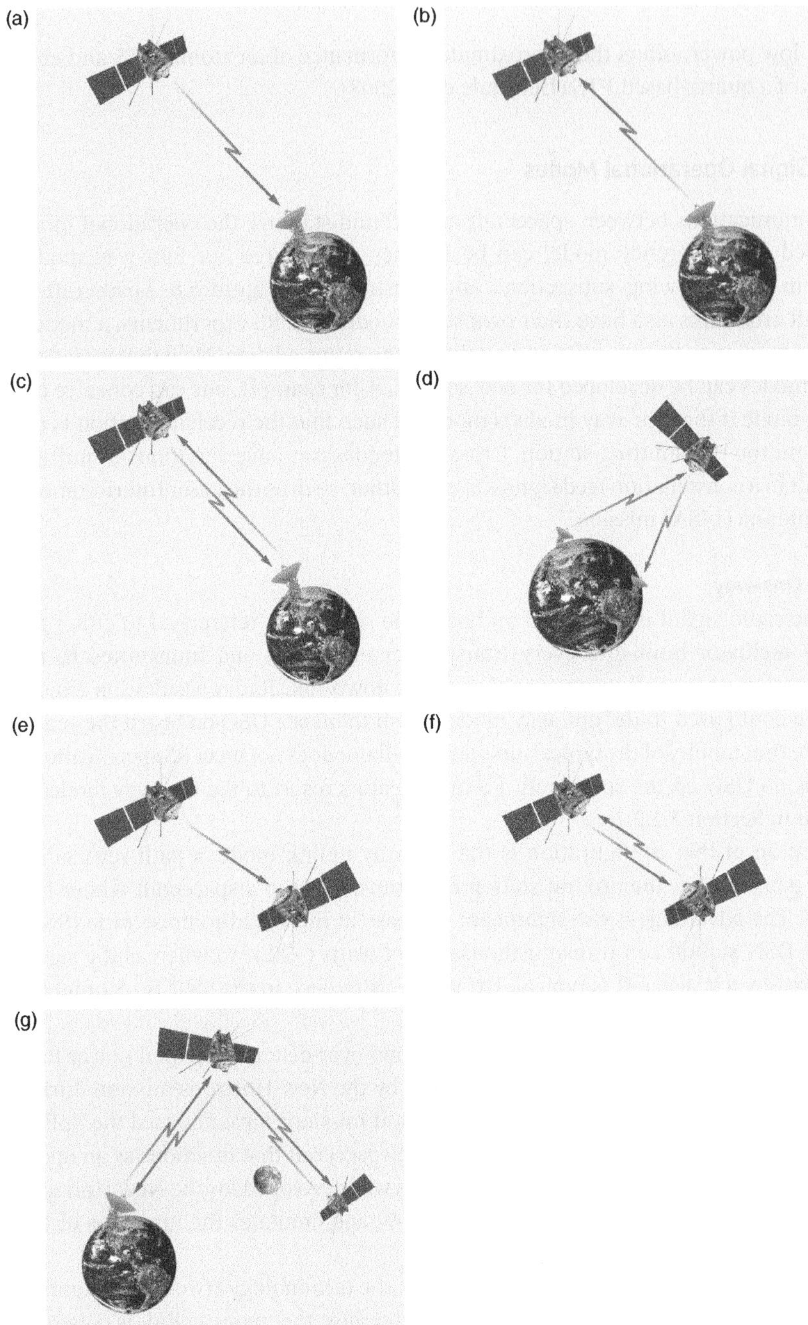

Figure 5.8 Illustration of link configuration modes for radio-science experiments. (a) One-way downlink from the spacecraft to a ground station referenced to an ultrastable oscillator on the spacecraft. (b) One-way uplink from the station to the spacecraft for onboard recording referenced to a USO. (c) Two-way links where a station transmits a signal to a spacecraft, which turns it around and transmits it back to the same station. (d) Three-way links where a station transmits a signal to a spacecraft, which turns it around and transmits it to a different ground station because the round-trip light was so long that the first station was no longer in view of the spacecraft and another station has come into view. (e) One-way crosslink where one spacecraft transmits a signal to another spacecraft. (f) Each spacecraft transmits a signal to the other spacecraft. (g) Four-way link exchange.

but commands or sequence updates still necessitated a transmission to the spacecraft. Because the one-way downlink is also called noncoherent mode, this unusual configuration is historically called "two-way noncoherent mode is on," or TWNC-ON, which can be confusing terminology when encountered in literature from that era.

5.2.2.2 Two-way

The two-way operational mode is attained when the radio signal is transmitted by the ground station, received by the spacecraft, and coherently retransmitted back to the same station. The downlink remains phase coherent with the uplink signal, which takes advantage of the superior quality FTS at the station, such as the hydrogen maser. As explained in Chapters 1 and 3, gravity experiments are configured in the two-way mode to benefit from the maser's stability.

Furthermore, the two-way mode is required for the ranging function, or precisely measuring the line-of-sight distance between a spacecraft and a ground station. Ranging requires transmission of ranging code that is received and transmitted back from the spacecraft to the station.

5.2.2.3 Three-way

In the three-way mode, the uplink signal is transmitted by the station, received by the spacecraft, and coherently retransmitted back to Earth, but to a different station. This happens due to the long RTLT over planetary distances where the transmitting station is no longer in view due to Earth's rotation and another station on a different continent comes into view of the same spacecraft. This mode also occurs when two or more stations are intended to receive the same spacecraft signal simultaneously where, regardless of RTLT, only one station is permitted to send the uplink, placing the other station in a three-way mode. Note that sometimes, the signal is received in either one-way downlink or three-way mode for a brief time period during the process of "station handover," the tracking transition from one station to another prior to settling into two-way mode.

5.2.2.4 Four-way

This rarely used mode describes a signal transmitted from the station to a spacecraft from which it is relayed to another spacecraft, then back to the first spacecraft and from there back to the ground station while maintaining phase coherency for precision Doppler measurements (see Figure 5.8(g)). The Japanese SELenological and ENgineering Explorer (SELENE) spacecraft applied this mode when using a subsatellite on the lunar farside, which had no direct LOS from a ground station, necessitating a coherent Doppler relay via the primary spacecraft; this radiometric configuration enabled measurement of the gravitational field on the farside.

The communication between the SELENE spacecraft to its subsatellite is a crosslink, as defined in the following section. The four-way mode and mix of ground-to-spacecraft and spacecraft-to-spacecraft modes opens the door for possible new innovative modes.

5.2.2.5 Crosslink Modes

When radio signals are exchanged between two spacecraft, they form one-way crosslink, two-way crosslink, dual one-way crosslinks, or dual two-way crosslinks, as shown in Figures 5.8(e) and (f). Dual one-way crosslinks are simultaneous one-way links (one-direction) from each spacecraft to the other. Similarly, dual two-way crosslinks are simultaneous two-way links from each spacecraft to the other for a total of four links. The Gravity Recovery and Climate Experiment (GRACE) and GRAIL missions used dual one-way crosslinks and produced a dual one-way ranging (DOWR) data product. Chapter 6 outlines proposals for future RS experiments with multiple small spacecraft utilizing crosslinks.

5.2.3 Reception Modes

Regardless of the FTS selection and operational modes described in the previous section, a signal can be received in one of two types of modes, closed-loop (CL) or open-loop (OL). Both types can be carried out simultaneously using independent receivers at the ground stations. The loop refers to the feedback technique used for signal detection and tracking in the CL receiver. Since there is no loop in OL reception, a better name is super-heterodyne receiver, but OL has been commonly used in the context of RS instrumentation.

Note that, although OLRs are often used for the one-way mode RS experiment (primary receivers for atmospheric radio occultations, for example), and CL receivers are often used in two-way mode (primary receiver for Doppler and range observations), there is no restriction to do so and one-way links can be received by a CL receiver, and vice versa.

Both CL and OL modes apply to ground station receivers as well as spacecraft transponders. Although the day-to-day operations of the spacecraft transponder are in the CL mode, the introduction of one-way uplink RO on New Horizons highlighted the rarely used OL mode reception of the transponder. The Electra radio, designed for proximity link communications between Mars landers and Mars orbiters, also has an OL mode (Satorius et al. 2011; Edwards et al. 2016).

5.2.3.1 Closed-loop Receivers

The DSN station's CL tracking receiver is used for all TT&C functions; it finds the carrier of the incoming spacecraft signal, locks onto it, and tracks it via a feedback loop, producing a time profile of the received frequency and signal amplitude, the latter provided by the automatic gain control (AGC) circuit.

The CL receiver has other designations in DSN documentation, such as the block V receiver (BVR) or downlink telemetry and tracking subsystem (DTT). It has two primary functions: telemetry extraction and Doppler and range radiometric data production for navigation and RS experiments. A phased-lock loop receiver detects the signal and precisely measures its phase in a loop bandwidth preselected to track the expected signal dynamics and SNR conditions. Telemetry demodulation can then be performed as detailed in DSN Document 810-005.

CL reception, however, has limitations that require an alternate method for cases of RS experiments or mission events of high dynamics in signal level and/or Doppler shift that exceed the CL receiver threshold. High dynamics can cause the receiver to drop lock, with subsequent loss of important data. These conditions typically take place during critical mission events, such as launch and initial acquisition, orbit insertion, landing, or other significant maneuvers, and are of interest to mission teams and, sometimes, to the RS community.

5.2.3.2 Open-loop Receivers

There are important experiments and specific signal conditions that require OL reception and processing at DSN stations. The OLR, also known as the radio science receiver (RSR), down-converts and records the incoming signal in a preselected bandwidth. There is no lock function or tracking threshold; instead, the OLR uses a prediction of the signal profile generated in advance of the experiment based on the known trajectory of the spacecraft. The "predict" file tunes a numerically controlled oscillator driving the down-converter that shifts the signal to a lower frequency (see Appendix 5A). This tuning compensates for the expected Doppler effects and keeps the recording bandwidth and output data file significantly smaller than if no tuning was applied.

The OLR captures the pre-detection signal in a user-selected bandwidth. Raw digital samples, complex in-phase and quadrature, or I&Q samples, of the signal voltage are produced for delivery to data users who apply their software tools to detect the signal from the OLR file and recover the frequency transmitted by the spacecraft for further analysis.

For RS experiments, the advantages of the OL reception method, over the CL receiver algorithm that provides real-time selection of the signal parameters, include

- Capturing the signal during conditions of low SNR.
- Capturing the signal during conditions of high signal dynamics in frequency.
- Capturing the signal during conditions of high signal dynamics in amplitude.
- Capturing multi-path signal conditions.
- Higher resolution data sampling.
- Preserving the information in the entire recording bandwidth.
- Allowing optimization and innovation in post-pass processing algorithms by users to maximize the scientific return.
- Capturing dual polarization channels, at stations capable of simultaneous dual polarization reception, for scientific post-processing.
- Enabling very-long-baseline interferometry (VLBI) and delta differential one-way ranging (DOR) observations.
- Allowing transport of a portable OLR (POLR; older version called portable RSR, PRSR) to facilities outside the DSN when needed.

The disadvantages of using OL data are

- Advanced software tools by each user group.
- Additional processing time and latency in delivery.
- Large data file sizes, requiring more resources for transfer and storage.

An example of the file size, a function of the recording bandwidth, is as follows: if a signal is captured in a 25-kHz bandwidth, it would have 25,000 samples per second multiplied by 2 for the I&Q components. In a 10-min (600 s) experiment, the total amount of data is 240 million bits (Mb), if each sample is 8 bits. Teams often record multiple channels (such as S- and X-band channels, right-hand and left-hand polarization channels, etc.) and the total data volume grow accordingly. For the same observation, a file derived from the CL receiver output containing the equivalent information of carrier frequency and AGC would be less than 1 Mb in size.

In this section, the description of OLRs has been limited to their use at ground stations; recently, SOLRs have been used on spacecraft; the radio-science experiment

(REX) on the New Horizons mission had the first operational use of a SOLR for an RS experiment during the Pluto encounter. JPL has also prototyped a flight processor called the Radio Atmospheric Sounding and Scattering Instrument (RASSI) for uplink as well as crosslink RS experiments (Navarro et al. 2010) and is currently addressing adding such capability to future deep space software-defined transponders such as the Iris radio and the UST.

The introduction of the SOLR (described further in Section 5.3.3) provides opportunities for additional scientific research in uplink (described in Chapter 2) and crosslink RS experiments (described in Chapter 6), as well as radiometry experiments (described in Chapter 1).

5.2.4 Signal Carrier Modulation Modes

A signal carrier uses various encoding and modulation techniques to deliver data. For deep space telecommunications engineers, encoding usually refers to error-correction code, which permits binary symbols that have been incorrectly detected in the receiving system to be corrected with the help of built-in redundancy in the encoded telemetry symbols.

Modulation is the conversion of information (telemetry) onto a signal downlink carrier for the purpose of delivery across a space communications channel from a spacecraft to a ground station, where there is a demodulator. The most common modulation scheme employed for NASA deep space missions is binary phase-shift keying (BPSK), where a modulation index is adjustable between 0 deg, a pure sinusoidal carrier with no telemetry modulation, and 90 deg, a suppressed carrier with all power placed in the data sidebands. Quadrature phase-shift keying (QPSK) is sometimes used and results in the carrier being completely suppressed.

Since RS experiments examine the properties of the signal carrier, the method selected for the telemetry modulation affects the signal's primary properties, such as variations in the signal carrier strength, one of the scientific observables. Mission teams use methods that place the carrier in one of three states and RS users sometimes influence this selection to optimize RS observations:

- Full carrier
- Residual carrier
- Suppressed carrier

A full carrier is an unmodulated carrier; thus, no telemetry or ranging is transmitted from the spacecraft, and the carrier signal is a pure sine wave (with noise) and used for RS experiments or navigation observations. Note that the small power penalty on the carrier due to ranging modulation is typically of negligible impact on RS experiments. When telemetry is transmitted, it can take partial power from the carrier, leaving a residual carrier, or take all the power from the carrier into the telemetry sidebands, leaving a suppressed carrier, as in the case of a modulation index of 90 deg.

Carrier modulation design criteria are discussed in Simon et al. (1996) who describe the considered trade-off between suppressed or unsuppressed carrier given the available total power, P_t, the power going into the carrier, P_c, and the power going into the data P_d, when telemetry phase modulates the downlink carrier (no ranging modulation):

$$P_t = P_c + P_d. \tag{5.7}$$

The telemetry modulation index determines the division of link power between the residual carrier and the data sidebands. The residual-carrier power is used in the CL receiver to achieve carrier synchronization. The data-sideband power is used in the demodulation and decoding process to recover the telemetry. An optimum modulation index that balances these competing needs may generally be estimated; this optimum index depends on the loop bandwidth of the CL receiver, the data rate, and the modulation/coding scheme in use.

Although the ideal configuration for RS experiments is full carrier, especially for RO experiments that experience signal attenuation in the medium under study (atmosphere of Venus or rings of Saturn, for example), the more common case of residual carrier configuration is acceptable for many RS experiments above an SNR threshold, such as gravity investigations.

Furthermore, RS experiments are not designed to be conducted in the case of a suppressed carrier, although navigation radiometric observations in this mode are possible due to the design of the DSN tracking receiver. However, if there is an opportunistic RS observation under suppressed carrier conditions, there are techniques to recover the signal carrier phase in the data post-processing by the users; see Asmar et al. (2013a) for a method of extracting RS data when a coded suppressed carrier modulation is used in deep space communications, such as BPSK.

5.3 Radio Science Technologies

USOs and KaTs are the most common additions used to supplement the telecommunication system specifically for RS purposes. A USO is required for RO and DWEs as well as crosslink gravity experiments, and a KaT is required for precision radiometric experiments, such as gravity fields for internal structure studies or testing the theories of general relativity. A KaT must be accompanied by compatible spacecraft antenna and microwave components. In addition, a SOLR, similar to the REX instrument on the New Horizons mission or comparable instruments, are likely to be used on future missions with uplink or crosslink RO experiments; see Section 5.3.3 and Chapter 6 for discussion on selected future technologies.

Accelerometers are also required on some missions to calibrate precision Doppler observations and remove the effect of nongravitational forces. GRACE and BepiColombo are examples of missions with accelerometers that measure the motion of the spacecraft due to external forces or outgassing. Similarly, thermometers on various spacecraft surfaces provide thermal data useful to high-precision RS observations; when an orbiter repeatedly crosses the terminator, for example, the heating–cooling cycle on spacecraft surfaces alters its dynamical behavior.

Furthermore, there are stand-alone science instruments and technologies, not described here, whose data complement RS observations and are used for correlation of data sets. For example, orbiters of rocky planets that measure high-resolution gravitational fields benefit from the presence of radars or altimeters that provide a comparable-resolution topography field, or other sources of topography or shape models, as described in Chapter 3. This was best demonstrated on MGS as well as

GRAIL and the Lunar Reconnaissance Orbiter (LRO) missions. On MGS, the Mars Orbiter Laser Altimeter (MOLA) described in Zuber et al. (1992), produced a high-resolution topography field that was correlated with the gravity field measured by the same mission (sometimes, also combined with historical data) leading to important discoveries (Zuber et al. 2000; McGovern et al. 2002). GRAIL did not carry an altimeter, but LRO was in operation in the same time period and carried the Lunar Orbiter Laser Altimeter (LOLA). Science teams used the GRAIL gravity data and LRO altimetry data to produce high-resolution correlations between the two data sets (Zuber et al. 2013b).

5.3.1 Spacecraft Ultrastable Oscillator

5.3.1.1 Purpose and Historical Development

The USO was developed for the purpose of carrying out planetary RO experiments in the one-way mode where there is a stable reference on each end of the link, spacecraft and ground station, or between two spacecraft. Reasons for carrying out the experiment in the one-way mode with a USO include:

- It eliminates the double propagation of the uplink and downlink signals in the same planetary medium under study. The double propagation requires more complex processing (Schinder et al. 2012; Hinson et al. 2008; Hinson et al. 2014; Tellmann et al. 2013), which is further complicated in cases of multi-path signal conditions in the planetary atmosphere (Fjeldbo et al. 1975; Lipa and Tyler 1979). The preferred one-way signal travels from the spacecraft through the medium once. This is typically a downlink configuration but, for the Pluto one-way uplink experiment, the direction was reversed and data were stored on the spacecraft then played back as telemetry files (Tyler et al. 2009).
- It eliminates the time needed by the transponder to lock-up on the uplink signal when the spacecraft is emerging from behind the occulting planetary body. Although the lock-up time can be short (a few seconds to a few minutes, depending on the atmospheric density, uplink SNR, sweep rate, etc.), this delay can be comparable to the length of the observation and cause the loss of the egress experiment in the lower atmospheric region. With a USO, the spacecraft can already be transmitting prior to the beginning of the occultation egress without loss of data.

Without a USO, the transponder's built-in auxiliary oscillator is not sufficiently stable to extract scientifically meaningful data (unless the aux osc stability improves significantly in the future). However, without a USO, an occultation experiment can be carried out in the two-way mode for ingress only (immersion). To clarify these issues, in a highly instructive 1975 paper on the Pioneer 10 RO of Jupiter's ionosphere at S-band (~2.1 GHz), Gunnar Fjeldbo (also known as Gunnar Lindal) and his team of RS pioneers discuss the use of an onboard auxiliary oscillator (referred to as the crystal oscillator, not thermally controlled) and summarize the state of the art at that time as well as the evolving thinking on the instrumentation due to the dilemma they faced prior to the invention of the USO for the Voyager program. The comment "considering to conduct the Pioneer 11 immersion measurements at Jupiter in the two-way mode" was addressed in Section 5.2.2.1. Fjeldbo et al. (1975) stated:

> In view of the unexpectedly large topside scale height and temperature revealed by the Pioneer 10 data, we are now considering to conduct the Pioneer 11 immersion measurements at Jupiter in the two-way mode. In this tracking mode, a stable Earth-based oscillator (either a hydrogen maser or a rubidium vapor oscillator) is used to generate a 2.1-GHz uplink carrier. The spacecraft's radio transponder coherently translates the frequency of the uplink signal to approximately 2.3 GHz and retransmits the new RF carrier back to the tracking station where the difference between the uplink and downlink frequencies can be measured. Since the stability of any one of the available ground-based frequency standards is at least an order of magnitude higher than the stability of the spacecraft crystal oscillator, the two-way measurements can be expected to help substantially reduce the current uncertainty in the topside plasma scale height and temperature. Of course, two-way tracking will not yield reliable data in the lower ionosphere where multipath on the uplink may cause the spacecraft's phase-locked loop receiver to change lock from one signal mode to another in an unpredictable manner. However, this disadvantage is not a serious problem since one-way emersion measurements can provide more data on the lower ionosphere.

Due to these challenges on Solar System missions, the first flight instrument developed specifically for RS enhancement was the USO, called the "single most dramatic improvement in radiometric performance" by G. L. Tyler (1987). The USO was developed for the Voyager atmospheric and ring RO experiments to provide a stable reference for the transponder as it transmits a one-way downlink signal to Earth.

The Voyager RS team leader at Stanford University charged a JPL team member, the late Gordon Wood, to partner with industry to develop a very stable clock small enough to fly on the Voyager spacecraft. Frequency Electronics Inc. (FEI) of Long Island, NY, was selected (its founder Martin Bloch and Gordon Wood coined the "ultra-stable" classification). Specifications were derived from the objectives of the Voyager investigations at the outer planets as well as the available technology.

After a prototype unit met the requirements, NASA procured five flight-qualified USOs. Voyager 1 and Voyager 2 carried one each, and the Galileo orbiter later carried USO serial 4 from that batch, later called the 10^{-12} class of stability. The Voyager USOs have since left our Solar System, although the Voyager 1 USO stopped functioning circa 1992 after successfully completing all its required scientific functions at planetary flybys. The Voyager 2 USO is still used for one-way communications at record distances and record longevity. USOs have been single-string science instruments while the transponders they are connected to (cross-strapped) are redundant.

After Voyager and Galileo, Cassini as well as orbiters at Mars, Venus, the Moon, and other planetary and Earth missions (listed in Asmar 2012b and updated in Table 5.1) have carried USOs from various manufacturers including FEI and the Johns Hopkins University Applied Physics Laboratory (APL); see example in Asmar et al. (2019). The post-Voyager USO models were as much as an order of magnitude better in their stability, the 10^{-13} class of stability. This was needed for Mars RO experiments where the atmosphere is significantly less dense than the outer planets. The first of these was flown on MGS, which, for a decade, carried thousands of atmospheric occultations with extremely valuable scientific results.

Table 5.1 Ultrastable oscillators on Solar System missions with approximate characteristics.

Deep space mission	Voyager Galileo	Galileo probe	MGS	MRO	Cassini
Maker	Frequency electronics	Frequency electronics	JHU-APL	Frequency electronic	JHU-APL
Year	1975	1975	1987	2002	1993
Crystal cut	AT	SC	SC		SC
Number of ovens	2	2	1	2	1
Mass (kg)	1.1	0.57	1.3	0.9	2
Dimensions (cm)	10.2 × 19.5	4.6 × 14	10.2 × 10.2 × 16.8	13 × 7.5 × 11.5	10.2 × 12.8 × 19.4
Power (W)	2.2	1.2	2.2		2.8
Resonator (MHz)	6.38	4.6	4.79		4.79
Nominal output (MHz)	19.1	23.117	19.144	76.7	114.917
Assigned channel	18/14	N/A	20	32	23
Bands/Links supported	S, X	1.387 GHz crosslink	X	X downlink UHF crosslink	S, X, Ka
Aging /24 h	5×10^{-11}		2×10^{-11}	5×10^{-11}	7×10^{-11}
Temperature (°C)	5×10^{-12}	3×10^{-12}	3×10^{-12}	5×10^{-12}	2×10^{-12}
Radiation (rad)	2×10^{-12}	2×10^{-13}	1×10^{-10}		1×10^{-10}
Magnetic field (Gauss)	5×10^{-12}	4×10^{-12}	8×10^{-13}		5×10^{-13}
Static acceleration (g)	1×10^{-9}	1×10^{-9}	3×10^{-9}		1×10^{-9}
Harmonic spurs (dBc)	−40		−60	−50	−60
Phase noise 1 Hz (dBc)	−100		−110	−90	−85
Phase noise 10 Hz	−108		−125	−110	−110
Allan deviation 1 s	3×10^{-11}		3×10^{-13}	3×10^{-12}	2×10^{-13}
Allan deviation 10 s	4×10^{-12}		1×10^{-13}	1×10^{-12}	1×10^{-13}
Allan deviation 100 s	1×10^{-12}		1×10^{-13}	2×10^{-12}	1×10^{-13}
Allan deviation 1000 s	1×10^{-12} 7×10^{-13}	5×10^{-10}	1×10^{-13}	3×10^{-12}	1×10^{-13}

(Continued)

Table 5.1 (Continued)

Huygens probe	New horizons	GRAIL	Rosetta VEX Akatsuki	LRO	JUICE (planned)
DASA, Germany	JHU-APL	JHU-APL	TimeTech GmbH	Symmetri-com	AccuBeat Israel
1995	1994	2010			
Rb	SC	SC			
2	1	1			
1.9	1.5	1.75	1.9	2.7	
17 × 14.9 × 11.8		19.4 × 10.2 × 11.8	17 × 13 × 16	22 × 11 × 15	12 × 12 × 12
9	3.5	3.2	5	2.9	
6835	5	4.832		5	4.79
10	30	57.98 38.66	38.2		56.4
N/A					
2 GHz crosslink	X uplink	S, X, Ka down/ crosslink	S, X X (Akatsuki)	S, Ka	X, Ka
2×10^{-9}	1×10^{-11}	5×10^{-11}		3×10^{-11}	
3×10^{-12}	1×10^{-12}	1×10^{-12}		2×10^{-12}	
2×10^{-12}					
4×10^{-12}					
−60	−60	−40			
−75	−110	−97		−127	
−95	−120	−115			
1×10^{-11}	2×10^{-13}	3×10^{-13}	1×10^{-12}	1.1×10^{-13}	5×10^{-13}
5×10^{-12}	1×10^{-13}	2×10^{-13}	3×10^{-13}	1.3×10^{-13}	5×10^{-13}
1×10^{-12}	1×10^{-13}	1×10^{-13}	3×10^{-13}	1.5×10^{-13}	5×10^{-13}
1×10^{-12}	2×10^{-13}	5×10^{-13}	1×10^{-12}	2×10^{-13}	5×10^{-13}

For the MRO mission, however, there was no NASA-selected RO investigation. However, a Voyager-class USO was included for the purpose of supporting the UHF relay communication system. According to Hinson et al. (2014), this USO was considered of "limited quality" for Mars RO and, as a result, the science observations strategy emphasized the two-way mode for occultation ingress. For the occultation egress, a coherent two-way link cannot be established quickly enough to prevent the loss of valuable observations of the lower neutral atmosphere. Similarly, the Mars Express (MEX) RO experiment (Patzold et al. 2016) and the Mars Atmosphere and Volatile Evolution (MAVEN) mission RS experiment (Withers et al. 2018) obtained electron density measurements via the two-way configuration at X-band (7–8 GHz).

In addition to planetary orbiters, two in-situ entry probes, the Galileo probe, delivered into the Jovian atmosphere by the Galileo orbiter, and the Huygens probe, delivered to Titan by the Cassini orbiter, also carried oscillators for RS purposes. The Galileo probe used a quartz oscillator that required many hours to reach its optimum stability, depending on its thermal environment, while the Huygens probe used a rubidium-based oscillator for the faster warm-up time required between turn-on and acquisition of scientific data; the rubidium oscillator can reach its required stability in tens of minutes. The DWE uses one oscillator at each end of the probe-to-orbiter link or spacecraft-to-spacecraft crosslink.

Probe oscillators were called a USO in the literature even though they do not meet the "ultra-stable" criterion of Section 5.3.1.2. As shown in Bird et al. (2005), the oscillator on the Huygens probe is called the transmitter USO (TUSO) and the oscillator on the orbiter as part of the probe relay package is called receiver USO (RUSO). The Huygens probe rubidium oscillator has an Allan deviation performance of $\sim 1 \times 10^{-10}$ at 100 s integration time (Asmar 1997); yet, it is labeled a USO in project documents and science publications. The same situation applied to the Galileo probe USO (different from the Galileo orbiter USO); the Galileo probe and Galileo orbiter USOs were both stress-compensated (SC) cut crystal oscillators, different from the Huygens rubidium USO. In the case of these planetary entry probes, the criterion of selecting quartz versus rubidium FTS was how quickly it can reach optimum stability due to the limited time of the experiment and addition to its power consumption requirement and warm-up time.

Due to an operational sequencing error, the RUSO was not turned on for the Huygens DWE relay package on the Cassini orbiter. The TUSO-based signal transmitted from Huygens was not received at the orbiter. It was, however, received at Earth-based radio telescopes (Bird et al. 2005). Radio telescopes had previously received the signal from the Galileo probe as an enhancement to the Galileo probe-to-orbiter link (Folkner et al. 1997).

Table 5.1 lists USOs on Solar System missions. Since the early USO development, other designs were introduced with lower volume, mass, and cost, but also with stability that does not match the original USO. The Mars Odyssey mission, for example, uses an oscillator for its UHF proximity communications equipment with Allan deviation of $\sim 1 \times 10^{-11}$ at 100 s (Barbieri et al. 2005); the designation of this oscillator was modified to a sufficiently stable oscillator (SSO). MRO also uses an oscillator, considered a USO, for its UHF proximity communications equipment with Allan deviation of $\sim 1 \times 10^{-12}$ at 100 s (Ely and Bell 2003).

Figure 5.7 shows the primary types of RS investigations along with an approximate range of frequency stability requirements, illustrating which ones require USOs. Fractional frequency stability, or Allan deviation, is shown for 100-s integration time as a representative value for the device.

For traditional planetary missions with radio links between spacecraft and a ground station, a USO is not used for gravity experiments except for rare occasions where the USO-referenced one-way data fill important gaps in the higher quality two-way data referenced to the station's atomic FTS. However, for crosslink missions, such as GRACE and GRAIL, and GRACE Follow-On that investigate high resolution gravitational fields, a state-of-the-art USO is used on each spacecraft for generating their DOWR data type, described in Klipstein et al. (2013) and Asmar et al. (2013b). In this system, the end-to-end stability of the crosslink instrumentation exceeds the stability limit of the USO due to the noise-cancellation technique described in Kim and Lee (2009) and Kim and Tapley (2002).

Klipstein et al. (2013) describe how direct correlation of the crosslink signals is accomplished by comparing the received signal to the transmitted signal and changing the LO arrangement in each spacecraft's instrument and how the USO phase noise is suppressed as follows:

> changes in the separation are correlated on the two orbiters; a longer light time for one means a longer light time for the other. USO phase noise, in contrast, causes opposite changes on the two sides. The local oscillator mixing scheme results in the phases on the two orbiters counter-rotating, so an advancing phase causes an immediate positive shift at the transmitting spacecraft, and a negative shift a light-time later for the receiving spacecraft. The sum of the signals from the two orbiters thus gives twice the distance change, suppressing USO phase noise.

5.3.1.2 Definitions, Performance, and Design Parameters

A USO is defined as an oscillator with stability expressed in terms of Allan deviation of $\sim 1 \times 10^{-12}$ to $\sim 1 \times 10^{-13}$ over an integration time of ~100 s; oscillator grades with worse stability should not be designated ultrastable (as noted in the previous section for entry probe oscillators). The Allan deviation (see Section 5.2.1) is the accepted measure of short-term stability and is specified, by convention, at four integration times: 1, 10, 100, and 1000 s, reporting fluctuations in output frequency as averaged over those selected periods. The stability at times longer than 1000 s is treated as drift and is typically quoted over 24-h periods (the Galileo probe had a short lifetime and specified the stability over 30 min).

For purposes of comparison of classes of oscillators, the 100-s integration time can act as a quick summary indicator of the quality since a quartz USO, which typically reaches its best stability near 100 s and then drifts at longer intervals (Prestage and Weaver 2007). The 100-s stability is sufficient to state, for example, that the Voyager USO has 100 s Allan deviation of $\sim 1 \times 10^{-12}$, the GRAIL USO has 100 s Allan deviation of $\sim 1 \times 10^{-13}$, the Mars Odyssey SSO has 100 s Allan deviation of $\sim 1 \times 10^{-11}$, and the

Huygens oscillator has 100 s Allan deviation of ~1×10^{-10}, where the approximation sign indicates 1–3 parts. The Voyager USO summary indicator also applies to the Galileo orbiter and MRO Electra oscillators and the GRAIL summary indicator also applies to MGS, Cassini, New Horizons, and GRACE.

The short time scale (one second or less to a few minutes) of the USO behavior characterizes typical processes associated with the crystal cut; the intermediate time scale is related to temperature fluctuations, and the long-term behavior is related to the aging of the crystal, which is the change in the oscillator's frequency due to internal, rather than environmental causes. A more complete technical specification on the USO performance includes the following, which are provided for the GRAIL USO as an example in Table 5.1:

- Allan deviation at four integration times
- Phase noise at various frequency offsets
- Daily aging rate
- Long-term drift

It is important to note the differences between accuracy, precision, and stability, best graphically illustrated in Figure 5.9 from Vig (2001). Accuracy refers to the closeness of measurements to a defined standard while precision refers to the closeness of measurements to each other. Stability is how well the device can statistically generate the same frequency when estimated over a given time period, called integration time. A device can be stable but not accurate and a corrective measure can be applied to remove an offset in the accuracy. This is why, for RS observation, precision is more important than accuracy since scientists establish a free-space baseline prior to an atmospheric radio occultation observation for example, and monitor relative changes in the frequency and amplitude, modeled or unmodeled, neglecting the absolute accuracy. Furthermore, RS teams periodically carry out activities for

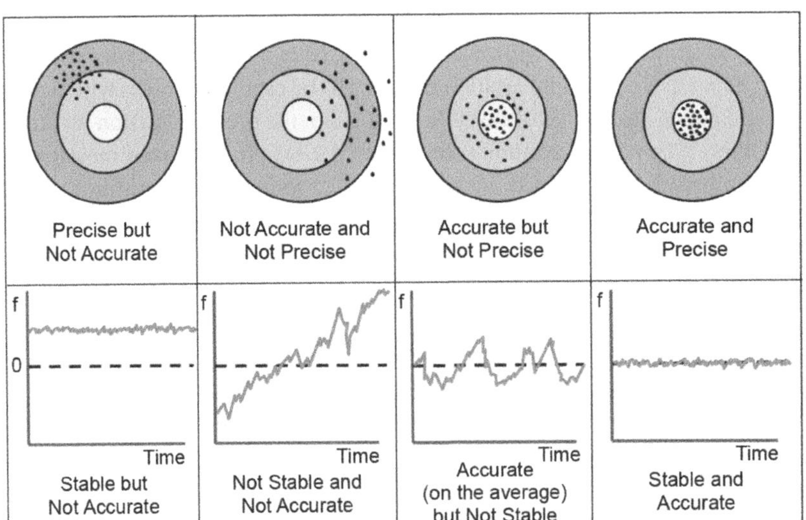

Figure 5.9 Illustration from Vig (2001) of examining frequency and timing instrumentation for precision, accuracy, and stability.

performance characterization of the USO long-term behavior; Figure 5.10 shows a typical long-term trend analyses from the Cassini USO over many years after the initial power turn-on.

For DWE purposes, due to the very rapid rotation of the giant planets, very small changes in probe trajectory will cause very large changes in the received Doppler frequency. To reduce this sensitivity to trajectory, frequency differences are used as the primary data input as opposed to absolute frequencies. The use of differential frequencies reduces the effect of absolute frequency uncertainty on the wind retriev al. This effect is not nearly as large on the more slowly rotating terrestrial planets. Therefore, for terrestrial planet Doppler wind measurements, absolute frequencies are the primary data input, and errors in the transmitted frequency translate directly into changes in the measured Doppler frequency. The effect of absolute transmitted frequency errors directly mimics and cannot be separated from the Doppler effect due to atmospheric winds (David H. Atkinson, private communications).

Terminology common in crystal oscillator (XO) literature includes the following devices:

- Voltage-controlled crystal oscillator (VCXO).
- Oven-controlled crystal oscillator (OCXO).
- Temperature-compensated crystal oscillator (TCXO).
- Oven-controlled voltage-controlled crystal oscillator (OCVCXO).
- Temperature-compensated voltage-controlled crystal oscillator (TCVCXO), sometimes called voltage-controlled temperature-compensated crystal oscillator (VCTCXO).

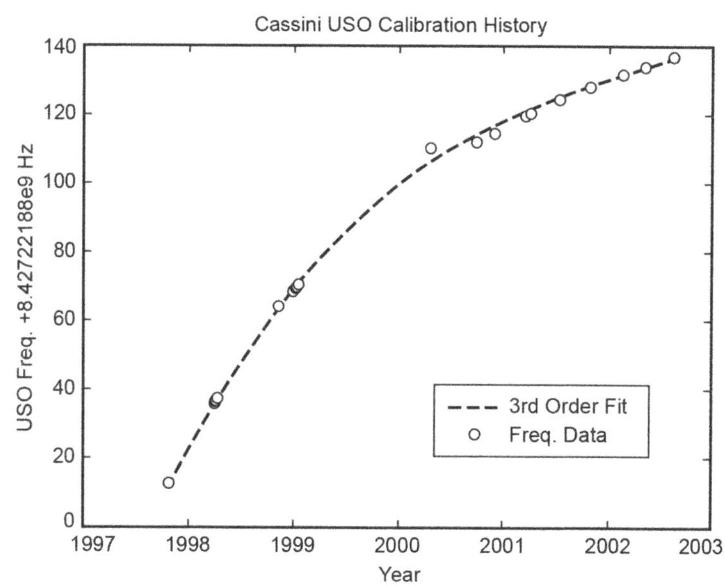

Figure 5.10 The frequency downlink at X-band (8.4 GHz) of the Cassini spacecraft when referenced to its onboard ultrastable oscillator. Data show drift and aging behavior of the precision quartz oscillator since launch with a line representing a third order fit to the data.

The fundamental limits on the stability of XO are discussed by Walls and Vig (1995) who point out that three major components contribute to instabilities: the crystal resonator, sustaining circuit, and the oven. Important factors include time from power turn-on, environmental temperature control during subsequent warm-up and the steady-state phase, pressure, humidity, acceleration, ionizing radiation, and the presence of a magnetic field.

Accelerations due to gravity, acoustic noise, vibration, or shock also affect the frequency stability. Walls and Vig (1995) reported that the best performance for quartz USO aging is 10^{-13} per day and Allan deviation fundamental limit of $\sim 1 \times 10^{-14}$ at 10 s due to the Johnson noise of the resonator and phonon scattering within the resonator. Crystal oscillators exposed to radiation experience frequency shifts that are attributed in part to the presence of impurity metallic ions within the raw quartz lattice. The overall performance of crystal oscillators depends on:

- Crystal cut
- Number of ovens
- Number of output channels
- Electronics
- Shielding
- Advanced features for frequency steering, etc.

Developers have refined the process of growing pure quartz crystals and cutting them with orientation that optimizes their performance. Voyager-class USOs use AT-cut crystals and dual ovens; they perform at typical Allan deviation of $\sim 1 \times 10^{-12}$ at 100 s. The GRAIL-class USOs use SC-cut crystals and single-oven; they perform at typical Allan deviation of $\sim 1 \times 10^{-13}$ at 100 s.

The AT-cut (orientation) was developed in the 1960s and was found to provide a very stable frequency output. A stress-compensated orientation, or SC-cut, was developed about a decade later. Kusters (1981) considers SC to be a generic name for a family of cuts that provide compensation for thermal transient and planar stress effects, which provide better short-term stability, among other advantages. So, the temperature-compensated AT-cut is oriented such that the temperature coefficients of the lattice have minimal impact on crystal performance, while the SC-cut minimizes the stress effects. AT remains the most popular cut in general use and SC, more expensive due to the double rotation manufacturing, fills a need in specialized applications that require higher stability and better aging performance.

5.3.1.3 Future Trends

The improvement of USO stability is the subject of considerable research. Breaking the 5×10^{-14} theoretical limit of quartz crystal oscillator 100 s Allan deviation can possibly be accomplished by using other material (Walls and Vig 1995). Furthermore, additional companies are producing USO, such as Microsemi (formerly known as Symmetricom) which produced the USO on the LRO, and Israel's AccuBeat, which produc the USO for ESA's JUICE mission.

Research facilities such as APL have prototyped a USO with a synthesizer to tune the output frequency in flight. USOs are currently designed to specific mission channel assignment and are not altered in flight. However, there are mission scenarios where in-flight modification of the USO output frequency can help mission operations and science planning. A synthesizer-equipped USO provides in-flight frequency control (Wallis et al. 2005). A synthesized USO will provide a frequency reference for transponders that is electronically adjustable to cover the entire deep-space communications band. This frequency agility allows in flight reassignment of the transponder frequencies.

Furthermore, a "disciplined" quartz USO can improve on the accuracy of output frequency (Weaver and Kantsiper 2007). With an increasing number of planetary spacecraft, coordination among spacecraft can require a local timekeeping system able to directly assist navigation, event synchronization, and trunk communication back to Earth. A disciplined quartz USO can have superior frequency stability extended to nearly 30 days of operating time via a Kalman estimator that optimally removes the drift of the free-running USO. The estimator is improved by periodically comparing the frequency of the USO to an Earth-based atomic FTS and transmitting correction parameters to the spacecraft USO.

Another trend is miniaturization. Spacecraft mass and power resources for small and low-cost missions require a significant reduction of mass and power without sacrificing the stability for science applications. The ~1×10^{-13}-quartz USO typically weighs ~1 kg and requires ~3 W of power in the steady state. Reduction can make USO's good candidates for future planetary SmallSat and CubeSat missions. Ideas based on an old design concept called "tactical BVA" (short for Boîtier à Vieillissement Amélioré or Enclosure with Improved Aging) still hold promise for miniaturization (Norton and Cloeren 1994).

In the era of CubeSats and small spacecraft, USOs can be difficult to accommodate since current USO volumes exceed 1 U. There are feasibility studies for reducing the size of oven-based quartz oscillators (Asmar et al. 2019). Quartz-based USOs require very precise thermal stability achieved via an enclosed oven system, signal processing, and accompanying electronics, and the thermal control mechanism is a prime candidate for possible miniaturization. Future trends in USO are further discussed in Chapter 6.

5.3.2 Spacecraft Ka-band Translator

The Ka-band translator (KaT or KaTS when including the power amplifier as a subsystem) is a stand-alone transponder/translator that is added on selected Solar System missions to provide an additional coherent link for scientific radiometric purposes. Signals at higher frequencies are less susceptible to noise from charged particles in the media of travel such as the interplanetary plasma and ionospheres.

Cassini pioneered the use of coherent Ka-band links in deep space; see historical figures in Appendix 5A of the first ever Ka-band signal during the early cruise phase of the mission. A team from the mission and the DSN developed detailed procedures that addressed the intricacies of prediction generation, antenna pointing, signal acquisition, etc., as discussed in the following sections.

To date, NASA's Cassini and Juno missions as well as ESA's BepiColombo mission carry a KaT; ESA's JUICE mission will also carry a KaT. The units on these missions were contributed by ASI and manufactured by TAS-I (Ciarcia et al. 2013). JPL has also developed transponders that can carry out the same functions (Pugh et al. 2017). The KaT is a transponder in that it receives a carrier signal from the ground station and transmits back a coherent signal at the frequencies shown in Chapter 1, Table 1.3, but is often called a translator when it handles the carrier signal only and not telemetry modulation. With ranging modulation, translator or transponder is in use interchangeably; a redesign of the Cassini-generation KaT, the BepiColombo KaT is identical to the Juno KaT but has an added board for providing ranging code at Ka-band, the first ever in deep space. Considered a science instrument on these missions, the KaT is not redundant, while the mission-critical X-band TT&C transponder is redundant.

The use of a KaT was not possible until a ground station was made Ka-band compatible. DSS-25 in Goldstone, California was upgraded with Ka-band transmission and reception capabilities in the 1990s to support investigations carried out during the Cassini cruise phase and was later refurbished for the Juno mission. The DSS-25 Ka-band system was later enhanced with ranging capabilities compatible with the BepiColombo mission. DSN stations at the other complexes are likely to be upgraded for Ka-band uplink to complement the existing Ka-band downlink capabilities. Furthermore, ESA's station in Argentina has been upgraded with Ka-band uplink and downlink capabilities.

The dispersive interplanetary plasma is the dominant noise source in Doppler and range measurements, most severely at S-band links, less at X-band links, and even less at Ka-band links. With the sensitivity inversely proportional to the square of the signal frequency, the improvement is about an order of magnitude from S- to X-band and from X- to Ka-band. As shown in Figure 1.8, for all frequencies, this effect is a function of the solar elongation angle (Sun–Earth–probe angle) and is at a minimum 180 deg (solar oppositions or inferior conjunctions), and a maximum at 0 deg (superior conjunctions). For the Ka-band link to provide such an advantage, it has to be Ka-band on the uplink as well as the downlink. A downlink-only Ka-band link derived from X-band uplink (on the Kepler mission, for example) is limited by the noise on the X-band uplink and provides only a small improvement on the downlink.

Using the Juno KaT as an example, Ciarcia et al. (2013) describe four functional modules shown in Appendix 5A:

- The receiver analog module carries out the low-noise amplification and down-converts the signal from RF to intermediate frequency (IF). It then digitizes the signal where a field-programmable gate array (FPGA) integrated circuit keeps the input power constant.
- The digital module uses an FPGA for another step of down-conversion from IF to baseband I&Q digital samples, processes the coherent turn-around conversion, and carries out the direct digital synthesis (DDS) functions for the uplink carrier tracking and the downlink signal generation.
- The transmitter analog module uses a phase-locked loop (PLL) function and a frequency multiplier (two cascaded $\times$ 2 multipliers) to set the downlink frequency to the correct value.
- The baseplate module contains the Ka-band SSPA of 2.5 W and includes two DC/DC converters for maximizing the power efficiency.

The use of Ka-band links also has operational challenges. The narrower Ka-band antenna beam-width necessitates more precise antenna pointing requirements. Future KaT capabilities are expected to be built into integrated SDRs, such as USTs.

5.3.3 Spacecraft Open-loop Receiver

Section 5.2.3.2 describes the advantages of having an OLR in the DSN for RS investigations. A SOLR is required when the link direction is reversed and the experiment is carried out as a one-way uplink experiment or for spacecraft-to-spacecraft crosslinks. The SOLR was first operated on New Horizon's REX instrument (Tyler et al. 2009).

The SOLR is generically a super-heterodyne receiver that captures and records the I&Q digital samples of the signal carrier received by the spacecraft. Next-generation SDRs, such as Iris and Electra's possible future version, as well as UST and UST-Lite can have this functionality built into the design.

The Electra SDR, so far only used at Mars, is another type of SOLR that provides relay service over the UHF frequency band of 390–450 MHz including proximity communications and radiometric Doppler observable from phase and power samples. It can collect and downlink one- and two-way UHF Doppler data when there is a link with another UHF transmitter or transceiver and can record the raw received UHF signal from another spacecraft in an OL mode and return the recorded spectrum to the Earth via a telemetry link.

5.3.4 Spacecraft Radio Science Beacon

GRAIL carried a transmitter called the Radio Science Beacon (RSB) to provide one-way X-band carrier-only signal to the ground for precision orbit determination and to support daily calibration of the USO frequency by the DSN. An RSB was driven by the onboard USO on each of the twin spacecraft (Klipstein et al. 2013). Future missions can benefit from this concept and possibly simplify or miniaturize the RSB for innovative RS investigations or calibration techniques.

5.3.5 Ground Water Vapor Radiometer

In the DSN, station DSS-25 at Goldstone is equipped with Ka-band uplink and downlink capabilities along with the X-band links specifically for RS investigations. Since this type of high-precision investigation requires additional calibrations, especially that of the Earth's troposphere, a custom-made advanced MCS was developed and installed next to the station (see Figure 5.3).

An AWVR determines the total water vapor content along a LOS by making measurements of the strength of microwave emissions from water molecules. Investigations that require more precision than the perturbation of the troposphere on the signal would require this RS instrument, especially at Ka-band links, which are less susceptible to charged-particles effect making the tropospheric effect more pronounced (Riley et al. 1990; Resch et al. 2002; Keihm et al. 2004; Tanner 1998; Bar-Sever et al. 2007). Chapter 6 describes a concept for possible future upgrade of the AWVR.

5.3.6 Ground Advanced Ranging Instrument

BepiColombo uses coherent Ka- and X-band Doppler links simultaneously along with the capability for the first ever precision ranging at Ka-band in deep space (Iess and Asmar 2007; Iess and Asmar 2018). The expected precision of the pseudo-noise (PN) ranging is under 20 cm; the average precision at X-band is ~1 m. Ranging at Ka-band requires compatible capabilities at a ground station and an advanced ranging instrument (ARI) was installed at DSS-25.

ARI takes advantage of dual Doppler calibration techniques centered on two-way coherent Ka-band tracking and applies differential Doppler to removing the effect of dispersive media out of the ranging data. The scheme reduces uncertainty in ranging accuracy by about one order of magnitude with simultaneous uplink and downlink at X- and Ka-bands, advanced ranging modulation, and the application of a constant calibration signal for internal station delay (Paik et al., 2018; Border et al., 2020).

Ranging at Ka-band with the spacecraft requires compatible capabilities at a ground station. With the introduction of spacecraft tracking at Ka-band (32–34 GHz) with Cassini and Juno in a two-way coherent mode, the Doppler (range rate) data quality improved by about one order of magnitude with the proper calibrations. A similar improvement in ranging was made with an ARI (Asmar et al. 2016) that also takes advantage of calibration techniques centered on two-way coherent Ka-band tracking. The scheme reduces uncertainty in ranging accuracy by about one order of magnitude with simultaneous uplink and downlink at X- and Ka-bands, advanced ranging modulation, and the application of a constant calibration signal for internal station delay. The improved ranging can be used for future RS experiments, such as the ones planned for the BepiColombo mission to examine possible violations of the theory of general relativity, improve the parameterized post-Newtonian formalism, improve measurements of solar oblateness, and characterize the solar corona. Advanced ranging instrumentation has been developed at the DSN. Results are summarized in Paik et al. 2018 and Section 6.4.3.

The calibration of the ground system ranging channel requires range calibration measurements via the use of a test translator to feed a frequency shifted replica of the uplink signal into the downlink signal path. The test translator will operate for the X/X, X/Ka, and Ka/Ka signal paths. In order for this calibration to be accurate, the delay of the test translator itself must be known. This is achieved via the use of a laboratory calibrated standard version of the test translator, known as a zero-delay device (ZDD). A total of six ranging channels will be supported in order to process the following ranging signals; this arrangement will allow real-time calibration of the ground system at the same time as performing ranging measurements with the spacecraft:

1. X/X calibration signal
2. X/X spacecraft signal
3. X/Ka calibration signal
4. X/Ka spacecraft signal
5. Ka/Ka calibration signal
6. Ka/Ka spacecraft signal

5.3.7 Ground Bethe Hole Coupler

In the era of the Galileo mission ground system development, the DSN developed a Bethe Hole coupler, a device that enables the Faraday Rotation experiment (Hartop 1987) (see Chapter 4). This experiment relied on the simultaneous reception of the right circular polarized (RCP) and left circular polarized (LCP) polarizations of the received signal while Galileo was designed to transmit a linearly polarized S-band signal (Howard et al. 1992) by design, which is usually difficult due to the loss of power; telecom engineers prefer a polarized signal over a linear signal. When the Galileo HGA failed to open, it was not possible to carry out this planned experiment and the couple installation of the Ground Bethe Hole coupler was abandoned (North Ham, private communications). Future missions can benefit from this concept and possibly simplify the implementation for innovative RS investigations or calibration techniques. Asmar and Bird (1996) demonstrated that a differential Faraday Rotation experiment was possible without this device, as shown in Chapter 4.

5.3.8 Ground Advanced Pointing Techniques

For X-band links, the DSN uses the conical scanning (conscan) antenna pointing technique (see Gudim et al. 1999). The antenna is scanned conically around the expected direction of the signal source in the plane of the sky. As the signal deviates from the optimum source, a receiver algorithm computes the location of the highest signal level and points the antenna toward it. This is typically carried out at the beginning of the pass and after spacecraft completing maneuvers that can change knowledge from previous predictions. Due to its disruptive impact on the phase of the signal, conscans are not carried out during RS observations.

For the narrower in beam-width Ka-band links, a monopulse technique is used to optimize the pointing. It compares the phase and gain of the nominal downlink with an error channel in a second receiver to determine elevation and cross-elevation corrections to the antenna pointing at the millidegree level. Periodic calibrations using a spacecraft signal are exercised during dedicated tracks to determine the required phase correction in advance of the RS observations. This is carried out by commanding the antenna to off-point by a few millidegrees; the monopulse system generates the needed correction and the commanded off-point angle determines the phase value to be applied during the pass and can be enabled throughout the DSN Ka-band pass.

For Ka-band links, where the antenna's beam width is narrower than X-band, beam aberration arises. A ground antenna points to a spacecraft moving within its narrow beam and transmits and receives in slightly different directions. Besso et al. (2007) and Imbriale et al. (1987) describe the aberration correction developed and applied by ground networks.

5.4 Operations and Experiment Planning

RS investigations require nuanced operational procedures and advanced planning for the flight and ground segments. As an example of advanced planning, the MRO gravity science team placed performance specifications on the X-band two-way coherent tracking link, as discussed in Section 5.1.2.3. A detailed "noise budget" study was carried out in the spacecraft design phase (of special concern was the presence on the HGA on a long boom) and the mission team concluded that no design changes were required. The in-flight radiometric observables confirmed the scientific quality of the observables.

On the other hand, no performance requirements were placed for Doppler noise on the smaller Mars Odyssey spacecraft beyond the navigation requirements since there was no official RS experiment. Opportunistic gravity observations were carried out in the orbital phase and the data also turned out to be of high scientific quality with proper planning (see Figure 1.9 in Chapter 1) and were incorporated in the aggregate Mars data sets for scientific analysis (Marty et al. 2009; Konopliv et al. 2016).

An example of operational planning to safeguard data quality is the timing of the frequent thruster firings on the Dawn spacecraft. Dawn used xenon ion thrusters for propulsion to achieve the velocity required to reach its target asteroids and maintain the necessary trajectories and orbits. Each thruster firing is dynamically disruptive to the radiometric observations for the gravity investigations of Vesta and Ceres, as are maneuvers to desaturate the momentum wheels. The RS measurements benefit from long uninterrupted data arcs, so careful sequencing of the competing events of gravity measurements, thruster firings, reaction wheel desaturation maneuvers, HGA pointing to Earth, and other payload requirements were carefully managed and eventually led to the successful investigations discussed in Chapter 3.

An example of complex operational procedures is the ground station noise temperature calibration procedures required prior to, during, and after bistatic radar observations. Since these surface scattering experiments are sensitive to changes in the received power level in the zeptowatt level (10^{-21} W), a very complex procedure, carefully rehearsed and implemented, requires the ground station personnel to insert a noise diode, point the antenna to zenith, and return to the planetary target several times throughout the observation. Station engineers provide the values for post-processing and analysis of the data (Simpson et al. 2009a and 2009b).

An example of operational experimental complexity is the Ka-band pointing procedures. As described in Asmar et al. (2017b), accurate pointing of the DSN antenna is required in order to optimize the power of the received narrow beam-width Ka-band signals. Two techniques are used to optimize the antenna pointing, conscan and monopulse, which are described in Section 5.3.8).

An example of a "time-share" between two types of RS experiment is the MGS mission that orbited Mars for about a decade. In its nearly circular orbit of approximately 2 h, gravity experiments were carried out in the portion of the orbit where the spacecraft was in view of Earth stations with the HGA pointed to Earth and the link in the two-way mode. On every occultation entry and exit, the configuration was altered to one-way mode for atmospheric occultations. With careful sequence planning, the arrangement was smooth and successfully led to a large and rich global data set for both types of experiments. Throughout this process, telemetry was available during the gravity observations and turned off during the RO observations.

A common activity on long-duration missions, such as Cassini, is negotiating when to turn off the telemetry modulation on the primary X-band communications links for the purpose of optimizing the RO experiments, but at the expense of communicating data packets to the mission control center. In most cases, radio occultation experiments are sensitive to the overall SNR and maximizing the carrier power boosts the quality of the observables.

5.5 Data Products

Key RS experiment data products are archived for formal experiments at NASA's Planetary Data System (PDS) along with higher-level data products resulting from the applying calibrations, processing, and further scientific analysis and interpretation.

The formulation for the observed and computed values of these data products and tracking observables, especially the light-time computation, are detailed in Moyer (2003). Thornton and Border (2003) is another authoritative reference on radiometric tracking techniques for deep space missions.

The names of the data files have changed over the years with the evolution of the DSN. For example, the radiometric tracking data files were named TRK-2-25 for the Galileo and Magellan era, TRK-2-18 for the Cassini and Mars missions, and TRK-2-34 (also known as the Tracking and Navigation File, TNF) for current products. In addition, a special ASCII archival file of tracking data, developed via international standards for exchange between agencies, is called the Tracking Data Message (TDM). Long-duration missions can span multiple file formats. Representative contents of a TNF file are shown in Asmar et al. (2014b).

5.5.1 Range Rate

Range rate is he Doppler data type, the time-tagged profiles of the signal frequency received by the DSN in one-, two-, or three-way modes that vary due to the modeled or unmodeled Doppler effect. The LOS velocity of the spacecraft relative to the station, including the motion of the station in inertial space due to Earth rotation, and other dynamical effects acting on the spacecraft are directly inferred from this observable.

The DSN's convention is that the Doppler shift is a positive quantity for a receding spacecraft (opposite the physicists' convention). Moyer (2003) describes the observables; the Doppler observable is defined in the DSN as

$$G * F_t - F_r + bias, \tag{5.8}$$

where G is the transponder ratio, F_t is the transmitted uplink frequency, F_r is the received downlink frequency, and *bias* is a constant (during a pass) of plus or minus 1 MHz.

Frequency is the derivative of phase so the accumulated carrier phase counts in a time interval are measured to generate the range rate at the station. In the one-way mode, the sky-frequency-level downlink carrier phase counts are measured while in the two-way or three-way coherent modes, both uplink and downlink phase counts are measured considering the transponder ratios listed in Table 1.3. The uplink carrier frequency is synthesized by the station's exciter using the highly stable hydrogen maser for reference with the stability preserved in the transponded downlink with the addition of the measurement errors described below in this section.

Range rate is a radiometric data type that is automatically generated for every DSN track and is the most common RS data product, used as the primary product for some experiments and secondary to the higher quality OL data, when available, for other experiments. A loss or interruption of the lock in the CL receiver produces gaps in the data. Although the motion of the spacecraft is best represented at its center of mass, the topocentric range rate is measured at the spacecraft antenna's electrical phase center; so, an additional correction is applied in the data processing phase. Some deep space missions request nonconstant uplink frequency.

An example of this was the test done on the Cassini mission after the Earth flyby to simulate the Huygens probe signal, which demonstrated that the receiver on the

Cassini spacecraft to detect and record the Huygens signal was not correctly designed. So, the uplink signal is tuned to compensate for the Doppler effect on the uplink leg and is mostly a constant value near the channel center upon arrival at the spacecraft. This reduces the stress on the spacecraft receiver's carrier loop (Taylor 2016).

Doppler residuals, the difference between the observed data points and the predicted model (based on a navigation solution), are available for real-time evaluation. They are called "pseudo residuals" to distinguish them from the better quality post-pass residuals with more complete calibration. Data points are available at the rate of 10 points per second; however, the uplink points are available from an uplink processor at one point per second. The quality of Doppler data, or Doppler noise, is in units of velocity with typical values of 0.1 mm/s at S-band at 60-s cycle count, and 0.03 mm/s at X-band for the same integration interval, and better at Ka-band (see DSN document 810-005 module 202B).

DSN Doppler tracking is described in Module 202 of the DSN Telecommunications Link Design Handbook, Document 810-005. This document specifies the measurement error due to thermal noise for two-way Doppler to be

$$\sigma_V = \frac{c}{2\sqrt{2}\pi f_C T}\sqrt{\frac{1}{\rho_L} + \frac{G^2 B_L}{P_C / N_0 |_{U/L}}}, \tag{5.9}$$

where T is the measurement integration time (s); f_C is the downlink carrier frequency (Hz); c is the speed of light in vacuum (mm/s); G is the transponding ratio; B_L is the one-sided, noise equivalent, loop bandwidth of downlink carrier loop (Hz); $P_C / N_0 |_{U/L}$ is the uplink carrier power to noise spectral density ratio (Hz); and ρ_L is the downlink carrier loop signal-to-noise ratio.

Note that σV has the dimensions of velocity as expected. This relationship accounts for thermal noise only, and does not include the effect of phase scintillations due to the signal passing through the interplanetary medium or solar corona, an effect that actually dominates the thermal noise at SEP angles during or near solar conjunctions. The magnitude of this effect depends strongly on the solar activity, and is dispersive so affects different wavelengths of the radio links (S-, X-, or Ka-bands) differently as shown in Figure 1.8. Shorter wavelengths are less susceptible to solar noise, so Ka-band performs better than X-band, which performs better than S-band. This also applies to the dispersive effects of the ionosphere.

5.5.2 Range

A deep-space range or distance measurement is fundamentally a measurement of phase delay. From the measured phase delay, a time delay is inferred. Since the time delay is always changing, it is essential to have a time tag associated with the range measurement (and the inferred time delay). For a two-way range measurement, the two-way delay is the time required for a signal arriving at the station's receiver at the time indicated by the measurement time tag, to propagate from the station to the spacecraft and back.

For a three-way range measurement, the three-way delay is the time required for a signal arriving at the receiving station, at the time indicated by the measurement time tag, to propagate from the transmitting station to the spacecraft and then to the receiving station. Range cannot be calculated directly and accurately from the time delays because the range changes significantly over the course of the signal travel time.

The two-way and three-way time delays calculated for deep space missions are useful in the orbit determination process. These calculated delays assist in the improvement of trajectory models; and so, indirectly, the delays assist in the estimation of range as a function of time (Berner et al. 2007). Range (or ranging) results in a measurement of the line-of-sight distance between the station and spacecraft that can vary, for deep space missions, from 300 km to more than 5 billion km or more.

The ranging instrumentation measures the round-trip phase delay of the returned ranging signal from which the RTLT is determined. It accommodates the constant uplink frequency mode or the mode where the uplink frequency varies with time. The quality of this product is currently approximately 1 m at X-band, 1 sigma (Thornton and Border 2003). Since this is a precise measurement, the signals delay due the electronics (the finite time it takes for the electronics to turn the signal around) at the station and the spacecraft are calibrated out of the data, along with other atmospheric corrections.

The spacecraft delays are measured for each spacecraft prior to launch. For example, Doody (2010) reports that the signal takes 420 ns ± 9 ns in the Cassini spacecraft hardware chain; if ignored, this would be an error of over 120 m. Border and Paik (2009) and Border et al. (2020) describe the process of calibrating the range delay in the station equipment and explore methods of improving it. There are two types of ranging to examine (the older sequential ranging system and newer pseudo-noise ranging) and a third data type for plane-of-sky position approximation.

Note that when considering an end-to-end range measurement at a fundamental level, a baseband ranging signal is generated at the station's exciter and a baseband ranging signal, after demodulation of the carrier, is collected at the station's receiver. The phase delay of that baseband ranging signal is measured through the system, with a time-tag applied. From the measured phase delay of the ranging signal, corrected with the help of the ambiguity-resolving components, the time delay is inferred, where the indicated time delay is for a ranging signal that arrived at the station's receiver at the time identified by the time-tag. The actual range is not calculated since the range changes during the time of flight of the ranging signal.

Within this system, there are filters and media effects that significantly delay both the carrier and the ranging signal that modulates the carrier. When considering the effect of a filter and the media on the ranging signal, the group delay of the filter is used because the ranging signal is modulated onto the carrier while passing through the filter and media. Range is a wideband signal that propagates as a group velocity.

5.5.2.1 Sequential Ranging

The sequential ranging signal is a sequence of signal components, where each signal component is periodic. The frequency of the signal component is halved at each new step in the sequence. The purpose of the highest frequency component, called the range clock, is to set the measurement precision. The purpose of all other components in the sequence is to resolve the phase-delay ambiguity. Sequential ranging is described in Module 203 of the DSN Telecommunications Link Design Handbook, Document 810-005.

5.5.2.2 Pseudo-noise Ranging

Pseudo-noise ranging is carried out with a ranging signal that is a logical combination of a range clock and several PN codes. Any deep-space transponder that has a turnaround ranging channel can support either sequential ranging or PN ranging. The performance of these two signaling schemes is comparable when used with a turnaround ranging channel. The important performance advantage of PN ranging arises when the transponder has a regenerative ranging channel. A regenerative ranging channel detects the PN ranging signal arriving on the uplink and creates a relatively clean copy of it for modulation onto the downlink carrier. In principle, regenerative ranging could also be done with a sequential ranging signal, but there are logistical difficulties associated with this approach. Pseudo-noise and regenerative ranging is described in Module 214 of the DSN Telecommunications Link Design Handbook, Document 810-005.

5.5.3 Delta Differential One-way Ranging (Delta-DOR)

For certain mission applications, especially targeting a landing site, it is also important to know the spacecraft position in the plane of the sky, which cannot be computed from the line-of-sight ranging data. Another method is needed to address this gap. This is differential one-way ranging (known as delta-DOR), a navigation capability that also benefits RS in that the experiments' data quality improves from optimizing the navigation solution.

Delta-DOR is derived from the more general VLBI interferometric measurements where two or more tracking stations on different continents are used to determine the position and velocity of the spacecraft in the plane of the sky (explained in detail in Thornton and Border (2003), and module 210 of DSN 810-005). Delta-DOR uses the angular location of a target spacecraft relative to a reference direction that is the direction of arrival of radio waves from a distant quasar with a well-known and catalogued direction; two baselines with orthogonal components are needed to measure both the right ascension and declination coordinates.

Delta-DOR measurements are relatively short, about 1 h, and are conducted using the east–west Goldstone-Madrid baseline, sensitive to right ascension information, or the Goldstone-Canberra DSN baseline, which is more sensitive to declination. Delta-DOR is being advanced through demonstrations of mixed baselines of DSN stations and stations of the European, Japanese, or Indian space agencies. This measurement requires that the spacecraft is capable of generating a one-way downlink carrier modulated by a set of continuous-wave tones (e.g., the delta-DOR tone in the SDST is at approximately 19.1 MHz from the carrier).

5.5.4 Differenced Range Versus Integrated Doppler

Differenced Range versus Integrated Doppler (DRVID) products are historical data products no longer in common usage but referred to in older RS literature. When X-band was introduced as a telecommunications downlink, initially to complement the S-band downlink, as in the case of the Voyager spacecraft that is capable of both S-band uplink and simultaneous S- and X-bands downlinks, it became apparent that dispersive effects from the presence of low-density solar plasma in the radio signal's propagation path could be isolated by differencing the effects in two links at different wavelengths. But since the two links are both downlinks, differencing the S- and X-band Doppler profiles would isolate the effects of charged particles on the downlink only.

Another technique was devised to isolate the effect on the uplink. According to Green (1980), JPL's Tracking Systems Analytic Calibration (TSAC) team provided calibrations for spacecraft radiometric data based on dual frequency (S- and X-band) Doppler data based on modeling the charged-particle effects on the uplink portion of the round-trip path using the downlink only information in the S/X data and certain assumptions about the spatial and temporal distribution of the contributing charged-particle environment.

The plasma in the propagation path causes a group velocity decrease and a phase velocity increase. Since the DSN ranging system measures the group velocity, the plasma mimics an increase in the apparent distance, and since the DSN range-rate (Doppler) system measures the phase velocity, the plasma causes an apparent Doppler decrease in the DSN convention. The apparent increase in range is equal in magnitude to the apparent decrease in range rate over the same timer interval, and this magnitude is proportional to the plasma density and inversely proportional to the square of the signal frequency, that is, S-band is more susceptible than X-band. In the DRVID method, the apparent increase from differencing two consecutive range measurements is compared to the apparent decrease from differencing two consecutive range-rate measurements leads to deducing the change in charged particles along the round-trip propagation path.

5.5.5 Open-loop Receiver (Radio Science Receiver)

The OLR data product, second most common to RS experiments after the range-rate data product, provides time-tagged digital samples of the spectrum containing the signal carrier, as described in reception mode in Section 5.2.2.2. The down conversion process is described in module 209 of the DSN 810-5 document and the data format, which depends on the version and date, is described in module "0159-Science" of the DSN 820-013 and is also available on the PDS website. The flexibility provided by the OL functionality, which preserves the information in the entire preselected bandwidth and centers on the signal carrier via a tuning prediction file, enables the users to capture the signal during conditions of low SNR, capturing the signal during conditions of high signal dynamics in amplitude or frequency and capturing multi-path signal conditions.

The tuning prediction is based on the latest navigation solution of the spacecraft trajectory and removes the first order Doppler effects, which, in turn, allows the use of narrower bandwidth; this lowers the size of the output files to manageable levels and improves the real-time visibility of the signal. The predicted frequency profile is preserved by the RSR and added to the frequency residual in the post-processing to reconstruct the sky frequency prior to down-conversion and tuning as shown in Asmar et al. (2014b) for the current generation receiver; the sky frequency is the actual received RF (e.g., 8.4 GHz) before the signal is down-converted to an IF within digital processors in real time.

The current RSR uses the Standard Formatted Data Unit (SFDU) with output samples at a single rate up to four channels at different sampling rates. Sample resolution and rate can vary from 16 bits at 1 kHz to 1 bit at 16 MH. Each RSR can capture data from a different band and polarization pair.

Because these are raw samples, the user needs to apply additional processing tools before the data are converted to physical parameters and interpreted for scientific analysis. This includes the following steps: signal detection via fast Fourier transform (FFT) or digital PLL code (or a combination, depending on the signal characteristics), scaling the carrier frequency from the recorded bandwidth (IF level) to the sky frequency (RF level), removing a model of the trajectory provided by the navigation team, applying atmospheric calibrations, applying known dynamical effects from the spacecraft telemetry, and prepping the resulting residuals, where the scientific content of interest resides for the next level of analysis.

Paik and Asmar (2010) describe three types of algorithms commonly used for the signal detection, FFT, PLL, and optimized FFT, referred to as spectral analysis, suitable for the case of high-frequency dynamics. In order to improve the accuracy of the estimated frequency, optimization can be applied to the FFT. Tortora et al. (2002) applied the Brent method for solar conjunction RS experiments. Lipa and Tyler (1979) use another optimization for atmospheric analysis based on a mathematical model of the rectangular window and a least-squares approximation to estimate optimum values of the frequency and amplitude in each FFT time interval.

5.5.6 Media Calibration

Three types of media affect the propagation path of signals from the spacecraft to the station and vice versa, not counting the atmosphere of the planet under study in the case of radio occultation experiments: the interplanetary plasma, the Earth's ionosphere, and the Earth's troposphere. The two layers of Earth's atmosphere (ionosphere and neutral atmosphere) have different characteristics since the ionosphere contains charged particles with a dispersive effect, as is also the case for the interplanetary plasma, while the troposphere is neutral and contains water vapor from the weather system that causes different path delay effects.

The standard data products are the DSN's media calibration data, where tropospheric (TRO) files and ionospheric (ION) files contain estimates of the radio delay in the Earth's troposphere and ionosphere, respectively, for use in calibrating their effects on the measured phase of the uplink and downlink carrier signals. These files are described by the 820-013, Deep Space Mission System (DSMS) External Interface Specification, JPL-16765, TRK-2-23: Media Calibration Interface document.

A more specialized data product is the AWVR data developed specifically for RS experiments at Ka-band, initially developed for the Cassini coherent Ka-band experiments in the cruise phase and also used by Juno's RS experiments. While the data quality at X-band reaches a limit comparable to the tropospheric contributions, as described in Asmar et al. (2005b), the Ka-band links overcome the bulk of the error contributions of dispersive media and benefit from additional corrections and calibrations of the noise contributed by the troposphere.

There is a stationary WVR at each DSN complex that measures the water emission in the atmosphere and provides data to analysts and modelers. At selected DSN stations, an advanced version measures emission at 20.7 and 31.4 GHz along the LOS to the spacecraft by pointing to the spacecraft and moving along with the DSN station. The AWVR, an element of a bigger Advanced Media Calibration (AMC) system, is part of DSN operations and is remotely operated from JPL.

Bar-Sever et al. (2007) describe the use of a global positioning system (GPS) for media calibration. The delay of radio signals due to the variations in the refractive index in the Earth's troposphere requires tropospheric calibration systems, an automated GPS-based system that calibrates the zenith tropospheric delays at all times and can be mapped to any LOS. Centimeter-level or better accuracy is reported as well as wet troposphere in short baseline interferometry that can be calibrated such that the Allan deviation of phase residuals, a unitless measure of the average fractional frequency deviation, is better than 2×10^{-15} on time scales of ~2000 to ~10,000 s.

5.5.7 Spacecraft Trajectory

The proper and most accurate reconstruction of the spacecraft trajectory is required for RS analysis. JPL navigation teams commonly produce the Orbit Data File (ODF), an abstracted version of the TNF (TRK-2-34), which is the fundamental input to orbit determination processes containing the line-of-sight Doppler and range, rather than raw phase. The ODF format is described by the TRK-2-18 document, 820-013, DSMS External Interface Specification, JPL D-16765, TRK-2-18: Tracking System Interfaces: ODF Interface.

A processed product is the SPK, a kernel of the SPICE library. According to Acton (1996), JPL's Navigation and Ancillary Information Facility (NAIF) has primary responsibility for design and implementation of the SPICE ancillary information system, supporting a wide range of space science mission design, observation planning and data analysis functions. NAIF also archives SPICE data produced by flight projects and distributes them with associated software tools to NASA researchers. The SPK contains the ephemeris information used for scientific analysis. Other space agencies use an equivalent file called the Orbit Ephemeris Message (OEM) files. The SPK is described in Spacecraft and Planetary Ephemeris Kernel NAIF Document No. 367.

5.5.8 Calibration Data Sets

When available, calibration can encompass advanced techniques, such as differencing multiple radio links, applying data from the advanced WVRs, using data from spacecraft accelerometers, and using noise-temperature measurements for precision bistatic radar RS experiments. In addition to orbital and geometrical considerations, the limits of experiment sensitivity are based on the frequency/phase stability of the link, its SNR, amplitude stability, spectral purity, calibration of nongravitational forces, calibration of the intervening media, and accuracy of the trajectory reconstruction. Measured observables require proper calibrations to correct for deterministic error sources.

Various calibration files are provided by the DSN for the ground-based elements of the RS instrumentation and by the flight project for the spacecraft elements.

- **Earth orientation parameters (EOPs)**: Data in the DSN EOP file describe the motion of the Earth in inertial space in terms of the orientation of its rotation axis and the angle through which the Earth has rotated on its spin axis since some reference epoch. The EOP file is used to calculate the Doppler contribution to a measurement from Earth rotation. The file and its contents are described by the TRK-2-21 document, 820-013, DSMS External Interface Specification, TRK-2-21: DSN Tracking System Earth Orientation Parameter Data Interface.

- **DSN monitor data**: The DSN generates a real-time stream of status and performance data from its own systems "monitor" data; the data may be used for anomaly resolution, performance validation, and calibration of other data. Some RS data sets include edited versions of this stream. Parameters may include azimuth and elevation angles of the antenna, monopulse correction values, carrier frequency, wind speed and direction, lock status of the CL PLL, receiver system noise temperature, and carrier signal-to-noise ratio. The data are stored in ASCII tables with labels, which describe both the format and content.
- **Spacecraft engineering telemetry (TLM)**: The spacecraft also generates a real-time stream of status and performance data from onboard systems, which can be downlinked in parallel with or separately from science data. These data may be used for anomaly resolution, performance validation, and calibration of other data. Parameters of interest to RS include lock status of spacecraft receivers, power output of spacecraft transmitters, and physical temperatures at various points on or within the spacecraft. Some Cassini data sets include edited versions of this data stream. Files are ASCII tables, which are accompanied by labels describing both the format and content of the tables.
- **C-Kernel (CKF)**: The navigation teams provide a file of reconstructed spacecraft inertial orientation that is available as a NAIF C-Kernel to determine HGA pointing, described by SPICE C-Matrix Kernel, NAIF Document No. 370.

Appendix 5A

Spacecraft Telecommunications System and Radio Science Flight Instrument for Several Deep Space Missions

Figures 5A.1-5A.7 illustrate block diagrams of the functional design of spacecraft telecommunication systems and RS flight instrumentation for several deep space missions for the purpose of historical archiving and comparison of the evolution over spacecraft generations.

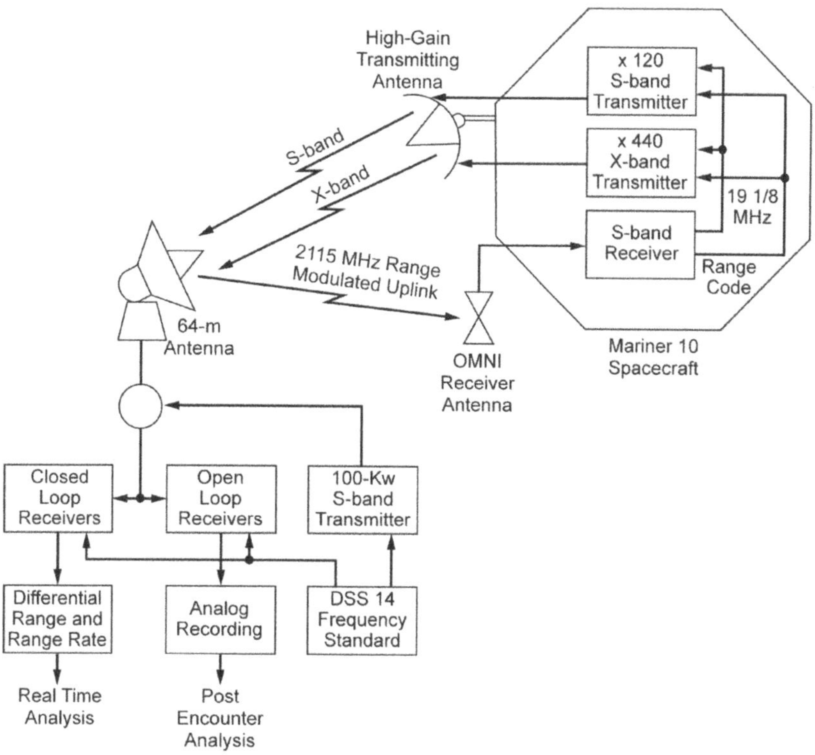

Figure 5A.1 Mariner 10 telecommunications system and RS flight instrument (based on Howard et al. 1974b).

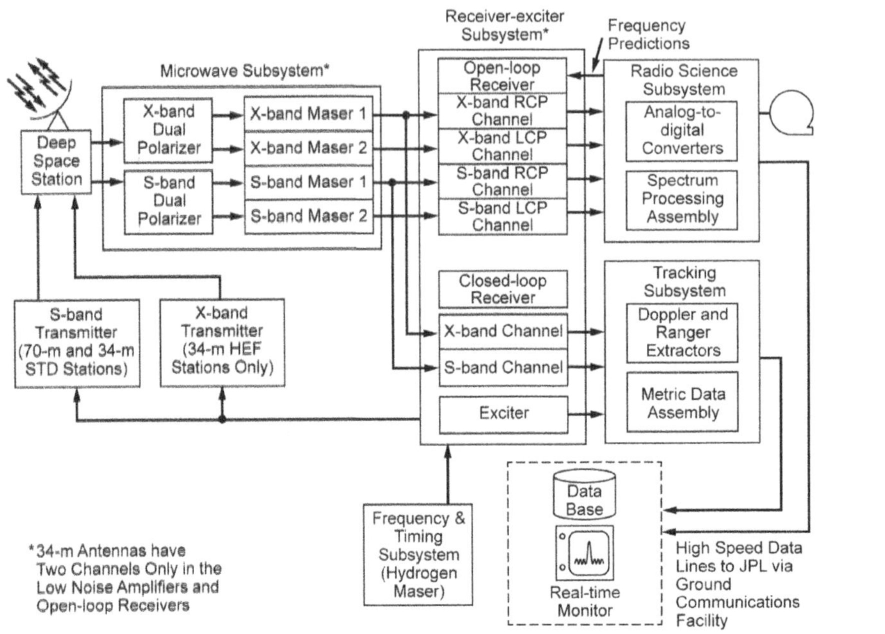

Figure 5A.2 Galileo telecommunications system and RS flight instrument (Howard et al. 1992).

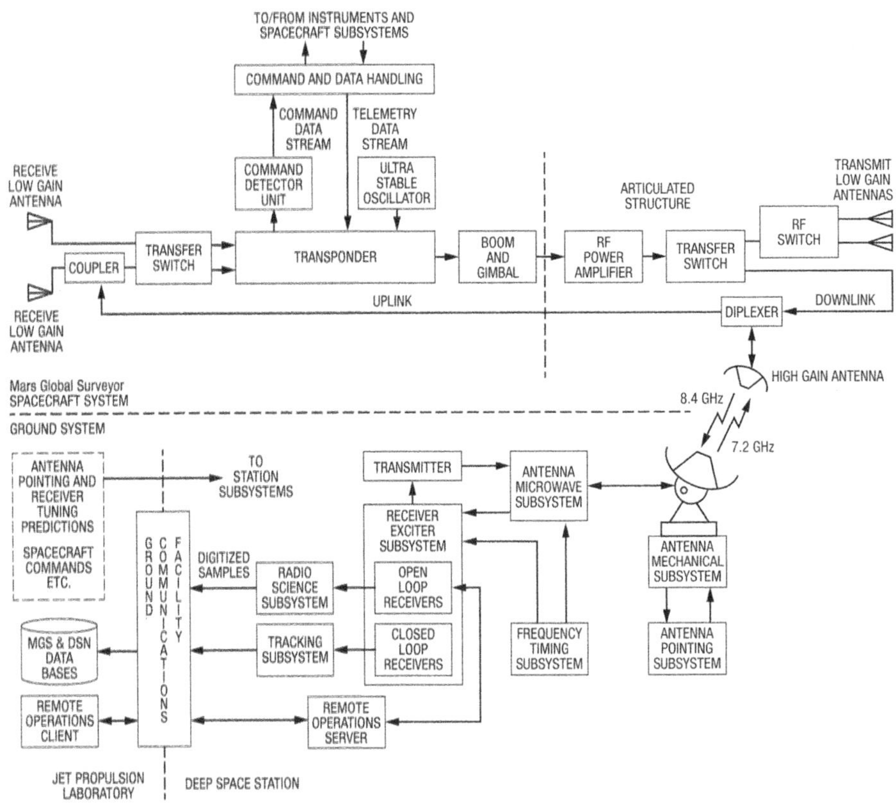

Figure 5A.3 Mars Global Surveyor telecommunications system and RS flight instrument (Tyler et al. 2001).

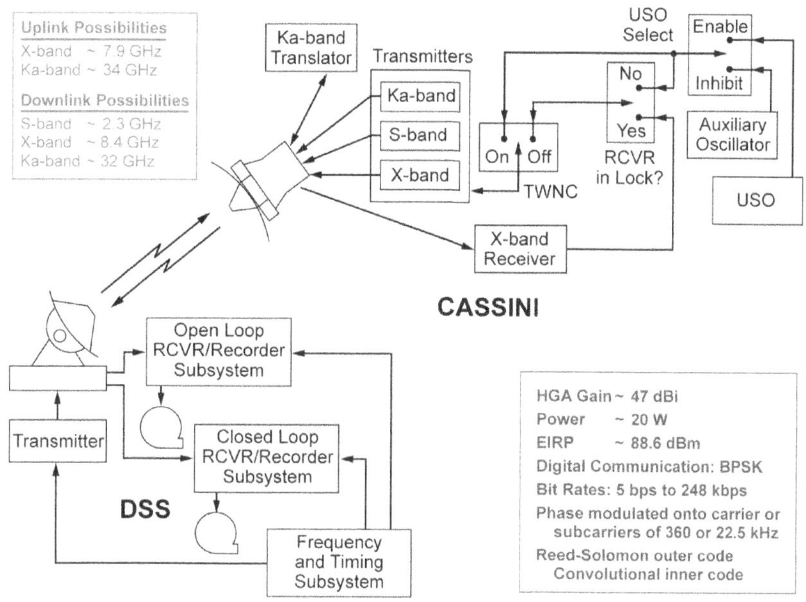

Figure 5A.4 Cassini telecommunications system and RS flight instrument (based on Kliore et al. 2004).

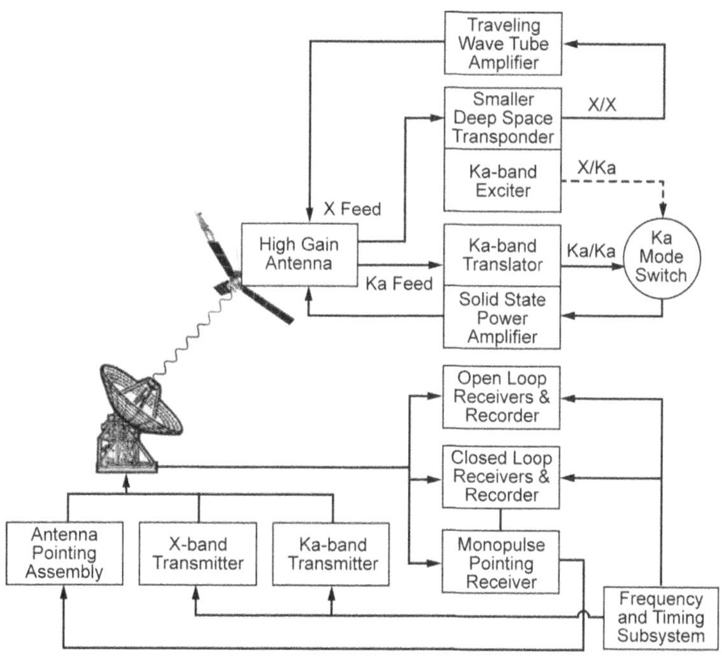

Figure 5A.5 Juno telecommunications system and RS flight instrument (based on Asmar et al. 2017b).

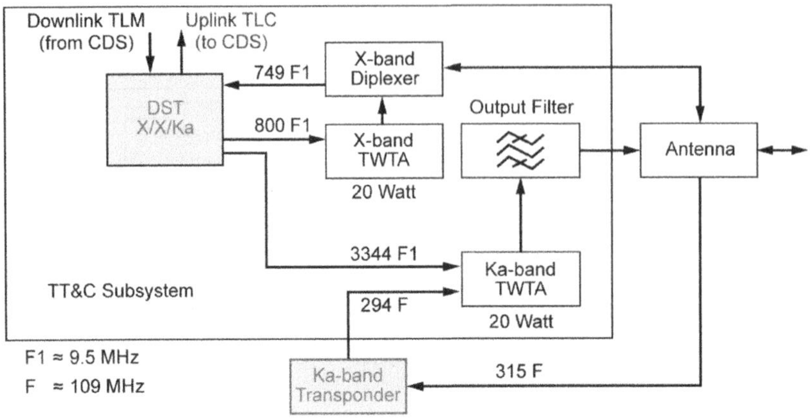

Figure 5A.6 Illustration of instrument for the BepiColombo radio science experiments (Iess and Boscagli 2001), highlighting two transponders onboard the spacecraft, a deep space transponder (DST) for communications functions and a Ka-band Transponder specifically for RS investigations.

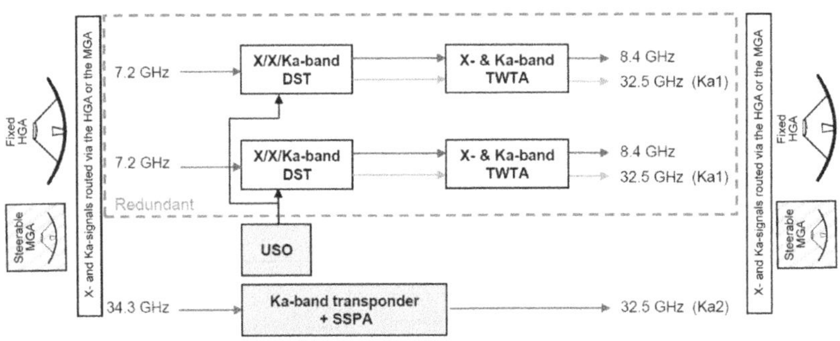

Figure 5A.7 Notional illustration of the RS instrument for the JUpiter ICy moons Explorer (JUICE) mission.

6

Future Directions in Radio Science Investigations and Technologies

In Collaboration with J. W. Armstrong and R. A. Preston

6.1 Fundamental Questions toward a Future Exploration Roadmap

Numerous scientific discoveries have resulted from nearly six decades of radio science (RS) investigations throughout the Solar System. The field continues to thrive due to innovative developments in many relevant areas, including:

- Expanding scientific objectives, such as high temporal and spatial resolution of atmospheric structure, high-resolution gravitational fields for internal structure of rocky and icy bodies, high-precision determination of relativistic parameters.
- Formulating new mission concepts, such as planetary small spacecraft, spacecraft constellations, links between multiple planets, entry probes, aerial vehicles, same beam interferometry.
- Advancing technologies, such as optical links, 20-cm-level ranging, RS-capable software-defined radios (SDRs), space-borne atomic clocks.
- Improving experimental methods and processing algorithms, such as time-delay local noise cancellation (TDLC) techniques, advanced signal detection software.
- Fostering strategic partnerships among space agencies, academia, and the private sector for contributed instruments, joint publications, special conference sessions, and cross-support among ground networks.

Future investigations will benefit from the infusion of new technologies developed to expressly enhance RS or synergistically overlap. Toward a roadmap for continued implementation, RS white papers submitted to the Decadal Surveys of the National Academies of Sciences, Engineering, and Medicine (reproduced in Appendices 6A and 6B) are used for guidance. In the past, important accomplishments were achieved based on the first white paper, such as an order-of-magnitude improvement in the quality of range and range-rate observables, shorter wavelength links, multiple links, and advanced calibration techniques.

Since the development of experimental capabilities is science-driven, RS investigations along with their associated observables and limits of sensitivity are grouped in Table 1.1 (in Chapter 1) and a general-purpose traceability matrix for functional and

Radio Science Techniques for Deep Space Exploration, First Edition. Sami W. Asmar.

performance requirements on instrumentation is provided in Table 1.2. Specific requirements are target-specific, particular to the selected space mission, and are based on science traceability matrices developed by proposal teams based on the state of the science, experimental objectives, mission design, state of the instrumental technology, cost considerations, etc. Requirements driving the RS design and cost of the investigation typically include:

- Performance of the end-to-end radio system (e.g., fractional frequency stability).
- Dynamical stability of the spacecraft.
- Required calibration techniques.
- Any special instrumentation (e.g., dual polarizations).

The future of RS for Solar System exploration is best defined by examining the following sets of questions.

6.1.1 Fundamental Questions about the Utility of RS Techniques

The interplay between technological and scientific advances can be explored by formulating fundamental questions about the utility of RS techniques:

1. What scientific opportunities can be made possible uniquely with increasing RS sensitivity?
2. Is one of the best future utilities of RS techniques the ability to probe regions that are otherwise inaccessible or are very risky with other techniques?
3. Will the rapid buildup of data, such as the case of future crosslink atmospheric occultations that can provide much higher spatial and temporal resolution, lead to insights not otherwise possible?
4. In what areas is RS still far from the fundamental limits of what can be measured (e.g., aspects of post-Newtonian parameters)?
5. Where do incremental improvements lie versus giant leaps that can influence our weltanschauung. For example, if our ability to make Doppler frequency measurements was improved by many orders of magnitude, would we truly learn a lot more about planetary interiors?
6. What is the full utility of small spacecraft for RS propagation, gravitation, surface characterization, and in-situ Doppler wind investigations?
7. What is the full utility of relay spacecraft assets for RS propagation and gravitation investigations?
8. What science would be enabled with swarms of small spacecraft?
9. What role would RS techniques have in the concept of postage-stamp spacecraft to be sent to other stars?
10. What is the cost-effectiveness of various strategies for improving RS sensitivity relative to the scientific gain?
11. Should funds be better spent analyzing data already acquired but not analyzed?
12. How could a drag-free "orbiting Deep Space Network (DSN)" be developed, and how could perturbations from Earth's gravity field and other factors be mitigated?
13. How can instrumentation detect planetary seismic/volcanic/geophysical (surface/subsurface) events with effects apparent in disturbances of planetary atmospheres?

6.1.2 Possible Triggers for Specific Innovations for Future Investigations

The following questions are triggers for possible specific innovations for future techniques, methods, and investigations.

1. What science is enabled as a result of flying atomic space clocks?
2. What science is enabled as a result of using miniaturized ultrastable oscillators (USOs)?
3. What science is enabled from newer USO technologies including molecular oscillators?
4. What additional science is enabled if all future planetary spacecraft have gone through a process of "gravity cleanliness" where the spacecraft properties relative to nongravitational forces are understood and calibrated for solar radiation pressure, outgassing, etc. (as was done with Gravity Recovery and Interior Laboratory (GRAIL), for example)?
5. What science is enabled if there are numerous relay satellites for planetary telemetry?
6. What science is enabled from observation of two or more spacecraft in the same beam?
7. What science is enabled via spacecraft at different planets with radio/laser links?
8. What science is enabled from arraying multiple large stations for significant increase in the downlink signal-to-noise ratio (SNR)? Can the resulting SNR from an array significantly enable experiments that would otherwise fall below the threshold of receiver sensitivity at more distant targets or due to atmospheric absorption and scintillation?
9. What science is enabled from uplink array technology (employing N antennas transmitting at X-band could produce signals that add coherently when received at the spacecraft providing a power gain of N_2 compared to a single antenna)?
10. What science is enabled from array elements along very long baselines, with enhanced spacecraft plane-of-sky interferometric radiometric observables?
11. What science is enabled from significant advances in quantum technologies, especially for clocks, gradiometers, and accelerometers?
12. What science is enabled from significant advances in DSN operation in the Multiple Spacecraft per Antenna (MSPA) mode?
13. What science is enabled from significant advances in polarimetry (e.g., measurement of Faraday rotation, detection of precipitation, etc.)?
14. What science is enabled from significant advances in balloons, sail planes, and other aerial vehicles?

6.1.3 Possible Synergies with Other Fields

A number of questions arise about the development of RS science goals, techniques, and technology that address the connections to other fields:

1. How can the synergy between RS and navigation techniques further advance the areas of Solar System ephemerides and relativistic gravity?
2. How can the synergy with onboard spacecraft instruments be further optimized for cross-correlations, such as gravity science and planetary shape models from radar/laser altimetry, imaging, or occultations, and the use of accelerometers?

3. How can the technological synergy between future optical link science and optical imaging instruments benefit the science?
4. How can the synergy with the other DSN science fields, radio astronomy, very-long-baseline interferometry (VLBI), and planetary radar advance the science, e.g., wide-band open-loop receivers (OLRs) and processing algorithms and overlapping DSN instrumentation? Can VLBI techniques be used to enhance radio occultations (ROs) or gravity science?
5. How can quasar occultations of typical planetary targets in the Solar System enhance research conducted by spacecraft radio occultations?
6. How can the synergy with the Earth science community further lessons learned as was the case with Gravity Recovery and Climate Experiment (GRACE) and global positioning system (GPS) occultation networks?
7. Is there synergy with techniques from spectroscopy, astrometry, or interferometry instrumentation and signal processing?
8. Is there synergy with the Search for Extra-Terrestrial Intelligence (SETI) and other nontraditional science in the DSN and elsewhere?

6.1.4 Examining Relevant Methodologies

Several questions arise about the future of RS that involve possible methodologies that could be developed or investigated:

1. What are the RS tradeoffs between bigger antennas and arrays of smaller antennas?
2. How could a drag-free “orbiting DSN” be developed, and how could perturbations from Earth’s gravity field and other factors be mitigated?
3. How would the RS data quality be impacted if future spacecraft modulation and coding methods were optimized to significantly increase the telemetry return at the expense of the carrier SNR?
4. How can the current capabilities for polarization-based science (e.g., bistatic radar and Faraday rotation measurements) be advanced and can specially designed ground instrumentation for the capture of both signal circular polarizations and sensitive power measurements lead to additional breakthroughs (e.g., precipitation measurements)?
5. Would breakthroughs in DSN scheduling, such as shorter passes-per-spacecraft due to improved media calibrations, quicker navigation solutions, and onboard recorder telemetering high-rate data enable additional science in the gaps between scheduled passes?
6. What is the utility of the long-lived of missions, some of which go into hibernation periods, and resulting unique investigations?
7. Can there be additional observations of background sky between DSN spacecraft tracks? And can there be RS application of unused DSN bandwidth?
8. How can the field benefit from artificial intelligence and smart spacecraft or “sciencecraft” (Hayati et al. 1997; Chien et al. 2001), as well as artificial intelligence applied to the DSN?
9. What is the utility of investigating anomalies, such as the Pioneer anomaly, which is now mostly understood (Turyshev et al. 2012), or the Earth flyby (an observed change in speed of some spacecraft during their flyby of Earth on their way to planetary targets; Anderson et al. 2008)?

To anticipate possible rapid formulation and implementation on future missions, efforts are underway to address many of these questions by examination and development to flight-readiness of potential science-enabling technologies by the international RS community, aided by strategic investments at NASA's Jet Propulsion Laboratory (JPL) (Asmar et al. 2019) and other research centers.

6.2 Science-enabling Technologies: Constellations of Small Spacecraft

A key technology priority for the near future is the development of small spacecraft that can perform precision RS observations similar to those of large spacecraft via links to either the Earth or other spacecraft. Also called SmallSats, CubeSats, or other names emerging for various size classes, small spacecraft have the advantage of low-cost which makes them ideal for:

- Multiplicity—a constellation (two or more) of small spacecraft can make scientific contributions by carrying out RS investigations that require:
  - formation-flying with crosslink, such as the GRACE or GRAIL planetary gravity missions
  - concepts such the Laser Interferometer Space Antenna (LISA) mission with 3-arm precision configuration
  - proposed concepts such Trilogy with planet-to-planet spacecraft-to-spacecraft laser crosslinks (Smith et al. 2012; Zuber et al. 2017)
  - interferometry or same-beam signal reception at ground stations to cancel out common noise sources that can be planetary orbiters or surface landers
  - atmospheric entry probes where the atmosphere can be sampled simultaneous at multiple locations
- Riskier endeavors—they can make scientific contributions by entering regions considered dangerous for more expensive larger spacecraft and carry out gravity investigations, for example, with links to a mother spacecraft;
- Technology demonstrations—they can carry new radio or optical elements or systems to gain flight experience prior to planning applications for RS investigations on future missions, a process useful for raising the technology readiness level (TRL), a metric for spacecraft components' technological maturity for flight on deep space missions;
- Educational tools—in addition to carrying out RS investigations, they can provide a platform for innovative trajectories, instrumentation, or other tools that would not be practical on more expensive larger missions.

Small spacecraft have been used extensively in the Earth environment, but not flown in deep space, until the Mars Cube One (MarCO) historical mission (Asmar and Matousek 2016). The resulting knowledge provided a leap in the TRL which quickly became a preferred baseline for follow-on proposals for more advanced radio instrumentation for various RS investigations. Future investigations will benefit greatly from the transition of a single large spacecraft exploring Solar System targets to small orbiters, entry probes, atmospheric balloons, other atmospheric vehicles, or surface landers that need to be robust to function for a reasonable lifetime in the target environment and placed in suitable trajectories or orbits. Small spacecraft have common functions, such as a capable radio system that fits the reduced mass, volume, and power resources. The maturation of multi-band (two or more wavelengths) radio systems can provide breakthrough science advances, especially while improving the spatial and temporal coverage of measurements.

This science-enabling technology requires advances in miniaturized radios, such as the next-generation SDRs (see Pugh et al. 2017). The Iris SDR used for the MarCO provided a successful interplanetary flight experience to further advance science-based capabilities, such as phase stability, simultaneous open-loop recordings at multiple wavelengths, and agility to swap the transmit/receive frequencies. MarCO also carried innovative ultrahigh frequency (UHF) and X-band antennas described in Hodges et al. (2016). A proposed new SDR-based instrument for RS investigations leverages the latest digital space-grade components to achieve the reduced size resources while providing flagship mission–quality RS capabilities and performance. It can be highly configurable to support single- and dual-band one-way and two-way links for both crosslink and traditional investigations using a reference from an external USO in addition to an internal oscillator.

Section 5.3.3 describes a spacecraft open-loop receiver (SOLR) as analogous to the science-based open-loop receiver (OLR) in the DSN used for RS investigations. A SOLR is required when the link direction is reversed and the experiment is carried out as a one-way uplink experiment as was done for the New Horizons mission RO experiment at Pluto with the Radio Science Experiment (REX) instrument (see Tyler et al. 2009) or for spacecraft-to-spacecraft crosslinks. The SOLR is generically a super-heterodyne receiver that captures and records the in-phase and quadrature (I&Q) digital samples of the signal carrier received by the spacecraft. The next-generation SDR proposed to have enhanced function as a RS instrument can have this capability built into the design.

USOs have been critical instrumentation for RO experiments since Voyager, the first Solar System exploration program to develop this science payload expressly for the RO investigations. As noted in Section 5.3.1, it is used for the downlink or uplink one-way mode links between spacecraft and ground stations or between spacecraft. Using the USO eliminates the double propagation effect of the uplink and downlink signals in the same planetary medium under study and also eliminates the time needed by the transponder to lock on the uplink signal when the spacecraft is emerging from behind the occulting planetary body.

Without an USO, the transponder's built-in auxiliary oscillator is not sufficiently stable for RO experiments, which, in that case, can be carried out in the two-way mode for occultation ingress only and with more complex signal processing. Small spacecraft cannot accommodate the mass and power of the current top-of-the-line flight USO (Cassini class with Allan deviation of $\sim 2 \times 10^{-13}$ at ~100 s integration time (Asmar 2012b)), so maturing smaller devices with comparable stability based on newer technologies is a goal for future development.

6.2.1 Constellations for Investigations of Atmospheric Structure and Dynamics

Compared to the traditional RO with spacecraft-to-Earth or, in the case of New Horizons at Pluto, Earth-to-spacecraft links, simulations have shown that substantial increases in global coverage as well as spatial and temporal resolution are possible using crosslinks among a constellation of small spacecraft, see Figure 6.1. This is akin to GPS-based RO constellations at Earth described in Chapter 2. At rocky planets, for example, near-global coverage can be accomplished in a couple of weeks with two spacecraft in optimized orbits as compared to many years via the traditional downlink to Earth (see Figure 6.2).

Figure 6.1 Illustration of a CubeSat Titan network for atmospheric crosslink occultations.

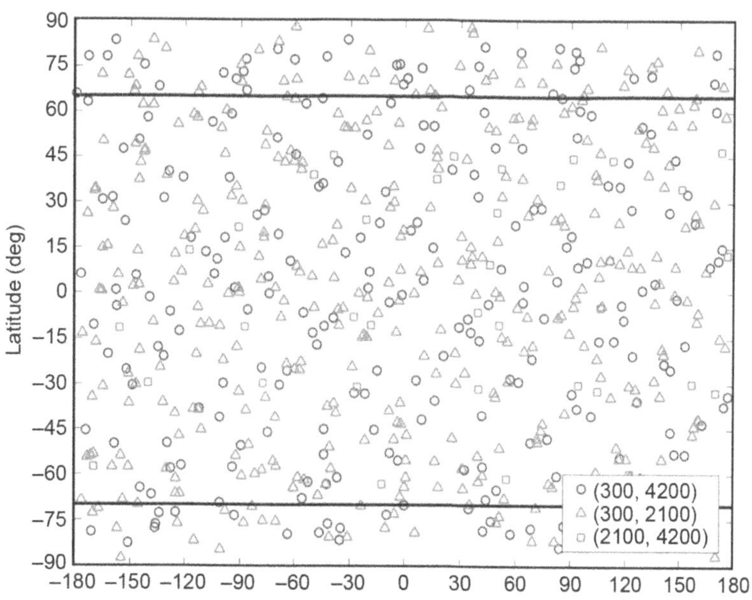

Figure 6.2 Simulation results showing number of occultations from Mars crosslink in 1 week using three CubeSats at three altitudes with pairwise occultation locations. Solid lines near poles show the traditional Mars-to-Earth occultation coverage in the same week.

To plan the future of crosslinks with planetary missions, a demonstration was carried out at Mars, as described in 2007 (see Ao et al. 2015; Asmar et al. 2016). Mars Odyssey and the Mars Reconnaissance Orbiter (MRO) are equipped with UHF links to communicate with surface landers, and one orbiter can sometimes communicate with the lander while in view of the other orbiter. The Electra Payload (see Chapter 5), an SDR onboard MRO, received a UHF signal transmitted from Odyssey toward a rover on the Martian surface. The digitized signal data were later transmitted from MRO to the DSN. This demonstrated the possible science return as well as the instrumental requirements to design the next-generation science-based crosslink instrumentation. The absence of an USO on Mars Odyssey limited the scientific return for the neutral atmosphere but electron density profiles of the ionosphere were successfully retrieved.

6.2.2 Constellations for Investigations of Interior Structure and Dynamics

The application of precision gravitational field measurements via spacecraft-to-spacecraft Doppler crosslinks has been demonstrated to significantly benefit high-impact geophysical exploration of planetary interiors and monitoring of atmospheric and surface mass transport. The dual-spacecraft GRACE Follow-On mission (Tapley et al. 2004) continues to produce high-resolution monitoring of time-variable Earth gravity. The GRAIL mission (Zuber et al. 2013b; illustrated in Figure 6.3) produced high-resolution gravitational fields of the Moon, as discussed in Chapter 3. The Selenological and Engineering Explorer (SELENE) mission had a Doppler crosslink with secondary spacecraft (a subsatellite) on the lunar farside (see four-way Doppler in Chapter 5).

Figure 6.3 GRAIL crosslinks at the Moon as an illustration of the concept that can be applied at other planetary bodies in the future.

Applying this concept to other rocky, icy, or gaseous bodies would enable rapid high-resolution global coverage and provide stringent constraints on the thickness of ice and the characteristics of subsurface oceans at icy moons. Furthermore, monitoring of static and dynamic gravity, including seasonal mass transport in the atmosphere, such as carbon dioxide (CO_2) on Mars, for example, would considerably improve our understanding of the Martian interior and climate. Discoveries akin to buried empty lunar lava tubes detected with GRAIL's data (Evans et al. 2016) are possible at other terrestrial planets; lava tubes are known visually to exist on Mars (Greeley 1973), but their underground network is largely unknown. Another important benefit of high-resolution gravity mapping is providing constraints on the thickness of ice and the characteristics of possible subsurface oceans at icy moons (Iess et al. 2014).

Planet-to-planet spacecraft-to-spacecraft crosslinks have also been proposed. The Trilogy concept suggests placing three spacecraft in orbit around Mars, Venus, and the Moon at the same time with laser links among them. This could provide insight into understanding planetary motion and fundamental physics. Due to changes in solar mass resulting from hydrogen fusion in the core, as well as the solar wind, the scale of the Solar System is slowly increasing as with additional changes possible if there is a secular variation of the gravitational constant, G. Measurements between planets could provide an accurate estimate of the Solar System expansion rate (Smith et al. 2012; Zuber et al. 2017).

6.2.3 Constellations for Simultaneous and Differential Measurements

The concept of using multiple spacecraft per antenna (sometimes called per aperture, MSPA) as well as interferometry can be realized through deployment of constellations. When more than one spacecraft is in the same beam of the ground station, such as multiple spacecraft at Mars, they can be simultaneously tracked by the same station using different receivers, saving on antenna resources and enabling interesting scientific opportunities. VLBI and delta differential one-way range (DOR), described in Chapter 5, are examples of interferometry techniques in which multiple radio signals can be received at one station or one signal received at multiple stations, typically on different continents for maximum separation, or other combinations for increased spatial resolution of specific measurements. Developments are underway to implement multiple uplinks per antenna (MUPA) at the DSN, a capability for which spacecraft transponders need to be programmable to receive one uplink channel and coherently return the signal at separated channels to eliminate interference.

From these concepts, differential interferometry can be implemented for planetary rotational state and tidal deformations. Tracking multiple orbiting or landed spacecraft from one or more Earth stations suppresses most of the common types of noise on the signal(s), such as atmospheric delays and clock drift, as illustrated in Figure 6.4. This enhances the sensitivity to the desired science goals, such as gravitational fields, rotational states, tidal deformations, and winds. This technique, sometimes referred to as same-beam differential interferometry (SBI), has benefits that were recognized by Counselman et al. (1972) and advanced by Preston (1974), Edwards et al. (1992), Goossens et al. (2011), and others. Gregnanin et al. (2012) describe the technological assessment of SBI for a Ka-band transponder for the BepiColombo RS experiments.

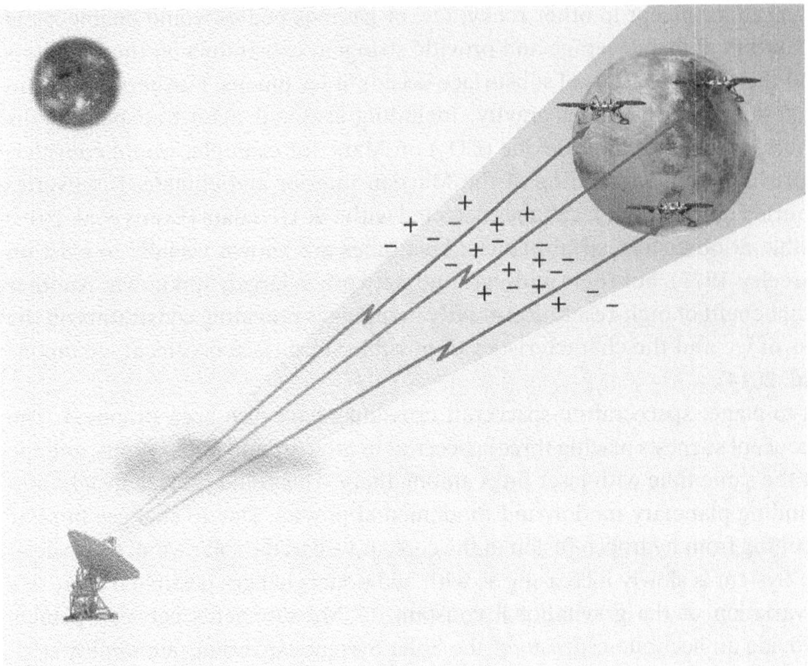

Figure 6.4 Artist's conception of interferometry technique for planetary rotational/tidal deformations.

In a manner similar to the lunar laser ranging techniques investigating lunar orbital dynamics, physical libration, and tidal deformation (Dickey et al. 1994), a proposal by P. Bender (1994) of a microwave interferometer involves a network of three widely separated landers coherently and simultaneously tracked at Ka-band by a single antenna on Earth. Folkner and Border (1992) describe orbiter–orbiter and orbiter–lander tracking using SBI. The technique was applied to examine librations (Apollo Lunar Surface Experiments Package (ALSEP) mission) and gravity field (SELENE mission) of the Moon, studying winds on Venus (Pioneer probes and Vega balloons), and observing Huygens' descent on Titan. New uses have been proposed for the Moon, Mars, and Jupiter.

6.2.4 Constellations of Entry Probes and Atmospheric Vehicles

Scientific goals to be pursued for future Solar System missions include new opportunities for profiling planetary winds via tracking entry probes, balloons, and other vehicles (see Figure 6.5). Radio links from probes descending through planetary atmospheres provide information on atmospheric dynamics, as demonstrated at Venus, Jupiter, and Titan. Links can be established with proximity spacecraft or ground stations and, if multiple links are possible, recovery of complete wind vectors can be made. See Atkinson et al. (1998) for the results of the Galileo probe experiment and Bird et al. (2005) for the results of the Huygens probe experiment. The Vega Venus Balloon Experiment is described in Sagdeev et al. (1986a, a1986b).

Figure 6.5 Concept illustration of a constellation of probes and balloons examining the winds at multiple locations in planetary atmospheres via Doppler tracking.

Furthermore, a constellation of small probes and/or balloons can provide a global assessment of atmospheric dynamics. Absorption on the probe's radio link can be used to infer the integrated abundance of ammonia (giant planets), sulfuric acid (Venus), or other absorbers. Similar techniques can be applied to atmospheric balloons as was done on the Vega mission at Venus. Atmospheric absorption on the probe's radio link can be used to infer the integrated abundance of ammonia as well as water and hydrogen sulfide along the probe signal ray path (Folkner et al. 1998). The effect examined by Doppler wind experiments (DWEs) exploits the Doppler shift in the direction of the line of sight caused by motion of the spacecraft due to wind. In the future, dual-link Doppler measurements, one link between the probe and a spacecraft, and a second link between the probe and a ground station or a secondary spacecraft (such as a dedicated SmallSat) would allow the derivation of two-dimensional wind vectors as illustrated in Figure 6.5. The links can be established with proximity spacecraft or, in some cases, directly to ground stations. Multiple small spacecraft descending through different parts of the atmosphere would provide a measurement of the profile of winds in multiple locations.

6.2.5 Constellations for Investigations of Planetary Surface

RS investigations have demonstrated via bistatic radar experiments that signal reflection from planetary surfaces and near-surface regions reveal their structure and composition. To date, investigations have been carried out at Venus, the Moon, Mars, Titan, comet 67P/Churyumov-Gerasimenko, Pluto, Vesta, and Kuiper Belt Object Arrokoth.

There is also significant experience with space-based reflectometry and scatterometry techniques used to study Earth's surface roughness and dielectric constant.

This type of RS investigation is possible between a flagship planetary mission and a small spacecraft or by a small constellation. Comparing the power between circularly polarized transmitted radio signals and oppositely polarized reflected signals provides information about icy moons' surface properties. It is speculated that mass flux could also be detected on Venus by a nadir-viewing spacecraft monitoring specific areas for changes between coherent and incoherent scattering that signify mass transport or surface terrain variations. Furthermore, Doppler delay maps from the differential travel time between transmitted and reflected signals could provide information on surface topography. Tracking-landed spacecraft can be used to study seasonal surface deposition (e.g., CO_2 on Mars) and finer scale surface and near-subsurface properties.

In a bistatic scattering experiment, a spacecraft downlink signal is reflected off a planetary surface and received on Earth or another spacecraft. Information can be deduced about the surface and near-subsurface electrical properties and roughness. Reversing the path via an uplink, as illustrated in Figure 6.6, provides higher SNR, as shown by New Horizons and Lunar Reconnaissance Orbiter (LRO); multiple simultaneous uplinks with frequency offset can improve the SNR and resolve subsurface degeneracies. A constellation of small spacecraft makes global characterization feasible and yields information at scales useful to the safety of planned landers. Scattering from subsurface structures provides added exploration opportunities where redundancies can be resolved using polarization, wavelength, and obliquity variations. Use of bistatic techniques with landed spacecraft can provide finer spatial resolution.

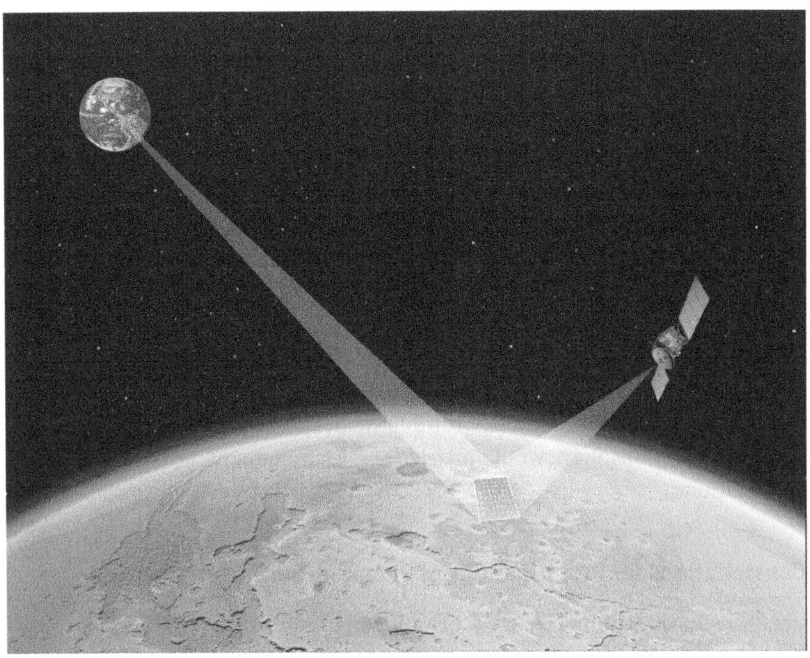

Figure 6.6 Illustration of uplink surface scattering (bistatic radar) study.

6.3 Science-enabling via Optical Links

Optical communication technologies are expected to be demonstrated on future deep space missions. Laser links can augment radio links for atmospheric propagation science and provide precision optimetric (equivalent to radiometric) observables to improve gravitational experiments and knowledge of Solar System ephemerides. The RS community has investigated opportunities for optical links including atmospheric occultations at optical wavelengths, as well as enhanced measurements of gravitational fields and interior structures, tests of fundamental theories of physics, and improved knowledge of Solar System ephemerides.

The DSN is planning to support communications with optical links starting with a technology demonstration on the NASA Psyche asteroid mission. Laser links can augment radio links for atmospheric propagation science and provide precision optimetrics (see Hemmati 2006; Asmar et al. 2019; Divsalar et al. 2014; Divsalar et al. 2016). Crosslink laser ranging is planned for the GRACE Follow-On mission (Klipstein et al. 2001; Sheard et al. 2012). The latter reference notes that the benefits of optical frequencies for interferometric technique stem from the fact that the basic unit of length in an optical system is 10,000 times smaller than typical radio wavelengths, measured in angstroms or microns instead of millimeters or meters, amounting to a ruler with finer markings.

There is also a dramatic reduction in diffraction effects from optical antennas and other structures near the receiver when compared to radio signals. Additionally, optical sources are by far more directional. Design choices have also been examined, including one-way versus two-way coherency modes, and actual in-flight demonstrations have been carried out (Divsalar et al. 2014). Buccino et al. (2016) describe the infusion of one-way optical ranging of LRO into the JPL software tools as precedent to optical link science and navigation. Divsalar et al. (2016) describe RS measurements in the presence of data on optical and radio frequency (RF) links. Hemmati (2006) discusses optical communications at planetary distances and the technologies required to reach this future capability, most of which would be synergistically used for optical link science. Divsalar et al. (2014) evaluate optimetric measurements in presence of data on optical links.

6.4 Science-enabling Calibration Techniques

The most accurate RS experiments to date were achieved during Cassini's gravitational wave observations. As detailed in Asmar et al. (2005b), the leading source of Doppler noise was the charged particles of the interplanetary plasma and Earth's ionosphere, calibrated by propagating spacecraft signals at two different wavelengths (e.g., X- and Ka-bands or S- and X-bands), and Earth's troposphere, which is partly calibrated via water vapor radiometers. The next noise source is unmodeled mechanical motion of the ground antenna's phase center. The following sections describe possible improvements to water vapor radiometry and attempt to characterize the antenna's mechanical noise.

6.4.1 Earth's Troposphere Water Vapor Radiometry

Another technique for radiometric data calibration is the calibration of the Earth tropospheric effects on the radio signal. Currently, a DSN-advanced water vapor radiometer (AWVR), which is adjacent to DSS-25 at Goldstone, provides brightness temperature measurements that can be converted into path delay data, as described in Chapter 5. Design and performance options for the next-generation radiometer include integrating the radiometer with the spacecraft tracking antenna as discussed in Border et al. (2018) and shown in Figure 6.7. The proposed improvements would have the most scientific impact in the determinations of the planetary interior structures reconstructed from gravity field measurements using the Allan deviation metric.

In the current system, a radiometer on a small antenna adjacent to the 34-m station makes measurements of the water vapor resulting in beam offset errors. In the proposed system, the radiometer is integrated into the station's electronics allowing water vapor calibration through the same beam. A design for this in-line (same optics for spacecraft Doppler and radiometer), cryo-cooled radiometer has been completed.

6.4.2 Antenna Mechanical Noise

The most sensitive Doppler fractional frequency stability achieved to date was $\approx 3 \times 10^{-15}$ Allan deviation at 1000 s (Asmar et al. 2005b). This used full calibration of the propagation media using X- and Ka-band uplink and downlink as well as tropospheric scintillation calibration instrumentation. With these calibrations applied, the next leading noise source is the unmodeled motion of the phase center of the ground antenna. This motion is due to intrinsic mechanical vibrations of the large DSN antennas and can be suppressed when two-way and three-way (receive-only) Doppler data from a smaller and stiffer antenna are suitably combined using a time-delay-correction procedure further enhancing radiometric observations for Solar System dynamics experiments. Armstrong et al. (2008) describe this technique along with the leading noise to be the time-dependent unmodeled motion of the ground antenna's phase center caused by antenna sag due to elevation angle changes, unmodeled subreflector motion, wind loading, and bulk motion of the antenna as it rolls over irregularities in its azimuth ring, etc. Using this technique, the mechanical noise in the combined Doppler observable can be reduced to that of the stiffer antenna and the overall data quality of the science observable would be significantly improved; recent analyses are found in Notaro et al. (2020, 2018).

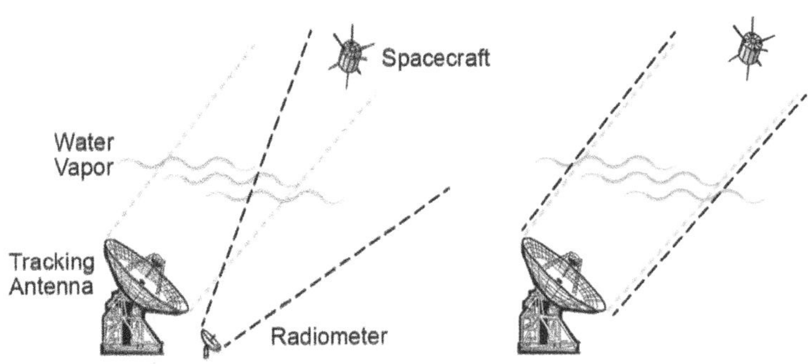

Figure 6.7 Concept of integrating the radiometer with the spacecraft tracking antenna (Jongeling et al. 2018).

Notaro et al. (2020) discuss Doppler tracking techniques for spacecraft navigation, planetary geodesy, and tests of the theory of general relativity and examine methods of reducing the noise in the radiometric observables. They elaborate on the TDLC technique introduced in Armstrong et al. (2008) as a linear combination of simultaneous Doppler data from a main antenna providing the two-way link and a smaller and stiffer antenna providing a three-way link via, as shown in Figure 6.8. This combination can considerably reduce the antenna mechanical noise. They report that "although X-band links are generally dominated by the highly-variable interplanetary plasma noise and therefore are not suitable for the TDLC, we found that, when local noises are particularly large at the two-way antenna, this technique may still lead to up to a factor-of-three noise reduction with respect to the two-way link."

6.4.3 Advanced Ranging

The European Space Agency's (ESA) BepiColombo mission to Mercury uses coherent Ka- and X-band Doppler links simultaneously. The mission also uses the first ever precision ranging at Ka-band in deep space with expected precision of the pseudo-noise (PN) ranging of under 20 cm; the average precision at X-band is ~1 m. These capabilities will be used to investigate Solar System dynamics, including planetary ephemerides, as well as tests of general relativity (Iess and Asmar 2007, Iess et al. 2021). This follows a previous test of aspects of the theories of relativity using Doppler tracking to improve the parameterized post-Newtonian formalism demonstrated via the Cassini mission RS instrumentation, as described in Bertotti et al. (2003).

Ranging at Ka-band with the spacecraft requires compatible capabilities at a ground station. With the introduction of spacecraft tracking at Ka-band (32–34 GHz) with Cassini and Juno in a two-way coherent mode, the Doppler (range-rate) data quality improved by about one order of magnitude with the proper calibrations. A similar improvement in ranging was made with an advanced ranging instrument (Asmar et al. 2016) that also takes advantage of calibration techniques centered on two-way coherent Ka-band tracking. The scheme, illustrated in Figure 6.9, reduces uncertainty in

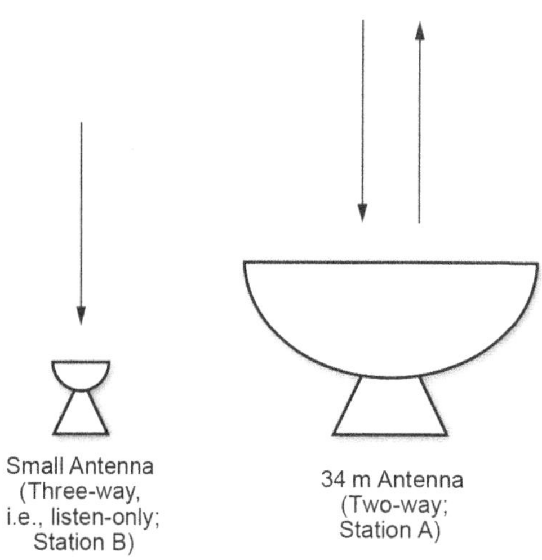

Figure 6.8 Concept of linear combination of simultaneous Doppler data from a main antenna providing the two-way link and a smaller and stiffer antenna providing a three-way link (Armstrong et al. 2008).

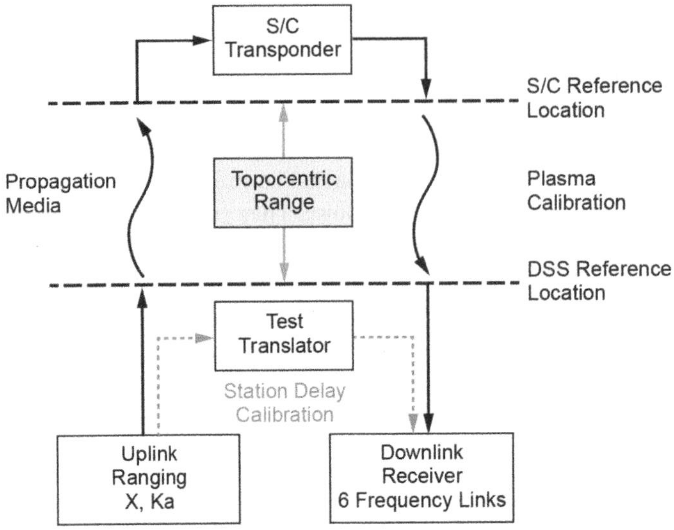

Figure 6.9 Concept of advanced deep space ranging at X- and Ka-band simultaneous uplink and downlink for full calibrations (Paik et al. 2018).

ranging accuracy by about one order of magnitude with simultaneous uplink and downlink at X- and Ka-bands, advanced ranging modulation, and the application of a constant calibration signal for internal station delay. The improved ranging can be used for future RS experiments, such as the ones planned for the BepiColombo mission, to examine possible violations of the theory of general relativity, improve the parameterized post-Newtonian formalism, improve measurements of solar oblateness, and characterize the solar corona. Advanced ranging instrumentation has been developed at the DSN (see Chapter 5 and Paik et al. 2018).

6.5 Summary

The past six decades have witnessed a vast set of discoveries achieved via RS methods. The ability to precisely measure electromagnetic properties of spacecraft radio signals provides unique method of obtaining new information about planetary atmospheres, ionospheres, rings, surfaces, shapes, and internal structure. Precision Doppler and ranging RS observables have been demonstrated for studies in fundamental physics and solar dynamics such as observing effects on signal passage through the Sun's gravitational field and solar wind, investigating gravitational waves, and monitoring planetary motion to study gravitational theory and solar mass loss.

Additional science discoveries could be enabled by developing new technologies and mission concepts such as deployment of low-cost small spacecraft with increasing radio capabilities that approximate those of full-scale spacecraft and can accept increased risk, for high spatial and temporal resolution of atmospheric and gravitational mapping. The future outlook for the field of RS is filled with exciting possibilities for more scientific discoveries.

Appendix 6A

The National Academies Planetary Science Decadal Survey, Radio Science Contribution, 2009

Planetary Radio Science: Investigations of Interiors, Surfaces, Atmospheres, Rings, and Environments

Sami W. Asmar[1] Kaare[2], Aksnes Roberto[3], Ambrosini Aseel Anabtawi[1], John D. Anderson[1], John W. Armstrong[1], David Atkinson[4], Jean-Pierre Barriot[5], Bruno Bertotti[6], Bruce G. Bills[1], Michael Bird[7], Veronique Dehant[8], Peter Edenhofer[9], F. Michael Flasar[10], William Folkner[1], Richard G. French[10], Hideo Hanada[10], Bernd Häusler[13], David P. Hinson[14], Luciano Iess[15], Özgür Karatekin[8], Arvydas J. Kliore[1], Alex S. Konopliv[1], Frank Lemoine[10], Ivan Linscott[14], Essam Marouf[16], Jean-Charles Marty[17], Koji Matsumoto[12], Hirotomo Noda[12], Kamal Oudrhiri[1], Meegyeong Paik[1], Ryan S. Park[1], Martin Pätzold[18], Robert Preston[1], Nicole Rappaport[1], Pascal Rosenblatt[8], Richard A. Simpson[18], David E. Smith[10], Suzanne Smrekar[1], Paul G. Steffes[19], Silvia Tellmann[18], Paolo Tortora[20], G. Leonard Tyler[14], Tim Van Hoolst,[8] Michael Watkins[1], James G. Williams[1], Paul Withers[21], Xiaoping Wu[1], Donald Yeomans[1], Dah-Ning Yuan Maria T. Zuber[22]

Author Institutions

1. Jet Propulsion Laboratory, California Institute of Technology
2. University of Oslo, Norway
3. Istituto di Radioastronomia, INAF, Italy
4. University of Idaho
5. Observatoire Géodésique de Tahiti
6. University of Pavia, Italy
7. University of Bonn, Germany
8. Royal Observatory of Belgium
9. University of Bochum, Germany
10. NASA Goddard Space Flight Center
11. Wellesley College

---

1 This work was carried out in part at the Jet Propulsion Laboratory, California Institute of Technology, under contract with the National Aeronautics and Space Administration.

12. National Astronomical Observatory, Japan
13. Universität der Bundeswehr München, Germany
14. Stanford University
15. University of Rome, Italy
16. San Jose State University
17. Centre National d'Etudes Spatiales, France
18. University of Cologne, Germany
19. Georgia Institute of Technology
20. University of Bologna, Italy
21. Boston University
22. Massachusetts Institute of Technology

6A.1 Summary

Scientists use radio links between spacecraft and Earth or between two spacecraft to examine changes in the phase/frequency, amplitude, line-width, and polarization, as well as round-trip light time of radio signals to investigate atmospheres and ionospheres, rings and tori, surfaces, shapes, gravitational fields, orbital motion, and dynamics of solar system bodies (e.g., planets, satellites, asteroids, and comets). Beyond planetary science, the tools and techniques of radio science (RS) on planetary missions are applied to investigations of the solar wind, corona, and magnetic field, as well as tests of fundamental physics including the theory of general relativity. RS techniques are also highly important in monitoring mission-critical maneuvers or spacecraft emergencies under conditions of low-signal level and/or high-signal dynamics.

RS will continue to be an important tool for solar system exploration in the coming decades. New scientific advances can be enabled or enhanced by recent developments in RS instrumentation and calibration techniques that improve the data quality by an order of magnitude over current technology. Investigations that will benefit from these improvements include missions to probe through the crust of icy satellites into the deep interior to detect subsurface water oceans. Studies of atmospheric occultations and surface scattering will be improved by uplink occultation instrumentation to overcome signal-to-noise ratio (SNR) limitations and by links between orbiters at the same planet to provide global coverage.

6A.2 Background

The broad sense of *Radio Science* is the use of propagating electromagnetic waves to explore the universe. RS observations can be passive, as in observations of naturally occurring emissions, or active, as in the results of probing with artificial signals. In planetary exploration, the term RS has been commonly used for about four decades to describe the scientific utilization of radio signals from spacecraft as opposed to natural celestial radio emissions. This field is distinguished from other sciences by its definition in terms of methods rather than by objects of study.

Current investigations include microwave radio occultations by planetary rings, ionospheres, and neutral atmospheres; propagation through the solar corona; reflection from surfaces; and distance/velocity measurements leading to determinations of mass and density distribution and construction of models for gravity fields and interior structure. These investigations can be grouped into two classes: *propagation* and *gravitation*, where the first investigates the effects of media on the signals and treats the effects of spacecraft motion as noise to be calibrated out of the data, and the second investigates the effects of forces on the spacecraft causing shifts in the signal and treats the effects of media as noise to be calibrated out of the data. Atmospheric balloons, descending probes, and atmospheric drag experiments constitute a hybrid category of RS investigations based on measuring perturbations in spacecraft position and velocity caused by the medium.

RS often benefits from collaborative fields such as the interactions between waves and charged particles, radar surface imaging, planetary topography, accelerometry, spacecraft navigation, very long baseline interferometry (VLBI) precision measurements of angular motion, and relativistic gravity. The discipline came of age in 1964 when the National Bureau of Standards renamed Section D of its *Journal of Research to Radio Science*; the journal has been published by the American Geophysical Union since 1969. In addition to *Radio Science*, planetary RS papers appear regularly in *Science*, *Nature*, *Journal of Geophysical Research*, *Icarus*, and others. RS data are preserved for future generations within the planetary data system since the measurements are self-calibrating and never outdated; as long as quality instrumentation is operated properly, measurements made today can be included in future analyses for very long periods of time.

6A.3 Historical Opportunities and Discoveries

With strong international collaboration, RS experiments have been conducted on almost all deep space missions in the past half-century. Selected examples of historical discoveries include:

- First estimate of Martian surface pressure
- Determination of ring structure and particle size distribution for Saturn and Uranus
- First measurement of surface pressure of Titan and Triton
- First detection of ionospheres of Titan and Triton
- Measurement of electron column density profile of the Io Plasma Torus
- Discovery of lunar mascons
- Determination of masses and modeling of bulk compositions of Jupiter, Saturn, Uranus, Neptune, Titan, Triton, Mars, Venus, and the Moon
- High-resolution gravitational fields of the Moon, Mars, and Venus
- Drag deceleration measurement in the comae of comets Halley and Grigg-Skjellerup
- Modeling large-scale coronal structure and electron densities in streamers and holes
- First evidence for the acceleration of coronal mass ejections far from the Sun

6A.4 Recent Opportunities and Discoveries

Since 1995, RS has had a significant role in subsequent missions and is credited with the following additional accomplishments:

- First detection of Martian core and bulk seasonal CO_2 deposition at poles (Mars Pathfinder)
- Vastly improved gravity field of Mars, with determination of Love number and correlation with topography (Mars Global Surveyor (MGS), Odyssey, Mars Reconnaissance Orbiter (MRO), and Mars Express)
- First detection of planetary gravity field variations other than Earth (MGS)
- First determination of the mass of Phobos (MGS)
- First nonspherical gravity field of an asteroid (Near-Earth Asteroid Rendezvous (NEAR) mission to Eros)
- High-accuracy profiling of Martian atmospheric structure from radio occultations (MGS)
- Detection of ionospheres on the Galilean Satellites (Galileo)
- Jovian deep atmospheric wind speeds and ammonia concentration (Galileo Probe)
- State-of-the-art tests of General Relativity and search for gravitational waves (Cassini)
- Improved gravity field of Mercury (Mercury: surface, space environment, geochemistry, and ranging (MESSENGER))
- Measurement of Mars upper atmospheric density from spacecraft drag (several orbiters)
- Surface characteristics of Venus, Moon, Mars, and Titan from bistatic radar experiments
- Detection of meteor layers in ionospheres of Mars and Venus (Mars Express, Venus Express)
- Modeling the mass distribution in the interiors of the large moons of Jupiter (Galileo)
- Height profile of winds on Titan with the Doppler Wind Experiment (Huygens)
- High-resolution gravitational field of the Earth via spacecraft-to-spacecraft links (Gravity Recovery and Climate Experiment (GRACE))
- Gravitational field of Saturn and its large satellites (Cassini)
- Atmospheric structure of Saturn and its large satellites (Cassini)
- Profiling of internal dynamics and of particle size distribution within Saturn's rings (Cassini)
- First direct gravity measurements in the lunar far-side (Selenological and Engineering Explorer (SELENE) (aka Kaguya))

6A.5 Future Opportunities

Future missions would include significant RS investigations such as:

- **Dawn:** Coherent X-band Doppler tracking will enable the measurement of the gravity field of Vesta and Ceres and constrain their interior structure.
- **Rosetta:** RS objectives include measurements of cometary nucleus mass, bulk density, and internal structure as well the composition and roughness of the nucleus surface.
- **Juno:** Coherent X- and Ka-band links will enable precise measurement of spacecraft motion during close polar orbits to determine the gravity field, distribution of mass, core characteristics, and convective motion in the deep atmosphere.

- **Gravity Recovery and Interior Laboratory (GRAIL):** Spacecraft-to-spacecraft radio links at Ka-band, timing synchronization at S-band, and X-band Doppler links to Earth6 will enable GRAIL to measure the lunar gravity field in unprecedented detail to be used to probe the Moon's interior from crust to core, reveal its subsurface structure, reconstruct its thermal history, and place limits on a possible inner core.
- **New Horizons:** High-precision uplink radio occultation will allow derivation of Pluto's atmospheric structure while Doppler tracking will constrain interior structure.
- **Exobiology on Mars (ExoMars):** Links from a Mars lander/rover or network would improve knowledge of the interior structure by determining the total moment of inertia, core moment of inertia, and state of the core in order to constrain the evolution of the planet Mars.
- **BepiColombo:** Order-of-magnitude improved Ka-Band Doppler and ranging will enable (1) investigating the interior structure of Mercury, (2) testing relativistic gravity and significantly improved general relativity post-Newtonian parameters, (3) testing any time variation of the gravitation constant to high accuracy, (4) determining the solar oblateness to high accuracy, and (5) characterizing the structure of the solar wind in and out of the solar ecliptic.
- **Jovian satellite orbiters:** Simultaneous coherent X- and Ka-band tracking data (two orders of magnitude more accurate relative to Galileo) with NASA's potential Jupiter-Europa mission and/or European Space Agency's (ESA) Jupiter-Ganymede mission. Both would yield detailed information about satellite interiors from gravity measurements, and together offer the possibility of spacecraft-to-spacecraft radio occultations of Jupiter's atmosphere and possible satellite surface scattering observations.
- **Saturnian satellite orbiter(s):** Similar internal structure objectives to the candidate Jovian missions and emphasis on Titan atmospheric investigations as well as surface scattering and rings.

The candidate Jovian orbiters illustrate the range of potential science. The Galilean satellites represent two classes of internally active bodies that likely maintain subsurface water layers beneath a modest ice layer above and in contact with a rocky ocean sea floor or between a thick outer ice shell and high-pressure ice phases below. RS teams would investigate (1) the deep interior structure: the radial density distribution of the deep satellite interiors would be inferred from improved measurements of moment of inertia and low-degree gravity field coefficients; the assumption of hydrostatic equilibrium would be tested by separate determination of the static components of the gravitational field coefficients J_2 and C_{22}; (2) crust and lithosphere: the gravitational signature of intrinsic density anomalies and regional topographic features would be inferred from higher degree gravity data in combination with altimetry, imposing constraints on the thermal evolutions of the moons; and (3) oceans: characterization of any global subsurface water ocean and the overlying ice shell which would require measuring the response of the shell to tidal forces exerted by Jupiter along the slightly eccentric orbit of each satellite using gravity and altimetry data. The dynamical tidal Love number k_2, characterizing the variation in the gravitational potential due to the satellite's time varying tidal distortion, is inferred from time-variable contributions to J_2, C_{22}, and S_{22} of the gravitational field while the Love number h_2, characterizing the radial displacement, would be determined from altimetry.

In addition to the gravity measurements by each candidate orbiter described above, when **two orbiters are present around Jupiter**, radio links between them could be used to (1) probe the neutral atmosphere of Jupiter to produce temperature-pressure profiles to new depth and profile the spatial and temporal variability of the atmospheric small-scale structure (e.g., waves and turbulence) as well as map the global distribution of ammonia; (2) probe the ionosphere of Jupiter to high accuracy in electron density; (3) profile the ionospheres of Io, Callisto, Europa, and Ganymede and search for potential tenuous neutral atmosphere around Europa and Ganymede; (4) examine Europa and Ganymede's surface properties via surface scattering; and (5) image surface emissivity with passive radiometry and dual spacecraft interferometry.

The **ExoMars** experiment will illustrate the utility of lander tracking and allow the determination of (1) planetary spin rate, direction of rotation pole, and polar motion; (2) precession and nutation for freely rotating bodies, or physical librations for synchronous bodies, which are sensitive to internal structure such as a core or a subsurface ocean; (3) changes in rotation rate and the direction of the rotation pole due to tides, changing polar caps, atmosphere changes, and an active interior; and (4) tidal displacements of the surface depend on Love numbers h_2 and l_2 that, like k_2, depend on the interior elastic properties and the size of any core or liquid layer. Knowledge of the state of Mars' core and its size is important for understanding the planet's evolution, deduced from the dynamics of its mantle and core. The evolution and the possibility of dynamo magnetic field generation in its core are highly dependent on the planet's ability to develop convection in the core and in the mantle; a core magnetodynamo is related either to a high thermal gradient in the liquid core (thermally driven) or to the growth of a solid inner core (chemically driven) or both. The state of the core depends on the percentage of light elements in the core and on the core temperature, which is related to the heat transport in the mantle.

Landing on Mars requires knowledge of likely atmospheric density during descent and landing. Radio occultation profiles provide very valuable information due to their absolute altitude scale, good vertical resolution, and well-understood uncertainties. Yet mission planners often find that few occultation profiles are available at the needed location, season, and time of day. Acquisition of radio occultation data whenever opportunities allow would provide a richer database than scheduling them as special events of interest only to a designated RS team. Furthermore, multiple spacecraft orbiting the red planet, if properly equipped for spacecraft-to-spacecraft links, can provide global coverage with links between them without scheduling ground resources.

6A.6 Technological Advances in Flight Instrumentation

The fundamental RS observation parameters are amplitude, frequency, phase, time, polarization, and direction of propagation of the wave. Measurement accuracy depends on the phase and amplitude stability of the equipment and SNR. Extraction of the highest quality science requires calibration of the intervening medium and nongravitational forces, as well as accurate reconstruction of position, attitude, and motion of the transmitter, receiver, and target. Precision time/frequency standards are essential; when possible, multiple links, water vapor radiometers, accelerometers, and noise power standards lead to accurate calibration of the data observables.

Prior to the Voyager mission, RS experiments were conducted using the spacecraft and ground telecommunications and tracking without modification to enhance the science. Voyager was the first deep space mission to be equipped with an Ultra Stable Oscillator (USO), added to provide a higher quality "clock" for atmospheric and ring occultations. Voyager was also the first mission to employ two coherently related wavelengths simultaneously, S- and X-bands, which improved the ionospheric occultations and enabled calibration of dispersive noise sources. Galileo was upgraded with elliptically polarized transmission to investigate Faraday rotation, but the payoff was not realized due to the deployment failure of the high gain antenna. Mars Global Surveyor used the next-generation USO, more stable by an order of magnitude than the USO on Voyager; Cassini, New Horizons, GRACE, GRAIL, and other missions later used this class of USO. European deep space missions such as Ulysses, Rosetta, Mars Express, and Venus Express also had the capability for simultaneous S- and X-band links. Rosetta and Venus Express carry USOs.

Current state of the art in RS instrumentation was achieved with Cassini, which pioneered the use of coherent Ka-band, the shortest of the three possible wavelengths in use by NASA's Deep Space Network (DSN). S-, X-, and Ka-bands (in decreasing susceptibility to interplanetary plasma noise) are available on Cassini. The Italian Space Agency contributed Ka-band radios to Cassini and Juno.

Future state of the art for RS starts with New Horizons, which is pioneering use of the uplink occultation configuration. The traditional downlink configuration (i.e., transmission from spacecraft to Earth) does not provide the necessary SNR to meet science objectives. By reversing the transmission direction, an improvement in sensitivity of about a thousand times can be achieved in probing Pluto's tenuous atmosphere at the cost of adding a miniaturized signal processor to the spacecraft radio.

6A.7 The Future of Flight Instrumentation

Most radio occultation and surface scattering experiments have been conducted in the downlink configuration with data acquisition on the Earth where the received SNR is limited primarily by the spacecraft transmitter power (~20 W), the antenna gain, and the Earth-spacecraft distance. Acquiring the data onboard the spacecraft while transmitting from the DSN at ~20 kW improves the SNR by a factor of 1000 (some loss results from the spacecraft receiver noise being higher than noise in ground receivers), overcoming any SNR limitations of the experiments.

6A.7.1 Crosslink Radio Science

Use of spacecraft-to-spacecraft links has many potential science applications. Crosslinks provide vastly improved global coverage with high SNR for atmospheric structure investigations, as has been shown with low Earth-orbiting receivers and GPS transmitters. Instrumentation to conduct the basic observations is included on selected planetary spacecraft; proof-of-concept demonstration has been conducted at Mars using an ultra-high frequency (UHF) link from Odyssey to MRO. The demonstration illustrated the possible science return as well as the equipment limitations, enabling a better design for the next-generation science-based instrument for crosslinks.

6A.7.2 Ka-band Transponders and Other Instrumentation

Since Doppler measurements used for interior structure determinations are limited by noise from fluctuations in the interplanetary plasma electron density, Cassini, Juno, BepiColombo, and the proposed Jupiter satellite orbiters employ the less susceptible Ka-band wavelength for up- and downlinks as well as multiple links, when possible, to isolate this dispersive effect. Wider band at Ka-band significantly improves ranging accuracy. Efficient and lightweight transmitters benefit both the RS and telecommunications users, especially for mass-limited landers. Routine inclusion of USOs in mission telecommunication systems would also be of utility to science, navigation, and communications.

6A.8 Ground Instrumentation

6A.8.1 NASA's Deep Space Network

Ka-band uplink capabilities have been installed at one DSN station (designated DSS 25) specifically to support RS experiments; Ka-band downlink capabilities exist at all complexes. This RS upgrade included advanced systems for pointing at Ka-band and media calibration, especially of the Earth troposphere with advanced water-vapor radiometers. Additional infrastructure improvements throughout the DSN, such as improved frequency and timing, general media calibrations, antenna pointing, antenna vibration reduction, and microwave amplitude and phase stability, also benefit RS.

A key component for RS DSN observations is the radio science receiver, which was developed (in predecessor design) to preserve the characteristics of the enhanced downlink signals from Voyager. It is a specially designed ultra-stable multichannel open-loop receiver independent of the tracking receivers, comprised a dedicated full-spectrum processor tuned by a prediction file. The current digital generation receives a down-converted signal in a preselected bandwidth appropriate to the specific RS experiment. The multi-mission radio science receiver is operated remotely from JPL by science teams and their staff for optimal configuration as well as quick reaction to changes; this increases the scope of RS activities, allows for improved feedback when needed, and lowers operational cost.

Precision distance measurements with ranging data are fundamental radio tracking observables. A concept for an advanced ranging instrument that utilizes wider bandwidths at Ka-band and open-loop receivers will, when implemented, improve the ranging resolution by an order of magnitude for science as well as navigation applications.

6A.8.2 Other Facilities

Science as well as missions have benefited from using radio telescopes external to the DSN, especially when the spacecraft transmit at non-DSN frequencies. The Greenbank Telescope, Usuda (Japan), Noto (Italy), Parkes and Narrabri (Australia), and other facilities often support RS activities, and a portable DSN radio science receiver has been developed for this purpose. It is anticipated that the new large Sardinia (Italy) radio telescope will be instrumented for full RS support of future missions. The Square-Kilometer Array (SKA) precursor arrays in Australia and South Africa offer expanded capability, and by ~2017, the SKA itself should be the most sensitive telescope in the world, eventually reaching the sensitivity of ~100 DSN 70-m antennas in ~2022, opening entirely new opportunities for RS and mission design (low power probes, probe telemetry direct to Earth).

6A.9 New Communications Architectures: Arrays and Optical Links

RS investigations are closely linked to communications architectures. As the latter advance, RS opportunities arise. Two future relevant architectures are arrays of ground stations and optical links. The future DSN would likely comprise arrays of many antennas, individually smaller than the 70-m antennas in use today but with an effective SNR comparable to, or better than, the existing large antennas. The new antennas would need to be equipped for multiple frequencies and dual polarization; a capability present today only at the 70-m antennas. In the array era, RS investigators would utilize the new facilities. As optical communications come into use on deep space missions, comparable link science would be conducted to investigate the same phenomena. The RS community is investigating the limitations and the opportunities for science using optical links.

6A.10 Conclusion and Goals

The future of RS will witness a leap in technical capabilities and subsequent scientific opportunities and discoveries. RS experiments flying on solar system missions will provide advances not only in planetary science but also in solar physics and relativity, often with the same instrument package. Design of instrumentation (e.g., coherent Ka-band transponders for Doppler and ranging along with compatible ground systems, on-board open-loop receivers and processors), advances in techniques (e.g., uplink occultations, spacecraft-to-spacecraft links), and improved calibration methods (e.g., media calibration, antenna mechanical noise reduction, etc.) will enable an order of magnitude improvement in data quality for the benefit of future missions. The RS community aims to carry the existing strong science capabilities into the next generation of flight and ground communications architectures, multi-mission low-cost infrastructures, and instrumentation with continued technical improvements.

Specific technical goals of the RS user community for the next decade include:

- Ensuring a proper replacement for the 70-m stations that meet RS requirements.
- Expanding Ka-band uplink capability to the other DSN complexes with accompanying pointing and media calibration systems in order to increase the sensitivity and experiment coverage.
- Increasing occultation experiment sensitivity by expanding the capability for uplink RS experiments using highly stable high-power transmitters, especially in the array era.

- Utilizing dual or multifrequency links on future missions in order to calibrate or measure the charged particle effects for propagation experiments.
- Adding dual polarization receiving systems to the 34-m stations.
- Developing advanced ultra-stable oscillators, transponders, on-board open-loop receivers and signal processors, and the technology for quiet and precisely pointed spacecraft.
- Developing spacecraft-to-spacecraft links for planetary orbiters, probes, and relay satellites.
- Placing spacecraft with RS instrumentation in permanent solar occultation orbits for solar and relativity investigations.
- Developing an advanced ranging instrument for dual X- and Ka-band wide-band coherent links utilizing open-loop ground receivers.
- Developing the next-generation DSN open-loop radio science receiver, including portable receivers for support at radio telescopes.
- Studying the performance of RS investigations for arrays and links in the optical spectrum.

Appendix 6B

The National Academies Planetary Science Decadal Survey, Radio Science Contribution

Solar System Interiors, Atmospheres, and Surfaces Investigations via Radio Links: Goals for the Next Decade

S. W. Asmar[1], R. A. Preston[1], P. Vergados[1], D. H. Atkinson[1], T. Andert[2], H. Ando[3], C.O. Ao[1], J. W. Armstrong[1], N. Ashby[4], J.-P. Barriot[5], P. M. Beauchamp[1], D. J. Bell[1], P. L. Bender[6], M. Di Benedetto[7], B. G. Bills[1], M. K. Bird[8], T. M. Bocanegra-Bahamon[1], G. K. Botteon[1], S. Bruinsma[9], D. R. Buccino[1], K. L. Cahoy[10], P. Cappuccio[7], R. K. Choudhary[11], V. Dehant[12], C. Dumoulin[13], D. Durante[7], C. D. Edwards[1], H. M. Elliott[1], T. A. Ely[1], A. I. Ermakov[14], F. Ferri, F. M. Flasar[16], R. G. French[17], A. Genova[7], S. J. Goossens[16,18], B. Häusler[2], R. Helled[19], D. P. Hinson[20], M. D. Hofstadter[1], L. Iess[7], T. Imamura[21], A.P. Jongeling[1], Ö. Karatekin[12], Y. Kaspi[22], M. M. Kobayashi[1], A. Komjathy[1], A. S. Konopliv [1], E. R. Kursinski[23], T. J. W. Lazio[1], S. Le Maistre[12], F. G. Lemoine[16], R. J. Lillis[14], I. R. Linscott[24], A.J. Mannucci[1], E.A. Marouf[25], J.-C. Marty[9], S.E. Matousek[1], K. Matsumoto[26], E. M. Mazarico[16], V. Notaro[7], M. Parisi[1], R. S. Park[1], M. Pätzold[8], G. G. Peytavi[2], M. P. Pugh[1], N. O. Rennó, P. Rosenblatt[13], D. Serra[28], R. A. Simpson[23], D. E. Smith[10], P. G. Steffes[29], B. D. Tapley[30], S. Tellmann[8], P. Tortora[31], S. G. Turyshev[1], T. Van Hoolst[12], A. K. Verma[32], M. M. Watkins[1], W. Williamson[1], M. A. Wieczorek[33], P. Withers[34], M. Yseboodt[12], N. Yu[1], M. Zannoni[31], M. T. Zuber[10]

Author Institutions

1. Jet Propulsion Laboratory, California Institute of Technology
2. Universität der Bundeswehr München, Germany
3. Kyoto Sangyo University, Japan
4. National Institute of Standards & Technology
5. Observatoire Géodésique de Tahiti
6. University of Colorado
7. Università di Roma (Sapienza), Italy
8. Universität zu Köln, Germany
9. Centre National d'Etudes Spatiales, France
10. Massachusetts Institute of Technology
11. Space Physics Lab., VSSC, Trivandrum, India

---

1 This work was carried out in part at the Jet Propulsion Laboratory, California Institute of Technology, under contract with the National Aeronautics and Space Administration. © 2020 Jet Propulsion Laboratory.

12. Observatoire Royal de Belgique
13. Université de Nantes, France
14. University of California, Berkeley
15. Università degli Studi di Padova, Italy
16. NASA/Goddard Space Flight Center
17. Wellesley College
18. University of Maryland
19. Universität Zürich, Switzerland
20. SETI Institute
21. The University of Tokyo, Japan
22. Weizmann Institute of Science, Israel
23. PlanetiQ, Inc.
24. Stanford University
25. San Jose State University
26. National Astronomical Observatory of Japan
27. University of Michigan
28. Università di Pisa, Italy
29. Georgia Institute of Technology
30. The University of Texas at Austin
31. Università di Bologna, Italy
32. University of California, Los Angeles
33. Observatoire de la Côte d'Azur, France
34. Boston University

6B.1 Summary

From Mercury to the outer reaches of the solar system, the past six decades have witnessed a vast set of discoveries utilizing Radio Science (RS) methods. For example, based on key gravitational evidence, subsurface oceans have been inferred at Titan, Enceladus, and Europa, where potential future missions may search for life.

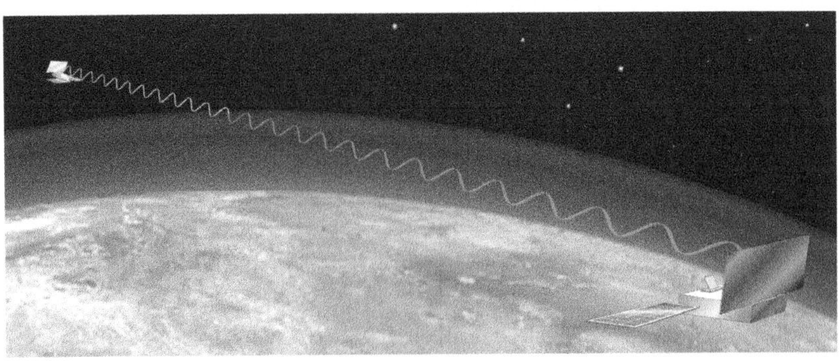

Illustration of two members of a constellation of small spacecraft at Venus with crosslinks for radio occultations that can lead to global coverage with high spatial and temporal resolutions.

The ability to precisely measure properties of spacecraft radio signals—frequency, phase, delay, amplitude, and polarization—provides unique leverage to extract new information about atmospheres, ionospheres, rings, surfaces, shapes, and the internal structure of solar system bodies (Asmar et al. 2019). In addition to planetary sciences, RS observables, such as precision Doppler and ranging, are critical to studies in fundamental physics and solar dynamics, such as observing effects on signal passage through the gravitational fields of the Sun and Jupiter, and monitoring planetary motion to study gravitational theory and solar mass loss (Armstrong et al. 2003; Genova et al. 2018; Smith et al. 2018).

RS remains a powerful and cost-effective tool for many investigations planned or conceived in the coming decade. Additional science discoveries could be enabled by developing new technologies and mission concepts such as:

- Deployment of small spacecraft missions for high spatial and temporal resolution of atmospheric and gravitational mapping (see illustration above),
- Novel instrumentation and calibration techniques to improve data quality by up to an order of magnitude over current levels, and
- Exploitation of uplink transmissions from Earth such as those used by New Horizons at Pluto, to improve sensitivity by orders of magnitude.

Recommendations for next decade: Scientific studies using spacecraft radio links have led to numerous discoveries. In the next decade, technological advances, and use of small spacecraft with increasing radio capabilities that approximate those of full-scale spacecraft, could enable many additional scientific breakthroughs in atmospheric dynamics, interior structures, and surface properties. Development of new techniques and technologies to reach these goals can enable all solar system missions to benefit from this enhanced low-cost science capability.

6B.2 Current Status of RS Investigations

To plan for the future, it's vital to understand RS current status. Major accomplishments include:

- **Interior structure:** Detailed gravity fields of Mercury, Venus, Mars, the Moon, Jupiter, and Saturn leading to refined models for their interior structures; fundamental insights into the interiors of Phobos, Eros, Vesta, Ceres, Lutetia, Bennu, cometary nuclei, Triton, and many Saturnian and Galilean satellites, leading to evidence for fluid layers in icy moons.
- **Atmospheric structure, dynamics, and composition:** Temperature–pressure profiles from Venus to Neptune, Pluto, Titan, and Triton; ionospheric electron densities; Jovian deep wind speeds; wind height profiles on Titan, Jupiter, and Venus; Venus temporal wind variations and gravity wave spectra, abundances of key molecules on Jupiter (ammonia) and Venus (sulfuric acid).
- **Planetary rings and tori:** Particle size distribution and internal dynamics of Saturn's rings and Jupiter's Io Plasma Torus; total ring masses measured with implications to age determination.

- **Surface properties:** Surface characteristics of Venus, the Moon, Mars, Titan, comet 67P/Churyumov-Gerasimenko, Pluto, Vesta, and Kuiper Belt object Arrokoth.

Ongoing RS experiments will continue in the next decade as Mars Atmosphere and Volatile EvolutioN (MAVEN) and Mars Express further investigate Mars' atmosphere, Akatsuki investigates Venus' atmosphere, Interior Exploration using Seismic Investigations, Geodesy and Heat Transport's (InSight) Rotation and Interior Structure Experiment (RISE) characterizes the Martian core, and Juno investigates Jupiter's interior structure, Jupiter's satellites, and the Io plasma torus. New investigations are also planned:

- ExoMars' Lander Radioscience (LaRa) will examine Mars' rotation and interior.
- ExoMars' Atmospheric Mars Entry & Landing Investigations & Analysis (AMELIA) will estimate wind velocity.
- Trace Gas Orbiter's (TGO) will investigate Mars' seasonal variations and CO_2 polar mass deposits.
- Trace Gas Orbiter and Mars Express will attempt crosslink atmospheric radio occultations (ROs).
- BepiColombo will investigate Mercury's interior structure.
- Europa Clipper will investigate the Jovian satellite's interior structure.
- Jupiter Icy Moons Explorer (JUICE) will study Galilean satellites' internal structure and ionospheres as well as Jupiter's atmosphere.
- Hera will measure Didymos' gravity via Juventas crosslink tracking.
- Psyche will investigate the interior structure of asteroid 16 Psyche.
- Lucy will determine the mass of five Trojan asteroids.
- Martian Moons eXploration (MMX) will investigate Phobos' interior structure.

6B.3 Key Science Goals for the Next Decade

Scientific goals to be pursued in the next decade include higher geographic and temporal resolution measurements of atmospheric and ionospheric structure from ROs, new opportunities for wind profiles from entry probes and balloons, more detailed interior structure from high-precision radiometric tracking, and increased exploration of surface electrical properties and roughness from scattering/bistatic radar. The emphasis is on improved measurement precision; use of small spacecraft individually, in pairs, or in constellations; and communication via crosslinks between spacecraft. Table 6B.1 lists selected science goals and enabling technology goals.

Atmospheres/Ionospheres: RS can obtain data key to scientific drivers centered on understanding the three-dimensional structure and variations of atmospheric composition and thermodynamics, the underlying processes that drive global dynamics such as thermal tides, the physical coupling of the surface-atmosphere system, including any ionospheric layer, and the global and local radiative balance and its relationship to atmospheric state and circulation.

RO refractivity measurements by small spacecraft constellations at Mars, Venus, and Titan potentially provide orders of magnitude improvements in the geographic

Table 6B.1 Radio science-based science and technology goals for the next decade.

Investigation	Representative science goals	Technology goals
Interiors	• Core and mantle size/density/composition • Interior structure models • Formation mechanism • Thermal history • Tidal and natural oscillations	• Spacecraft constellations crosslinks • Same antenna beam tracking • Advanced radiometric calibrations • Atomic clocks • Small spacecraft orbit insertion
Atmospheres, rings, and other media	• Temperature–pressure profiles • Neutral and ionospheric densities • Wave structures and origins • Shape models	• Spacecraft constellations crosslinks • Advanced software-defined radios • Miniaturized stable oscillators • Small spacecraft orbit insertion
Atmospheric winds	• Wind velocities	• Small spacecraft delivery/entry
Surfaces	• Electrical properties and roughness	• Uplink and crosslink instrumentation
Surface-atmosphere interactions	• Seasonal ice deposition on Mars • Geophysically induced waves	• Geographical and temporal resolution

and temporal resolution of the atmospheric vertical temperature, pressure, and density profiles (Mannucci et al. 1998; Lillis et al. 2020). Refined measurements of rotational dynamics improve knowledge of CO_2 seasonal movements. This can enhance research in Martian climate science by providing foundational data for global atmospheric modeling and meteorological information for atmospheric stability, geostrophic and isothermal winds, and geopotential height. Global observations of Titan's atmospheric structure would provide a leap in understanding its atmospheric dynamics. For Saturn, Uranus, and Neptune, detailed vertical variations of winds provide insights into the circulation, thermodynamics, and evolution of their atmospheres.

For Venus, detailed spatial and temporal characterization is possible for large-scale atmospheric wave structures, such as planetary waves, jet streams, and polar vortices. This improves our understanding of their formation, evolution, and role in global circulation, including spectral and structural view of lower atmosphere thermal tides, hypothesized to sustain super-rotation. By exploiting the strong coupling between the surface and dense atmosphere, long-term observations may allow detection of waves in the neutral or charged regions possibly generated by seismic events. RO absorptivity profiles could allow for the retrieval of chemical species of the sulfuric acid cycle for accurate chemical composition abundances over prolonged periods.

Interiors: Key drivers for planetary interiors center on modeling the density and composition, leading to an improved understanding of interior formation and evolution. RS provides key evidence toward these goals through high-resolution gravity fields and their time-variations due to tidal response and mass transport. Rotational dynamics provide additional clues about interiors such as core size via radio tracking of landed spacecraft or long-duration orbiters. Investigations of small-body interior structure through their gravity fields can reflect formation mechanisms.

Uranus and Neptune, whose interior structures are fundamentally different from Jupiter and Saturn, are potential targets of upcoming missions in the next decade. Searching for a possible liquid layer in Neptune's moon Triton is also a key RS goal. Icy giant gravity data via Doppler tracking of an orbiter, preferably with a secondary spacecraft (subsatellite), will be crucial to understanding their internal structure by significantly reducing the error bars on J_2 and J_4 and possibly detecting J_6. Current uncertainties don't allow unique compositional solutions (Bailey and Stevenson 2021; Helled et al. 2020). Hofstadter et al. (2017) noted interior structure as a future mission's highest priority. This understanding will provide clues to exoplanets' origin since Neptune-sized bodies are prevalent around other stars. Detecting giant planet normal modes when orbiters pass through their resonances offers a possible new RS approach to interior studies (Durante et al. 2017).

Surfaces: Signal reflections from planetary surfaces and near-surface regions reveal their structure and composition. Space-based reflectometry and scatterometry techniques are used to study Earth's surface roughness and dielectric constant. This is possible between a flagship mission and a small spacecraft or by a small constellation. Comparing the power between circularly polarized transmitted radio signals and oppositely polarized reflected signals provides information about icy moons' surface properties. Speculatively, mass flux could also be detected on Venus by a nadir-viewing spacecraft monitoring specific areas for changes between coherent and incoherent scattering that signify mass transport or surface terrain variations. Doppler delay maps from the differential travel time between transmitted and reflected signals could provide information on surface topography. Radio tracking of landed spacecraft can be used to study seasonal surface deposition (e.g., CO_2 on Mars) and finer-scale surface and near-subsurface properties.

6B.4 Radio Science Techniques for Achieving the Science Goals of the Next Decade

Spacecraft constellations for atmospheric structure and dynamics: Compared to spacecraft-to-Earth ROs, substantial increases in global coverage as well as spatial and temporal resolution are possible using crosslinks among a constellation of small spacecraft, similar to global positioning system (GPS)-based RO constellations at Earth. At Mars and Venus, for example, simulations show that near-global coverage can be accomplished in a couple of weeks with two spacecraft in optimized orbits as compared to many years via the traditional downlink to Earth. Such crosslinks (see crosslink demo carried out with Mars orbiters in Ao et al. 2015) can profile the atmosphere through clouds and aerosols and detect the signature of phenomena such as gravity waves, planetary waves, turbulence, and jet streams, and for Venus, possibly detect seismically induced waves, potentially opening a window into possible current tectonic activity.

Spacecraft constellations for interior structure and dynamics: Precision gravity experiments via spacecraft-to-spacecraft crosslinks have been used at Earth (Gravity Recovery and Climate Experiment (GRACE), Tapley et al. 2004) and the

Moon (Gravity Recovery and Interior Laboratory (GRAIL), Zuber et al. 2013b; Selenological and Engineering Explorer's (SELENE) farside subsatellite) for high-impact studies of planetary interiors and monitoring mass transport. Applying this concept to other rocky, icy, or gaseous bodies would enable rapid high-resolution global coverage and provide stringent constraints on the thickness of ice and the characteristics of subsurface oceans at icy moons.

Differential interferometry for planetary rotational state and tidal deformations: Tracking multiple landers or spacecraft from one or many Earth stations suppresses most common noise sources, enhancing the sensitivity to the desired science goals (e.g., gravitational field, rotational state, tidal deformations, and winds). The benefits of same-beam differential interferometry were recognized by Counselman et al. (1972) and advanced by Preston (1974), Edwards et al. (1992), Bender (1994) and others, and applied to examine librations (ALSEP) and gravity field (Selene) of the Moon, studying winds on Venus (Pioneer probes and Vega balloons), and observing Huygens' descent on Titan. New uses have been proposed for the Moon, Mars, and Jupiter.

Doppler wind experiments for measurement of atmospheric dynamics: Radio links from probes descending through planetary atmospheres provide information on atmospheric dynamics as demonstrated at Venus, Jupiter, and Titan. Links can be established with proximity spacecraft or ground stations, and, if multiple links are possible, recovery of complete wind vectors can be made. In some cases, ranging, very-long-baseline interferometry (VLBI) techniques, and onboard observations provide additional information. A constellation of small probes and/or balloons can provide a global assessment of atmospheric dynamics. Absorption on the probe's radio link can be used to infer the integrated abundance of ammonia (giant planets), sulfuric acid (Venus), or other absorbers. Similar techniques can be applied to atmospheric balloons, as was done on the Vega mission at Venus.

Scattering studies for surface properties: In a bistatic scattering experiment, a spacecraft downlink signal is reflected off a planetary surface and received on Earth or another spacecraft. Information can be deduced about the surface and near-subsurface electrical properties and roughness. Reversing the path via an uplink provides higher signal-to-noise ratio (SNR), as shown by New Horizons and the Lunar Reconnaissance Orbiter (LRO); multiple simultaneous uplinks with frequency offset can improve the SNR and resolve subsurface degeneracies. A constellation of small spacecraft makes global characterization feasible and yields information at scales useful to the safety of planned landers. Scattering from subsurface structures has added exploration opportunities where redundancies can be resolved using polarization, wavelength, and obliquity variations. Use of bistatic techniques with landed spacecraft can provide finer spatial resolution.

6B.5 Technology Development Needed in the Next Decade

The RS observational accuracy depends primarily on the phase and amplitude stability of the end-to-end system and the SNR of radio links. Furthermore, extracting high-quality science requires calibration of the intervening media and nongravitational forces, as well as accurate reconstruction of position, attitude, and motion of

Table 6B.2 Radio science technology methods and typical requirements for the next decade.

Investigation	Method	Requirements
Interiors	• Precision two-way Doppler link(s); one-way possible if atomic clock on board • Shorter wavelength less noisy (multiple wavelengths preferred)	• Allan deviation ~1 × 10–14 (1000 s) • Dynamical stable spacecraft • Calibration of intervening media and nongravitational forces
Atmospheres, rings, and other media	• Ultrastable links (two preferred) • One-way or dual one-way with an ultrastable oscillator (USO); two-way if no USO, limited to RO ingress	• Allan deviation ~1 × 10–13 (~100 s) • Phase noise ~ –90 dBc/Hz, 10 Hz offset • Amplitude stability ~0.1 dB at 60 s
Atmospheric winds	• Cross or one-way Doppler links with stable oscillators at both ends • Signal power if monitoring absorption	• Allan deviation ~1 × 10–11 (~100 s) • Amplitude stability: 0.1 dB at 60 s • Multiple links for better wind vectors
Surfaces	• One-way or crosslinks at good geometry • Signal power at dual polarizations • Multiple signals (for uplink case)	• Amplitude stability ~0.1 dB at 60 s • Allan deviation ~1 × 10–12 (~100 s) • Power measurement calibration
Fundamental physics	• Precision two-way Doppler link(s) • Shorter wavelength less noisy • Precision ranging	• Allan deviation ~1 × 10–14 (~1000 s) • Dynamical stability • Calibration of known noise sources
Solar Science	• Ultrastable link(s) • Dual links to isolate dispersive media • Ranging for coronal densities	• Allan deviation ~1 × 10^{-12} (~100 s) • Amplitude stability: 0.1 dB at 60 s • Dual polarization receivers

the transmitter-target-receiver system. Precision time and frequency standards are essential; multiple frequency links enable isolating dispersive media, water vapor radiometers enable calibrating Earth's tropospheric effects, and precise onboard accelerometers enable calibrations of nongravitational forces for optimized scientific results.

Table 6B.2 lists the technology methods needed to address challenges and meet the RS decadal goals, using general investigation cases, which will be refined as specific science objectives, planetary targets, and mission configurations are identified. The Allan deviation values represent fractional frequency instability for classes of clocks (ultra-stable oscillator (USO), for one-way mode or hydrogen maser for two-way mode). Dynamical stability is determined by thruster balance or timing of spacecraft maneuver or solar electric propulsion events that disturb the Doppler data quality, and precision power calibration is required for comparing small variations in the two polarizations of received signals.

To meet the new RS goals for the next decade, we have identified essential technology developments needed in radios for small spacecraft, stable oscillators, data calibration, and atomic clocks. Insertion of small spacecraft into planetary orbits is also a key technology area for future missions along with advances in optimizing all spacecraft orbits and attitude stability for precision measurements. Ground station resources for constellation missions are being considered by the deep space network (DSN). Finally, preparation for the coming age of optical links is discussed later in this section.

Small spacecraft enabled for radio science: The highest technology priority is the maturation of multiband crosslinks between robust small spacecraft in suitable planetary orbits. This requires advances in miniaturized radios, stable oscillators, and orbit insertion. A critical need is the development of software-defined radios (SDRs) suitable for small spacecraft. The Iris SDR used for the Mars Cube One (MarCO) mission (Asmar and Matousek 2016) provided experience for further advances toward science-quality radiometric and functional capabilities (e.g., simultaneous open-loop recordings at multiple wavelengths and agility to swap the transmit/receive frequencies). USOs have been critical instrumentation for RO experiments since the Voyager mission. Small spacecraft cannot accommodate the mass and power of the current top-of-the-line flight USOs (Allan deviation ~2×10^{-13} at ~100 s), such as those flown on Mars Global Surveyor (MGS), Cassini, GRACE/GRACE Follow-On, GRAIL, New Horizons, Venus Express (VEX), Rosetta, Akatsuki, and LRO. The goal is to mature smaller units with comparable stability based on newer technologies.

Advanced calibrations at the Deep Space Network: The most accurate RS experiments to date were achieved during Cassini's gravitational wave observations; further improvement is possible. A leading source of radiometric noise is Earth's troposphere, which is partly calibrated via water vapor radiometers. An order of magnitude precision improvement is possible by pointing along the same path as the signal arriving at the antenna, rather than being nearby. The next noise source is unmodeled mechanical motion of the ground antenna's phase center. A method (Armstrong et al. 2008) for suppressing this intrinsic noise is to combine the two-way data from the principal antenna with Doppler data received at a smaller stiffer antenna; early testing by Notaro et al. (2020) has shown promise. Another concept involves measuring the antenna surface motion with laser or microwave beams and combining those data with a mechanical antenna model in post-processing.

Enhanced planetary gravity via atomic clocks: Space-borne atomic clocks would revolutionize spacecraft tracking methods (Ely et al. 2014). Future one-way or crosslink observations enabled by these onboard clocks, as they get further improved in stability and reduced in size, could achieve performance comparable to two-way coherent links referenced to ground station atomic clocks. This could enable nearly continuous tracking using smaller ground stations, enhancing the science at lower cost. A miniaturized prototype Mercury-trapped-ion atomic clock (DSAC-1) with Allan deviation of ~2×10^{-14} (1000 s) is being tested in Earth orbit and DSAC-2 is offered to Discovery missions as a technology infusion. An even smaller version with comparable stability is under development.

Small spacecraft orbit insertion: A big challenge in planning small spacecraft missions is the method of insertion into orbits around solar system bodies. Small spacecraft could be carried to orbit by larger orbiters if suitable attachment systems were available. Placing conventional propulsion systems onto small spacecraft has thus far not been practical and more suitable systems are needed. This challenge needs to be solved through a dedicated technology or alternative insertion techniques in order to achieve the full potential of small spacecraft for RS experiments. Aerocapture and aerobraking technologies for small spacecraft can enable them to achieve orbit of some bodies, as discussed in Austin et al. (2020).

Enhanced science via optical links: Although the previous discussion has addressed the science possible with radio links, optical links are expected to be demonstrated on future deep space missions. Laser links can provide measurements in most areas already pioneered by RS but, in many cases, with new scientific capabilities and precision, occultations could better probe upper atmospheres and the Doppler measurements could provide improved gravity fields measurements. In the next decade, it is important to lay the technological framework for future optical link science.

References

Abbate, S.F., Armstrong, J.W., Asmar, S.W., Barbinis, E., Bertotti, B., Fleischman, D., Gatti, M., Goltz, G.L., Herrera, R.G., Iess, L., Lee, K.J., Ray, T.L., Tinto, M., Tortora, P., and Wahlquist, H. (2003). The Cassini gravitational wave experiment. In: Gravitational Wave Detection, (eds. M. Cruise and P. Saulson), 4856: 90–97. *Proceedings of the International Society for Optical Engineering (SPIE).*

Abbott, B.P., the LIGO Scientific Collaboration and Virgo Collaboration. (2016). Observation of gravitational waves for a binary black hole merger. *Physical Review Letters* 116 (6): 061102.

Acton, C.H. (1996). Ancillary data services of NASA's navigation and ancillary information facility. *Planetary and Space Science* 44 (1): 65–70.

Allan, D.W. (1966). Statistics of atomic frequency standards. *Proceedings of the IEEE* 54 (2): 221-230.

Allan, D.W. (2016). Historicity, strengths, and weaknesses of Allan variances and their general applications. *Gyroscopy and Navigation* 7 (1): 1–17.

Allen, C.W. (1947). Interpretation of electron densities from corona brightness. *Monthly Notices of the Royal Astronomical Society* 107 (5–6): 426–432.

Aleksandrov, Y.N., Alexandrov, Y.N., Vasilyev, M.B., Vyshlov, A.S., Dubrovin, V.M., Zaitsev, A.L., Kolosov, M.A., Savich, N.A., Samovol, V.A., Samoznaev, L.N., and Sidorenko, A.I. (1977). Some results of dual-frequency radio occultation exploration of the night-time ionosphere of Venus with satellites Venera 9 and 10. *Space Research* XVII: 663–666.

Ambrosini, R., Asmar, S.W., Bolli, P., and Flamini, E. (2011). The planned space science utilizations of the new Sardinia 64-m radio telescope. *Proceedings of the IEEE* 99 (5): 875–880. doi:10.1109/JPROC.2010.2063690.

Anderson, A.J. (1971). Probability of long period gravitational radiation. *Nature* 229: 547–548.

Anderson, J.D. (1974). Geodetic and dynamical properties of planets. *Eos, Transactions American Geophysical Union* 55 (5): 515–523.

Anderson, J.D., Armstrong, J.W., Campbell, J.K., Estabrook, F.B., Krisher, T.P., and Lau, E.L. (1992). Gravitation and celestial mechanics investigations with Galileo. *Space Science Reviews* 60 (1–4): 591–610.

Anderson, J.D., Armstrong, J.W., Estabrook, F.B., Hellings, R.W., Lau, E.K., and Wahlquist, H.D. (1984). Pioneer 10 search for gravitational waves – no evidence for coherent radiation from Geminga. *Nature* 308 (5955): 158–160.

Radio Science Techniques for Deep Space Exploration, First Edition. Sami W. Asmar.

Anderson, J.D., Campbell, J.K., Ekelund, J.E., Ellis, J. and Jordan, J.F. (2008). Anomalous orbital-energy changes observed during spacecraft flybys of Earth. *Physical Review Letters* 100 (9): 091102.

Anderson, J.D., Campbell, J.K., Jacobson, R.A., Sweetnam, D.N., Taylor, A.H., Prentice, A.J.R., and Tyler, G.L. (1987a). Radio science with Voyager 2 at Uranus: Results on masses and densities of the planet and five principal satellites. *Journal of Geophysical Research: Space Physics* 92 (A13): 14877–14883.

Anderson, J.D., Colombo, G., Esposito, P.B., Lau, E.L., and Trager, G.B. (1987b). The mass, gravity field, and ephemeris of Mercury. *Icarus* 71 (3): 337–349.

Anderson, J.D., Jacobson, R.A., Lau, E.L., Moore, W.B., Olsen, O., Schubert, G., and Thomas, P.C. (2001a). Shape, mean radius, gravity field and interior structure of Ganymede. *Bulletin of the American Astronomical Society* 33: 1101.

Anderson, J.D., Jacobson, R.A., Lau, E.L., Moore, W.B., and Schubert, G. (2001b). Io's gravity field and interior structure. *Journal of Geophysical Research: Planets* 106 (E12): 32963–32969.

Anderson, J.D., Jacobson, R.A., McElrath, T.P., Moore, W.B., Schubert, G., and Thomas, P.C. (2001c). Shape, mean radius, gravity field, and interior structure of Callisto. *Icarus* 153 (1): 157–161.

Anderson, J.D., Johnson, T.V., Schubert, G., Asmar, S.W., Jacobson, R.A., Johnston, D., Lau, E.L., Lewis, G., Moore, W., Taylor, A.H., Thomas, P.C., and Weinwurm, G. (2005). Amalthea's density is less than that of water. *Science* 308 (5726): 1291–1293. doi:10.1126/science.1110422.

Anderson, J.D., Krisher, T.P., Borutzki, S.E., Connally, M.J., Eshe, P.M., Hotz, H.B., Kinslow, S., Kursinski, E.R., Light, L.B., Matousek, S.E., and Moyd, K.I. (1987c). Radio range measurements of coronal electron densities at 13 and 3.6 centimeter wavelengths during the 1985 solar conjunction of Voyager 2. *The Astrophysical Journal* 323: L141–143.

Anderson, J.D., Laing, P.A., Lau, E.L., Liu, A.S., Nieto, M.M., and Turyshev, S.G. (2002). Study of the anomalous acceleration of Pioneer 10 and 11. *Physical Review D* 65 (8): 082004.

Anderson, J.D., Lau, E., Asmar, S.W., Bird, M.K., Clark, B.C., Giampieri, G., Gilliand, K., and Pätzold, M. (2004a). Stardust dynamic science at comet 81P/Wild 2. *Journal of Geophysical Research: Planets* 109: E12.

Anderson, J.D., Lau, E.L., Sjogren, W.L., Schubert, G., and Moore, W.B. (1996). Gravitational constraints on the internal structure of Ganymede. *Nature* 384 (6609): 541–543.

Anderson, J.D. and Mashoon, B. (1985). Pioneer 10 search for gravitational waves – limits on a possible isotropic cosmic background in the microhertz region. *Astrophysical Journal* 290: 445–448.

Anderson, J.D., Null, G.W., and Wong, S.K. (1974). Gravity results from Pioneer 10 Doppler data. *Journal of Geophysical Research* 79 (25): 3661–3664.

Anderson, J.D. and Schubert, G. (2007). Saturn's satellite Rhea is a homogeneous mix of rock and ice. *Geophysical Research Letters* 34 (2).

Anderson, J.D. and Schubert, G. (2010). Rhea's gravitational field and interior structure inferred from archival data files of the 2005 Cassini flyby. *Physics of the Earth and Planetary Interiors* 178 (3): 176–182.

Anderson, J.D., Schubert, G., Asmar, S.W., Jurgens, R.F., Lau, E.L., Moore, W.B., Slade, M.A., and Standish, E.M. (2001d). Mercury's global topography from radar ranging data. In: Workshop on Mercury: Space Environment, Surface, and Interior (ed. Mark Robinbson, G. Jeffrey Taylor), Chicago, IL: The Field Museum. (October 2001).

Anderson, J.D., Schubert, G., Jacobson, R.A., Lau, E.L., Moore, W.B., and Palguta, J.L. (2004b). Discovery of mass anomalies on Ganymede. *Science* 305 (5686): 989–991.

Anderson, J.D., Schubert, G., Jacobson, R.A., Lau, E.L., Moore, W.B., and Sjogren, W.L. (1998). Europa's differentiated internal structure: inferences from four Galileo encounters. *Science* 281 (5385): 2019–2022.

Anderson, J.D., Turyshev, S.G., Asmar, S.W., Bird, M.K., Konopliv, A.S., Krisher, T.P., Lau, E.L., Schubert, G., and Sjogren, W.L. (1997). Radio science investigations on a Mercury orbiter mission. *Planetary and Space Science* 45 (1): 21–29.

Anderson, J.D. and Warner, M.R. (1966). J. Kovalevsky Determination of the masses of the Moon and Venus, and the astronomical unit from the radio tracking of Mariner II. In: Trajectories of Artificial Celestial Bodies as Determined from Observations/Trajectoires Des Corps Celestes Artificiels Déterminées D'après Les Observations (ed. J. Kovalevsky), 216–246. Berlin: Springer.

Anderson, J.D., Campbell, J.K., Ekelund, J.E., Ellis, J. and Jordan, J.F. (2008). Anomalous orbital-energy changes observed during spacecraft flybys of Earth. *Physical Review Letters* 100 (9): 091102.

Andreev, V.E., Efimov, A.I., Samoznaev, L.N., Chashei, I.V., and Bird, M.K. (1997). Characteristics of coronal Alfvén waves deduced from Helios Faraday rotation measurements. *Solar Physics* 176 (2): 387–402.

Andreev, V.E. and Gavrik, A.L. (1993). Radio transmission measurements of the plasma shell of comet Halley. *Astronomy Letters* 19: 437–440.

Andrews-Hanna, J.C., Asmar, S.W., Head III, J.W., Wieczorek, M., Kiefer, W.S., Taylor, G.J., Konopliv, A.S., Lemoine, F.G., Mazarico, E.M., McGovern, P.J., Melosh, H.J., Neumann, G.A., Phillips, R.J., Smith, D.E., Solomon, S.C., Williams, J.G., and Zuber, M.T. (2013). Ancient igneous intrusions and the early expansion of the Moon revealed by GRAIL gravity gradiometry. *Science* 339 (6120): 675–678. doi:10.1126/science.1231753.

Ao, C.O., Edwards, C.D., Kahan, D.S., Pi, X., Asmar, S.W., and Mannucci, A.J. (2015). A first demonstration of Mars crosslink occultation measurements. *Radio Science* 50 (10): 997–1007.

Aoki, S., Guinot, B., Kaplan, G.K., Kinoshita, H., McCarthy, D.D., and Seidelmann, P.K. (1982). The new definition of Universal Time. *Astronomy and Astrophysics* 105: 359–361.

Aoki, S. and Kinoshita, H. (1983). Note on the relation between the equinox and Guinot's non-rotating origin. *Celestial Mechanics* 29: 335–360.

Argus, D.F. and Gordon, R.G. (1991). No-net-rotation model of current plate velocities incorporating plate motion model NUVEL-1. *Geophysics Research Letters* 18: 2039–2042.

Armstrong, J.W. (1993). Bispectral analysis of spacecraft radio scintillation. *Waves in Random Media* 3 (2): 63–70.

Armstrong, J.W. (2016). Low-frequency gravitational wave searches using spacecraft Doppler tracking. *Living Reviews in Relativity* 9 (1): 1–59. (updated 11 January 2016) http://www.livingreviews.org/lrr-2006-1.

Armstrong, J.W. and Coles, J.W. (1972). Analysis of three-station interplanetary scintillation. *Journal of Geophysical Research* 77: 4602–4610.

Armstrong, J.W., Coles, W.A., Kojima, M., and Rickett, B.J. (1986). Solar wind observations near the Sun. In: The Sun and Heliosphere in Three Dimensions, (ed. R.G. Marsden), 59–64. Dordrecht: Springer.

Armstrong, J.W., Coles, W.A., and Rickett, B.J. (2000). Radiowave scattering in the outer heliosphere. *Journal of Geophysical Research: Space Physics* 105 (A3): 5149–5156.

Armstrong, J.W., Estabrook, F.B., Asmar, S.W., Iess, L., and Tortora, P. (2008). Reducing antenna mechanical noise in precision spacecraft tracking. *Radio Science* 43 (3): 1–7. doi:10.1029/2007RS003766.

Armstrong, J.W., Estabrook, F.B., and Wahlquist, H.D. (1987). A search for sinusoidal gravitational radiation in the period range 30–2000 seconds. *Astrophysical Journal* 318: 536–541.

Armstrong, J.W., Iess, L., Tortora, P., and Bertotti, B. (2003). Stochastic gravitational wave background: upper limits in the 10^{-6}–10^{-3} Hz band. *Astrophysical Journal* 599 (2): 806–813.

Armstrong, J.W. and Sramek, R.A. (1982). Observations of tropospheric phase scintillations at 5 GHz on vertical paths. *Radio Science* 17 (6): 1579–1586.

Armstrong, J.W. and Woo, R. (1980). Starprobe Scintillation Study. Interoffice Memorandum 3331-80-070, Jet Propulsion Laboratory, Pasadena, CA: California Institute of Technology. 15 December 1980

Armstrong, J.W. and Woo, R. (1981). Solar wind motion and structure within 30 RO: spacecraft radio scintillation observations. *Astronomy and Astrophysics* 103: 415–421.

Aschwanden, M.J., Poland, A.I., and Rabin, D.M. (2001). The new solar corona. *Annual Review of Astronomy and Astrophysics* 39 (1): 175–210.

Ashby, N., Bender, P.L., and Wahr, J.M. (2007). Future gravitational physics tests from ranging to the BepiColombo Mercury planetary orbiter. *Physical Review D* 75 (2): 022001.

Asmar, S.W. (1996a). Through the solar wind, the DSN measures electron density outside the ecliptic plane. *DSN Science and Technology Newsletter* (6): (September).

Asmar, S.W. (1996b). Trends in performance and characteristics of quartz ultra-stable oscillators for radio science experiments on deep space missions. *Proceedings of Workshop on Scientific Applications of Clocks in Space*, Pasadena, CA: (November 1996).

Asmar, S.W. (1997). Characteristic trends of ultra-stable oscillators for radio science experiments. The Telecommunications and Data Acquisition Progress Report 42-129, Jet Propulsion Laboratory, Pasadena, CA: California Institute of Technology. 15 May 1997.

Asmar, S.W. (2005). Eavesdropping on Huygens. *The Planetary Report* (March/April 2005).

Asmar, S.W. (2010). Radio as a science tool. *Proceedings of the IEEE* 10: 1686–1687. doi:10.1109/JPROC.2010.2060247.

Asmar, S.W. (2011b). Spacecraft-to-spacecraft radio links instrumentation for planetary gravity, atmospheric and surface sciences. *8th International Planetary Probe Workshop*, Norfolk, VA (June 2011).

Asmar, S.W. (2012a). Communicating by Doppler detecting spacecraft dynamics during critical maneuvers. *12th SpaceOps Conference*, Stockholm, Sweden (June 2012).

Asmar, S.W. (2012b). Ultra-stable oscillators for probe radio science investigations. *International Planetary Probe Workshop*, Toulouse, France (June 2012).

Asmar, S.W. (2015a). Mars Relay CubeSat. *NASA Tech Briefs Magazine* (September 2015).

Asmar, S.W., Anderson, J.D., Castillo, J.C., Folkner, W.M., Konopliv, A.S., Marouf, E.A., Rappaport, N.J., Schubert, G., Spilker, T.R., Tyler, G.L., Watkins, M.M., and Yoder, C.F. (2003b). Radio science concepts for exploring the interior structures of Jupiter's icy moons. P12A-1054, *American Geophysical Union*, San Francisco, CA: (December 2003).

Asmar, S.W., Anderson, J.D., and Turychev, S. (2005a). Deep Space Network radio science performance for precision Doppler. *The Pioneer Anomaly Collaboration*, the International Space Science Institute, Bern, Switzerland (November 2005).

Asmar, S.W., Ao, C., Edwards, C., Kahan, D.S., Pi, X., Paik, M., and Mannucci, A.J. (2016). Demonstration of Mars crosslink occultation measurements for future small spacecraft constellations. In *IEEE Aerospace Conference*, pp. 1–6. (March 2016).

Asmar, S.W., Armstrong, J.W., Ashby, N., Bender, P., Bertotti, B., Folkner, W.M., Iess, L., Milani, A., Preston, R.A., Tortora, P., Turyshev, S.G., Williams, J.G., and Wu, X. (2009b). Radio science techniques for solar system tests of general relativity. *A White Paper submitted to the Fundamental Physical Sciences Panel of the 2009 Decadal Survey*.

Asmar, S.W., Armstrong, J.W., Iess, L., Rappaport, N.J., Somenzi, L., Tortora, P., and Zingoni, F. (2006a). The gravity fields and interior structure of the Saturnian satellites. *American Geophysical Union Fall*.

Asmar, S.W., Armstrong, J.W., Iess, L., and Tortora, P. (2005b). Spacecraft Doppler tracking: noise budget and achievable accuracy in precision radio science observations. *Radio Science* 40 (2): RS2001. doi:10.1029/2004RS003101.

Asmar, S.W., Armstrong, J.W., Iess, L., Tortora, P., and Bertotti, B. (2004a). Precision of radio science instrumentation for planetary exploration. ESA 3rd Tracking, Telemetry & Command Systems for Space Applications Workshop, Darmstadt, Germany: European Space Agency (September 2004).

Asmar, S.W. and Bird, M.K. (1996). Coronal faraday rotation observed with the Deep Space Network and the Magellan spacecraft. *American Geophysical Union Western Pacific Geophysics Meeting*, Brisbane, Australia: (July 1996).

Asmar, S.W., Bolton, S.J., Buccino, D.R., Cornish, T.P., Folkner, W.M., Formaro, R., Iess, L., Jongeling, L.A.P., Lewis, D.K., Mittskus, A.P., Mukai, R., and Simone, L. (2017b). The Juno gravity science instrument. *Space Science Reviews* 213 (1–4): 205–218.

Asmar, S.W., Divsalar, D., Oudrhiri, K., and Hamkins, J. (2013a). Radio science measurements with suppressed carrier. *2013 IEEE Aerospace Conference*, pp. 1–9. Big Sky, MT (2–9 March 2013).

Asmar, S.W., Folkner, W.M., Iess, L., and Tortora, P. (2009c). Galilean satellites gravity investigations and interior structure with future missions. *American Geophysical Union Fall* 2009: P51E-1172.

Asmar, S.W., Folkner, W.M., Johnston, D., Preston, R.A., Bird, M.K., Dutta-Roy, R., Dzierma, Y., Allison, M., Atkinson, D.H., Malecha, J., Edenhofer, P., Plettemeier, D., Tyler, G.L., Gurvits, L., Avruch, I., and Pogrebenko, S. (2005c). Initial results of the Huygens Doppler wind experiment. Third International Planetary Probe Workshop, Anavyssos (ed. BJ Bienstock), Greece: (June 2005).

Asmar, S.W., French, R.G., Marouf, E.A., Schinder, P., Armstrong, J.W., Tortora, P., Iess, L., Anabtawi, A., Kliore, A.J., Parisi, M., Zannoni, M., and Kahan, D. (2014b). *Cassini Radio Science User's Guide*. https://pdsatmospheres.nmsu.edu/data_and_services/atmospheres_data/JUNO/logs/Cassini Radio Science Users Guide - 3 may 2014.pdf (accessed 4 September 2014).

Asmar, S.W. and Herrera, R.G. (1995). Radio Science Handbook, Rev. 6, JPL Document D-7938, Jet Propulsion Laboratory, California Institute of Technology, Pasadena, CA.

Asmar, S.W., Iess, L., Folkner, W.M., Simone, L., and Bolton, S.J. (2014c). Advanced radio science instrumentation for the Juno, BepiColombo, and JUICE missions. #1130. *International Workshop on Instrumentation for Planetary Missions*, Greenbelt, MD: (4–7 November 2014).

Asmar, S.W. and Johnston, D. (2005). Open-loop tracking during critical mission phases. National Radio Science Meeting (URSI), Boulder, CO: (January 2005).

Asmar, S.W., Konopliv, A.S., Bills, B.G., Park, R.S., Raymond, C.A., Russell, C.T., Smith, D.E., and Zuber, M.T. (2009d). Simulations of Dawn gravity for Vesta and Ceres and implications for interior structure. *Division for Planetary Sciences Meeting Abstracts #41* (50-01).

Asmar, S.W., Konopliv, A.S., Park, R.S., Raymond, C.A., Bills, B.G., Gaskell, R., Russell, C.T., Smith, D.E., Toplis, M., and Zuber, M.T. (2012b). The gravity field of Vesta from Dawn. *European Planetary Science Conference*, Madrid, Spain.

Asmar, S.W., Konopliv, A.S., Watkins, M.M., Williams, J.G., Park, R.S., Kruizinga, G., Paik, M., Yuan, D.-N., Fahnestock, E.G., Strekalov, D., Harvey, N.E., Lu, W., Kahan, D.S., Oudrhiri, K., Smith, D.E., and Zuber, M.T. (2013b). The scientific measurement system of the Gravity Recovery and Interior Laboratory (GRAIL) mission. *Space Science Reviews* 178: 25–55. doi:10.1007/s11214-013-9962-0.

Asmar, S.W. and Kursinski, E.R. (1991). The role of clocks in operating deep space missions. *Proceedings of the 23rd Annual Precise Time and Time Interval Applications and Planning Meeting*, Pasadena, CA: (December 1991).

Asmar, S.W., Lazio, J., Atkinson, D.H., Bell, D.J., Border, J.S., Grudinin, I.S., Mannucci, A.J., Paik, M., and Preston, R.A. (2019). Future of planetary atmospheric, surface, and interior science using radio and laser links. *Radio Science* 54 (4): 365–377.

Asmar, S.W., Mannucci, A.J., Ao, C.O., Kobayashi, M.M., Lazio, J., Marinan, A.D., Preston, R.A., and Williamson, W. (2017d). Small spacecraft constellation concept for Mars atmospheric radio occultations. American Geophysical Union, New Orleans, LA.

Asmar, S.W. and Matousek, S.E. (2016). Mars Cube One (MarCO) shifting the paradigm in relay deep space operation. *14th International Conference on Space Operations*, p. 2483.

Asmar, S.W., Oudrhiri, K., Kahan, D.S., Esterhuizen, S., Sarkissian, J., Jackson, S., Preisig, B., Karatekin, Ö., and Griebel, H. (2013c). Curiosity's landing dynamics as observed at the CSIRO Parkes Radio Telescope. International Planetary Probe Workshop, San Jose, CA: (June 2013).

Asmar, S.W., Oudrhiri, K., Kurtik, S., and Weinstein-Weiss, S. (2014d). Communications during critical operations: preparing for InSight's landing on Mars. *13th SpaceOps Conference*, Pasadena, CA: (May 2014).

Asmar, S.W., Oudrhiri, K., and Weaver, G. (2013d). Ultra-stable oscillator reference spacecraft-to-spacecraft links for planetary exploration. *IEEE-UFFC Joint Symposia*, Prague, Czech Republic: (July 2013).

Asmar, S.W., Park, R.S., Bills, B.G., Gaskell, R., Konopliv, A.S., Raymond, C.A., Smith, D.E., Zuber, M.T., and Russell, C.T. (2013e). The gravity and rotation models of Vesta from the Dawn mission. European Geosciences Union General Assembly 2013, 13289. Vienna, Austria.

Asmar, S.W. and Renzetti, N.A. (1993). The Deep Space Network as an Instrument for Radio Science Research. Pasadena: JPL Publication, JPL document (ed. N. Renzetti), 80–93. Jet Propulsion Laboratory, California Institute of Technology.

Asmar, S.W. and Resch, R. (2011). Instrumentation of large radio telescopes for radio science investigations. Sardinia Radio Telescope: The Impact of Large Antennas on Radio Astronomy and Space Science (ed. N. D'Amico), Cagliari, Italy: (November 2001).

Asmar, S.W. and Schubert, G. (2002a). Admittance computations from lunar gravity and topography data. *The Moon Beyond 2002: Next Steps in Lunar Science and Exploration Conference*, Taos, NM: (12–14 September 2002).

Asmar, S.W. and Schubert, G. (2006). South pole Aitken compensation mechanisms based on gravity and topography data. 36th Committee on Space Research (COSPAR) Scientific Assembly, 3279. Beijing, China: (July 2006).

Asmar, S.W., Schubert, G., Konopliv, A.S., and Moore, W. (1999b). The lunar crustal thickness from analysis of the lunar prospector Gravity and Clementine topography datasets. New Views of the Moon II: Understanding the Moon through the Integration of Diverse Datasets, Flagstaff, AZ: (September 1999).

Asmar, S.W., Schubert, G., and Nimmo, F. (2003d). Lunar laboratory for admittance calculations (more on estimations of lunar elastic thickness. *Division of Planetary Sciences 35th Meeting*, Monterey, CA: (September 2003).

Atkinson, D.H., Ingersoll, A.P., and Seiff, A. (1997). Deep winds on Jupiter as measured by the Galileo probe. *Nature* 388 (6643), 649–650.

Atkinson, D.H., Pollack, J.B., and Seiff, A. (1998). The Galileo probe Doppler Wind Experiment: measurement of the deep zonal winds on Jupiter. *Journal of Geophysical Research: Planets* 103 (E10): 22911–22928.

Austin, A., Afonso, G., Albert, S., Ali, H., Alunni, A., Arnold, J., Bailet, G., Beauchamp, P., Cassell, A., Cutts, J., and Deshmukh, R. (2020). Enabling and Enhancing Science Exploration Across the Solar System: Aerocapture Technology for SmallSat to Flagship Missions. In International Planetary Probe Workshop Webinar Series.

Bailey, E. and Stevenson, D.J. (2021). Thermodynamically Governed Interior Models of Uranus and Neptune. *The Planetary Science Journal* 2 (2): 64.

Balmino, G., Lambeck, K., & Kaula, W. M. (1973). A spherical harmonic analysis of the Earth's topography. *Journal of Geophysical Research* 78 (2): 478–481.

Banerdt, W.B., Smrekar, S.E., Banfield, D., Giardini, D., Golombek, M., Johnson, C.L., Lognonné, P., Spiga, A., Spohn, T., Perrin, C. and Stähler, S.C., 2020. Initial results from the InSight mission on Mars. *Nature Geoscience* 13 (3): 183–189.

Barbieri, A.J., Butman, S., Danos, M.J., Greenberg, E., Ilott, P.A., Kazz, G.J., Torgerson, J.L., Vaisnys, A., Adams, W.R., Johnson, C.E. and Dapore, M. (2005, March). Development and flight performance of CCSDS Proximity-1 on Odyssey and the Mars exploration rovers. In: *2005 IEEE Aerospace Conference* 1444–1454. IEEE.

Barriot, J.P. and Balmino, G. (1992). Estimation of local planetary gravity fields using line of sight gravity data and an integral operator. *Icarus* 99 (1): 202–224.

Bar-Sever, Y.E., Jacobs, C.S., Keihm, S., Lanyi, G.E., Naudet, C.J., Rosenberger, H.W., Runge, T.F., Tanner, A.B. and Vigue-Rodi, Y. (2007). Atmospheric media calibration for the deep space network. *Proceedings of the IEEE* 95 (11): 2180–2192.

Baumbach, S. (1937). Coronal densities from white hght echpse observations. *Astronomische Nachrichten* 263: 121.

Belton, M.J.S., Chapman, C.R., Thomas, P.C., Davies, M.E., Greenberg, R., Klaasen, K., Byrnes, D., D'Amario, L., Synnott, S., Johnson, T.V., McEwenstar, A., Merline, W.J., Davis, D.R., Petit, J.-M., Storrs, A., Veverka, J., and Zellner, B. (1994). Bulk density of Asteroid 243 Ida from the orbit of its satellite Dactyl. *Nature* 374: 785–788.

Belton, M.J.S., Chapman, C.R., Thomas, P.C., Davies, M.E., Greenberg, R., Klaasen, K., Byrnes, D., D'amario, L., Synnott, S., Johnson, T.V. and McEwen, A. (1995). Bulk density of asteroid 243 Ida from the orbit of its satellite Dactyl. *Nature* 374 (6525): 785–788.

Bender, E. (1973). An equation of state for predicting vapour–liquid equilibria of the system N_2–Ar–O_2. *Cryogenics* 13: 11–18.

Bender, P.L. (1994). Proposed microwave transponders for early lunar robotic landers. *Advances in Space Research* 14: 233–242.

Bender, P.L., Ashby, N., Vincent, M.A., and Wahr, J.M. (1989). Conceptual design for a Mercury relativity satellite. *Advances in Space Research* 9 (9): 113–116.

Berner, J.B. and Bryant, S.H. (2002). Operations comparison of deep space ranging types: sequential tone vs. pseudo-noise. *IEEE Aerospace Conference* 3: 3–1313.

Berner, J.B., Bryant, S.H., and Kinman, P.W. (2007). Range measurement as practiced in the Deep Space Network. *Proceedings of the IEEE* 95 (11): 2202–2214.

Bertotti, B., Ambrosini, R., Asmar, S.W., Brenkle, J.P., Giampieri, G., Iess, L., Messeri, A., and Wahlquist, H.D. (1992). The gravitational wave experiment. *Astronomy and Astrophysics Supplement Series* 92: 431–440.

Bertotti, B., Ambrosini, R., Asmar, S.W., Comoretto, G., Giampieri, G., Iess, L., Koyama, Y., Messeri, A., Vecchio, A., and Wahlquist, H.D. (1995b). Ulysses gravitational wave experiment. *Astronomy and Astrophysics* 296: 13–25.

Bertotti, B., Comoretto, G., and Iess, L. (1993). Doppler tracking of spacecraft with multifrequency links. *Astronomy and Astrophysics* 269: 608–616.

Bertotti, B., Iess, L., and Tortora, P. (2003). A test of general relativity using radio links with the Cassini spacecraft. *Nature* 425: 374–376.

Bertotti, B. and Giampieri, G. (1998). Solar coronal plasma in Doppler measurements. *Solar Physics* 178 (1): 85–107.

Besserer, J., Nimmo, F., Wieczorek, M.A., Weber, R.C., Kiefer, W.S., McGovern, P.J., Andrews-Hanna, J.C., Smith, D.E. and Zuber, M.T. (2014). GRAIL gravity constraints on the vertical and lateral density structure of the lunar crust. *Geophysical Research Letters* 41 (16): 5771–5777.

Besso, P., Bozzi, M., Formaggi, M., Pasian, M. and Perregrini, L. (2007, October). Design of K-band inductive dichroic mirrors for upgrading ESA deep-space antenna DSA2. In: *2007 European Microwave Conference* 485–488. IEEE.

Bierman, G.J. (1977). Factorization Methods for Discrete Sequential Estimation. New York: Academic Press.

Bills, B.G., Asmar, S.W., Konopliv, A.S., Park, R.S., and Raymond, C.A. (2014). Harmonic and statistical analyses of the gravity and topography of Asteroid 4 Vesta. *Icarus* 240: 161–173.

Bills, B.G. and Ermakov, A.I. (2019). Simple models of error spectra for planetary gravitational potentials as obtained from a variety of measurement configurations. *Planetary and Space Science* 179: 104744.

Biondi, R., Steiner, A.K., Kirchengast, G., Brenot, H., and Rieckh, T. (2017). Supporting the detection and monitoring of volcanic clouds: A promising new application of Global Navigation Satellite System radio occultation. *Advances in Space Research*, 60 (12): 2707–2722.

Bird, M.K. (1982). Coronal sounding with occulted spacecraft signals. *Space Science Reviews* 33: 99–126.

Bird, M.K., Allison, M., Asmar, S.W., Atkinson, D.H., Avruch, I.M., Dutta-Roy, R., Dzierma, Y., Edenhofer, P., Folkner, W.M., Gurvits, L.I., and Johnston, D.V. (2005). The vertical profile of winds on Titan. *Nature* 438 (7069): 800–802.

Bird, M.K., Asmar, S.W., Brenkle, J.P., Edenhofer, P., Funke, O., Pätzold, M., and Volland, H. (1992a). Ulysses radio occultation observations of the Io plasma torus during the Jupiter encounter. *Science* 257: 1531–1535.

Bird, M.K., Asmar, S.W., Brenkle, J.P., Edenhofer, P., Pätzold, M., and Volland, H. (1992b). The coronal-sounding experiment. *Astronomy and Astrophysics Supplement Series* 92 (2): 425–430.

Bird, M.K., Asmar, S.W., Brenkle, J.P., Edenhofer, P., Pätzold, M., and Volland, H. (1994). The coronal electron density distribution determined from dual-frequency ranging measurements during the 1991 solar conjunction of the Ulysses spacecraft. *Astrophysical Journal* 426: 373–381.

Bird, M.K., Asmar, S.W., Edenhofer, P., Funke, O., Pätzold, M., and Volland, H. (1993). The structure of Jupiter's Io plasma inferred from Ulysses radio occultation observations. *Planetary and Space Science* 41 (11/12): 999–1010.

Bird, M.K., Dutta-Roy, R., Asmar, S.W., and Rebold, T.A. (1997). Detection of Titan's ionosphere from Voyager 1 radio science occultation observations. *Icarus* 130: 426–436.

Bird, M.K., Dutta-Roy, R., Heyl, M., Allison, M., Asmar, S.W., Folkner, W.M., Preston, R.A., Atkinson, D.H., Edenhofer, P., Plettemeier, D., Wohlmuth, R., Iess, L., and Tyler, G.L. (2002). The Huygens Doppler wind experiment, Titan winds derived from probe radio frequency measurements. *Space Science Reviews* 104.

Bird, M.K. and Edenhofer, P. (1990). Remote sensing observations of the solar corona. In: Physics of the Inner Heliosphere, (ed. R. Schwenn and E. Marsch), 13–97. Berlin, Germany: Springer-Verlag.

Bird, M. K., Linscott, I. R., Tyler, G. L., Hinson, D. P., Pätzold, M., Summers, M. E., ... & Team, T. N. H. S. (2019). Radio thermal emission from Pluto and Charon during the New Horizons encounter. *Icarus* 322: 192–209.

Bird, M.K., Pätzold, M., Edenhofer, P., Asmar, S.W., and McElrath, T.P. (1996b). Coronal radio sounding with Ulysses: solar wind electron density near 0.1 AU during the 1995 conjunction. *Astronomy & Astrophysics* 316: 441–448.

Bird, M.K., Volland, H., Holland, R.A., Koomen, M.J., Michels, D.J., Sheeley, N.R., Jr., Armstrong, J.W., Seidel, B.L., Stelzried, C.T., and Woo, R. (1985). White-light and radio sounding observations of coronal transients. *Solar Physics* 98: 341–368. doi:10.1007/BF0015246.

Bird, M.K., Volland, H., Stelzried, C.T., Levy, G.S., and Seidel, B.L. (1977). Faraday rotation transients observed during solar occultation of the Helios spacecraft. In: Study of Traveling Interplanetary Phenomena, (ed. D. Reidel), 63–74. Holland: Dordrecht.

Biswas, A., Srinivasan, M., Piazzolla, S. and Hoppe, D. (2018, February). Deep space optical communications. In: Free-Space Laser Communication and Atmospheric Propagation XXX Vol. 10524. 105240U. International Society for Optics and Photonics.

Black, M.T., Zuber, M.T., and McAdoo, D.C. (1991). Comparison of observed and predicted gravity profiles over Aphrodite Terra, Venus. *Journal of Geophysical Research: Solid Earth* 96 (B1): 301–315.

Bolton, S.J. (2010). The Juno mission. *Proceedings of the International Astronomical Union* 6 (S269): 92–100.

Bolton, S.J., Levin, S., and Bagenal, F. (2017). Juno's first glimpse of Jupiter's complexity. *Geophysical Research Letters* 44 (15): 7663–7667.

Bolton, S.J., Lunine, J., Stevenson, D., Connerney, J.E.P., Levin, S., Owen, T.C., Bagenal, F., Gautier, D., Ingersoll, A.P., Orton, F.S., Guillot, T., Hubbard, W., Bloxham, J., Coradini, A., Stephens, S.K., Mokashi, P., Thorne, R., and Thorpe, R. (2010). The Juno mission. *Space Science Reviews* 213 (1–4): 5–37.

Border, J.S. and Paik, M. (2009). Station Delay Calibration for Ranging Measurements. The Interplanetary Network Progress Report (ed. J. Hamkins), 42–177. Pasadena: Jet Propulsion Laboratory, California Institute of Technology. 15 May 2009.

Border, J.S., Paik, M., Lee, C., Shin, D.K. and Volk, C.P. (November 15, 2020) Demonstration of Advanced Ranging Instrument. IPN Progress Report. 42–223.

Boucher, C., Altamimi, Z., and Duhem, L. (1994). Results and analysis of the ITRF93. *IERS Technical note* 18: 1–313.

Bourgois, G., Coles, W.A., Daigne, G., Silen, J., Turunen, T., and Williams, P.J. (1985). Measurements of the solar wind velocity with EISCAT. *Astronomy and Astrophysics* 144 (2): 452–462.

Buccino, D. R., Helled, R., Parisi, M., Hubbard, W. B., & Folkner, W. M. (2020). Updated equipotential shapes of Jupiter and Saturn using Juno and Cassini grand finale gravity science measurements. Journal of Geophysical Research: *Planets* 125 (8): e2019JE006354.

Buccino, D.R., Seubert, J.A., Asmar, S.W., and Park, R.S. (2016). Optical Ranging Measurement with a Lunar Orbiter: limitations and Potential. *Journal of Spacecraft and Rockets* 53 (3): 457–463. https://doi.org/10.2514/1.A33415.

Caetta, J., Asmar, S.W., Abbate, S.F., Connally, M., and Goltz, G. (1998). Remote Operations of the Deep Space Network Radio Science Subsystem. Telecommunications and Mission Operations Progress Report 42-134, Pasadena: Jet Propulsion Laboratory, California Institute of Technology. 15 August 1998.

Cahoy, K., Beerer, I., Marinan, A., Asmar, S.W., Withers, P., and Moore, L. (2012). iROCC: interplanetary radio occultation CubeSat constellation. iCubeSat: First Interplanetary CubeSat Workshop, Cambridge, MA: (May 2012).

Cain, D.L. (1960). Refraction Corrections. Inter-Office Memorandum, Jet Propulsion Laboratory, California Institute of Technology, Pasadena, CA, 6 July 1960.

Cain, D.L. (1964). Refraction of a Radio Signal During Occultation due to a Planetary Atmosphere. Inter-Office Memorandum, Jet Propulsion Laboratory, California Institute of Technology, Pasadena, CA, IOM #312-404, 16 March 1964.

Cain, D.L. and Hamilton, T.W. (1964). Determination of tracking station locations by Doppler and range measurements to an Earth satellite. Technical Report 32-534 Jet Propulsion Laboratory, Pasadena, CA: California Institute of Technology.

Cain, D.L., Kliore, A.J., and Levy, G.S. (1966). The Mariner 4 occultation experiment: summary of data and reduction methods. In *3rd and 4th Aerospace Sciences Meeting*, p. 148.

Cain, D.L., Kliore, A.J., and Seidel, B.L., M.J. and Sykes. (1972). The shape of Mars from the Mariner 9 occultations. *Icarus* 17 (2): 517–524.

Callahan, P.S. (1973). Prediction of Tropospheric Wet Component Range Error from Surface Measurements, JPL Technical Report 32–156, XVIII: 41–46, Pasadena, CA: Jet Propulsion Laboratory, California Institute of Technology. http://ipnpr.jpl.nasa.gov/progress_report2/XVIII/XVIIIF.PDF (accessed September 23, 2016).

Campbell, J.K. and Anderson, J.D. (1989). Gravity field of the Saturnian system from Pioneer and Voyager tracking data. *The Astronomical Journal* 97: 1485–1495.

Campbell, J.K. and Synnott, S.P. (1985). Gravity field of the Jovian system from Pioneer and Voyager tracking data. *The Astronomical Journal* 90: 364–372.

Castillo-Rogez, J.C. and McCord, T.B. (2010). Ceres' evolution and present state constrained by shape data. *Icarus* 205 (2): 443–459.

Chao, C.C. (1972). The Tropospheric Calibration Model for Mariner Mars 1971. In Tracking System Analytic Calibration Activities for the Mariner Mars 1971, Technical Report 32-1587, Pasadena, CA: Jet Propulsion Laboratory, California Institute of Technology.

Chappaz, L., Sood, R., Melosh, H.J., Howell, K.C., Blair, D.M., Milbury, C. and Zuber, M.T. (2017). Evidence of large empty lava tubes on the Moon using GRAIL gravity. *Geophysical Research Letters* 44 (1): 105–112.

Chen, C.C., Shambayati, S., Makovsky, A., Taylor, F.H., Herman, M., Zingales, S., Nuckolls, C. and Siemsen, K. (2000). Small Deep Space Transponder (SDST) DS1 Technology Validation Report. *JPL document.*

Cheng, Y., Johnson, A.E., Matthies, L.H. and Olson, C.F. (2003). Optical landmark detection for spacecraft navigation.

Chien, S., Sherwod, R., Burl, M., Knight, R., Rabideau, G., Engelhardt, B., Castano, R., Stough, T., Roden, J. and Davies, A. (2001). The Techsat-21 Autonomous Sciencecraft Constellation.

Choudhary, R.K., Ambili, K.M., Choudhury, S., Dhanya, M.B., and Bhardwaj, A. (2016). On the origin of the ionosphere at the Moon using results from Chandrayaan-1 S-band radio occultation experiment and a photochemical model. *Geophysical Research Letters* 43 (19).

Ciarcia, S., Simone, L., Gelfusa, D., Colucci, P., De Angelis, G., Argentieri, F., Iess, L. and Formaro, R. (2013). MORE and Juno Ka-band transponder design, performance, qualification and in-flight validation. *Image* 15: F1.

Cimino, J.B., Elachi, C., Kliore, A.J., McCleese, D.L., and Patel, I.R. (1980). Polar cloud structure of Venus as derived from the Pioneer Venus orbiter. *Journal of Geophysical Research: Space Physics* 85 (A13): 8082–8088.

Coles, W.A. (1978). Interplanetary scintillation. *Space Science Reviews* 21: 411–425.

Coles, W.A. (1993). Scintillation in the Solar Wind (IPS). In: Wave Propagation in Random Media, (ed. V.I. Tatarskii, A. Ishimaru and V.U. Zavorotny), 156–168. SPIE/IOP Publishing.

Coles, W.A. and Harmon, J.K. (1989). Propagation observations of the solar wind near the Sun. *Astrophysical Journal* 337: 1023–1034.

Coles, W.A., Liu, W., Harmon, J.K., and Martin, C.L. (1991). The solar wind density spectrum near the Sun: results from Voyager radio measurements. *Journal of Geophysical Research* 96: 1745–1755.

Coles, W.A., Rickett, B.J., Rumsey, V.H., Kaufman, J.J., Turley, D.G., Ananthakrishnan, S., Armstrong, J.W., Harmon, J.K., Scott, S.L., and Sime, D.G. (1980). Solar cycle changes in the polar solar wind. *Nature* 286: 239–241. doi:10.1038/286239a0.

Corlies, P., Hayes, A.G., Birch, S.P.D., Lorenz, R., Stiles, B.W., Kirk, R., Poggiali, V., Zebker, H. and Iess, L. (2017). Titan's topography and shape at the end of the Cassini mission. *Geophysical Research Letters* 44 (23): 11–754.

Davies, R.W. (1974). Issues in gravitational wave detection with space missions. In *Gravitational Waves and Radiation*, 33–35. *Proceedings of the International Conference*, Universite de Paris VII, vol. 220 of *Colloques Internationaus du CNRS*, Paris, France: CNRS.

Deep Space Mission System (DSMS). External Interface Specification, 820-013, JPL D-16765, Jet Propulsion Laboratory, California Institute of Technology, Pasadena, CA.

Counselman, C.C., Hinteregger, H.F., and Shapiro, I.I. (1972). Astronomical applications of differential interferometry. *Science*, 178 (4061): 607–608.

Counselman III, C.C., Gourevitch, S.A., King, R.W., Loriot, G.B. and Ginsberg, E.S., 1980. Zonal and meridional circulation of the lower atmosphere of Venus determined by radio interferometry. *Journal of Geophysical Research: Space Physics* 85 (A13): 8026–8030.

Danzmann, K. (2000). LISA mission overview. *Advances in Space Research* 25 (6): 1129–1136.

Damour, T. and Nordtvedt, K. (1993). General relativity as a cosmological attractor of tensor-scalar theories. *Physical Review Letters* 70 (15): 2217.

Deep Space Mission System (DSMS). Telecommunications Link Design Handbook, JPL Document 810-005. Pasadena, CA: Jet Propulsion Laboratory, California Institute of Technology.

Dehant, V., Hinderer, J., Legros, H. and Lefftz, M. (1993). Analytical approach to the computation of the Earth, the outer core and the inner core rotational motions. *Physics of the Earth and Planetary Interiors* 76 (3–4): 259–282.

Dehant, V., Folkner, W.M., Renotte, E., Orban, D., Asmar, S.W., Balmino, G., Barriot, J.-P., Benoist, J., Biancale, R., Biele, J., Budnik, F., Burger, S., De Viron, O., Häusler, B., Karatekin, Ö., Le Maistre, S., Lognonné, P., Menvielle, M., Mitrovic, M., Pätzold, M., Rivoldini, A., Rosenblatt, P., Schubert, G., Spohn, T., Thomassen, L., Tortora, P., Van Hoolst, T., Witasse, O., and Yseboodt, M. (2009). Lander radioscience for obtaining the rotation and orientation of Mars. *Planetary and Space Science* 57 (8–9): 1050–1067. doi:10.1016/j.pss.2008.08.009.

Dehant, V. and Mathews, P.M. (2015). Precession, nutation and wobble of the Earth. Cambridge University Press.

DeNatale, J.F., Borwick, R.L., Stupar, P.A., Newgard, R.A. and Zhu, M. (2008). Compact, low-power atomic time and frequency standards. THOUSAND OAKS CA: TELEDYNE SCIENTIFIC COMPANY.

Dennison, P.A. and Hewish, A. (1967). The Solar Wind Outside the Plane of the Ecliptic. *Nature* 213: 343–346. doi:10.1038/213343a0.

Dermott, S.F. (1979). Shapes and gravitational moments of satellites and asteroids. *Icarus* 37 (3): 575–586.

Dickey, J.O., Bender, P.L., Faller, J.E., Newhall, X.X., Ricklefs, R.L., Ries, J.G., Shelus, P.J., Veillet, C., Whipple, A.L., Wiant, J.R. and Williams, J.G. (1994). Lunar laser ranging: A continuing legacy of the Apollo program. *Science* 265 (5171): 482–490.

Divsalar, D., Asmar, S.W., Farr, W., Dolinar, S., and Vilnrotter, V. (2016). An optical receiver for science measurements and data detection. pp. 1–9. *IEEE Aerospace Conference*, March: 2016.

Divsalar, D., Asmar, S.W., Moision, B., and Oudrhiri, K. (2014). Radio science measurements in presence of data on optical and RF links. *IEEE Aerospace Conference*, pp. 1–10. (March 2014).

Divsalar, D., Asmar, S.W., and Oudrhiri, K. (2015). Radio science measurements using phase modulated optical links. #2256, *IEEE Aerospace Conference.*

Duncan, C. B. et al. (2014). Iris Transponder – Communications and Navigation for Deep Space. in Small Satellite Conf., Logan, UT.

Dunham, D.W. (1981). Recently-observed planetary occultations. *Occultation Newsletter, International Occultation Timing Association (IOTA)* 2: 139–143.

Durante, D., Guillot, T., and Iess, L. (2017). The effect of Jupiter oscillations on Juno gravity measurements. *Icarus* 282: 174–182.

Durante, D., Hemingway, D.J., Racioppa, P., Iess, L. and Stevenson, D.J. (2019). Titan's gravity field and interior structure after Cassini. *Icarus* 326: 123–132.

Durante, D., Parisi, M., Serra, D., Zannoni, M., Notaro, V., Racioppa, P., ... & Bolton, S. J. (2020). Jupiter's gravity field halfway through the Juno mission. *Geophysical Research Letters* 47 (4): e2019GL086572.

Doody, D. (2010). Deep space craft: An overview of interplanetary flight.

Doody, D., 2011. Basics of Space Flight: Black and White Edition. Blüroof Press.

Drinkwater, M.R., Haagmans, R., Muzi, D., Popescu, A., Floberghagen, R., Kern, M. and Fehringer, M. (2006). November. The GOCE gravity mission: ESA's first core Earth explorer. In Proceedings of the 3rd international GOCE user workshop, 6–8. Noordwijk, The Netherlands: European Space Agency.

Eanes, R.J., Schutz, B., and Tapley, B. (1983). Earth and ocean tide effects on Lageos and Starlette. In: Proceeding 9th International Symposium on Earth Tides, (ed. J.T. Kuo), 239–249. New York City, NY: (17–22 August 1981). Stuttgart, Germany: E. Sckweizerbart'sche Verlagabuchhandlung.

Edwards, C.D., Bell, D.J., Biswas, A., Cheung, K.M. and Lock, R.E. (2016, March). Proximity link design and performance options for a Mars areostationary relay satellite. In: *2016 IEEE Aerospace Conference* 1–10. IEEE.

Edwards, P.C., Hajj, G., Ao, C., Romans, L., Callas, J., Mannucci, A., Asmar, S.W., and Kahan, D.S. (2008). *A Chapman-Layers Ionospheric Model for Mars*. JPL Publication 08-24, Jet Propulsion Laboratory, California Institute of Technology, Pasadena, CA, August 2008.

Edwards, C., Rogstad, D., Fort, D., White, L., and Iijima, B. (1992). A demonstration of real-time connected element interferometry for spacecraft navigation. *Astrodynamics* 1991: 1179–1192.

Edwards Jr, C., Stelzried, C., Deutsch, L., & Swanson, L. (1998). NASA's Deep Space Telecommunications Roadmap.

Efimov, A.I., Armand, N.A., Lukanina, L.A., Samoznaev, L.N., Chashei, I.V., and Bird, M.K. (2009). Investigation of coronal mass ejections by the two-position radio sounding method. *Geomagnetism and Aeronomy* 49: 1165–1169.

Efimov, A.I., Bird, M.K., Rudash, V.K., Andreev, V.E., Chashei, I.V., Plettemeier, D., and Edenhofer, P. (2005). Solar wind velocity measurements near the Sun using Ulysses radio amplitude correlations at two frequencies. *Advances in Space Research* 35: 2189–2194.

Efimov, A.I., Iakovlev, O.I., Razmanov, V.M. and Shtrykov, V.K., 1977. Turbulence of plasma near the sun and the solar-wind velocity derived from the Venera 10 radio occultation experiment. *Soviet Astronomy Letters*, 3, pp.172–174.

Elachi, C., Allison, M.D., Borgarelli, L., Encrenaz, P., Im, E., Janssen, M.A., Johnson, W.T.K., Kirk, R.L., Lorenz, R.D., Lunine, J.I. and Muhleman, D.O. (2004). Radar: the Cassini Titan radar mapper. *The Cassini-Huygens Mission* 71–110.

Elkins-Tanton, L.T., Asphaug, E., Bell, J., Bercovici, D., Bills, B.G., Binzel, R.P., Bottke, W.F., Goldsten, J., Jaumann, R., Jun, I., Lawrence, D.J., Marchi, S., Oh, D., Park, R., Peplowski, P.N., Polanskey, C.A., Prettyman, T.H., Raymond, C.A., Russell, C.T., Scheinberg, A., Weiss, B.P., Wenkert, D.D., Wieczorek, M., and Zuber, M.T. (2015). The discovery science of Asteroid (16) Psyche. *46th Lunar and Planetary Science Conference* 1832: 1632.

Ellis, J. (1980). Large Scale State Estimation Algorithms for DSN Tracking Station Location Determination. *Journal of the Astronautical Sciences* 28: 15–30.

Elliot, J.L. and Olkin, C.B. (1996). Probing planetary atmospheres with stellar occultations. *Annual Review of Earth and Planetary Science* 24 (1): 89–123.

Ely, T.A. and Bell, D. (2003). Sensitivity of Mars surface navigation to timing errors. AAS, 03–527.

Ely, T.A., Murphy, D., Seubert, J., Bell, J. and Kuang, D. (2014, January). Expected performance of the deep space atomic clock mission. In AAS/AIAA Space Flight Mechanics Meeting.

Ely, T.A., Murphy, D., Seubert, J., Bell, J., and Kuang, D. (2014). Expected performance of the Deep Space Atomic Clock mission. *AAS/AIAA Space Flight Mechanics Meeting* (January 2014.

Ermakov, A.I., Fu, R.R., Castillo-Rogez, J.C., Raymond, C.A., Park, R.S., Preusker, F., Russell, C.T., Smith, D.E. and Zuber, M.T. (2017). Constraints on Ceres' internal structure and evolution from its shape and gravity measured by the Dawn spacecraft. *Journal of Geophysical Research: Planets* 122 (11): 2267–2293.

Ermakov, A.I., Park, R.S. and Bills, B.G. (2018). Power laws of topography and gravity spectra of the solar system bodies. *Journal of Geophysical Research: Planets* 123 (8): 2038–2064.

Eshleman, V.R., and Manning, L.A. (1954). Radio communication by scattering from meteoric ionization. *Proceedings of the IRE* 42 (3): 530–536.

Eshleman, V.R., Fjeldbo, G., Anderson, J.D., Kliore, A.J., and Dyce, R.B. (1968). Venus: lower atmosphere not measured. *Science* 162: 661–665.

Eshleman, V.R., Tyler, G.L., Anderson, J.D., Fjeldbo, G., Wood, G.S., and Croft, T.A. (1977). Radio science investigations with Voyager. *Space Science Reviews* 21: 207–232.

Esposito, P.B., Anderson, J.D., and Ng, A.T.Y. (1976). Experimental determination of Mercury's mass and oblateness. *COSPAR Meeting*.

Esposito, P.B., Banerdt, W.B., Lindal, G.F., Sjogren, W.L., Slade, M.A., Bills, B.G., Smith, D.E. and Balmino, G. (1992). Gravity and tophography. *Mars* 209–248.

Estabrook, F.B. (1978). Gravitational Wave Detection with the Solar Probe: II. The Doppler Tracking Method. A Close-Up of the Sun. In *Proceedings of the Solar Probe Science Workshop*, (eds. M. Neugebauer and R.W. Davies) 78–70. JPL Publication 441-449, Jet Propulsion Laboratory, California Institute of Technology, Pasadena, CA.

Estabrook, F.B., Hellings, R.W., Wahlquist, H.D., and Wolff, R.S. (1979). Gravitational radiation detection with spacecraft Doppler tracking: limiting sensitivities and prospective missions. In: Sources of Gravitational Radiation, (ed. L.L. Smarr), 37–48. In *Proceeding of the Battelle Seattle Workshop* (24 July–4 August 1978). Cambridge, UK; New York: Cambridge University Press.

Estabrook, F.B. and Wahlquist, H.D. (1975). Response of Doppler spacecraft tracking to gravitational radiation. *General Relativity and Gravitation* 6: 439–447. doi:10.1007/BF00762449.

Evans, A.J., Soderblom, J.M., Andrews-Hanna, J.C., Solomon, S.C., and Zuber, M.T. (2016). Identification of buried lunar impact craters from GRAIL data and implications for the nearside Maria. *Geophysical Research Letters* 43 (6): 2445–2455.

Ferri, F., Karatekin, Ö., Lewis, S.R., Forget, F., Aboudan, A., Colombatti, G., Bettanini, C., Debei, S., Van Hove, B., Dehant, V., and Harri, A.M. (2019). Exomars atmospheric mars entry and landing investigations and analysis (amelia). *Space Science Reviews* 215(1): 8.

Finocchiaro, S., Iess, L., Folkner, W.M., and Asmar, S.W. (2011). The determination of Jupiter's angular momentum from the lense-thirring precession of the Juno spacecraft. *American Geophysical Union Conference Fall Meeting. Abstract* 1: 1620.

Fjeldbo, G. (1964). Bistatic Radar Methods for Studying Planetary Ionospheres and Surfaces. SU-64-025. Stanford, CA: Stanford Electronics Laboratories.

Fjeldbo, G., Fjeldbo, W.C., and Eshleman, V.R. (1966). Models for the atmosphere of Mars based on the Mariner 4 occultation experiment. *Journal of Geophysical Research* 71 (9): 2307–2316.

Fjeldbo, G. and Eshleman, V.R. (1968). The atmosphere of Mars analyzed by integral inversion of Mariner IV occultation data. *Planetary and Space Science* 16: 1035–1059.

Fjeldbo, G. and Eshleman, V.R. (1969). Atmosphere of Venus as studied with the Mariner 5 dual radio-frequency occultation experiment. *Radio Science* 4: 879–897.

Fjeldbo, G., Kliore, A.J., and Eshleman, V.R. (1971). The Neutral Atmosphere of Venus as Studied with the Mariner V Radio Occultation Experiments. *The Astronomical Journal* 76: 123–140.

Fjeldbo, G., Kliore, A.J., Seidel, B., Sweetnam, D., and Woiceshyn, P. (1976). The Pioneer 11 Radio Occultation Measurements of the Jovian Ionosphere. In: Jupiter Studies of the Interior, Atmosphere, Magnetosphere and Satellites (JSIA), (ed. T. Gehrels), 238–246. Tucson: The University of Arizona Press.

Fjeldbo, G., Seiden, B., Sweetnam, D. and Howard, T. (1975). The Mariner 10 radio occultation measurements of the ionosphere of Venus. *Journal of Atmospheric Sciences* 32 (6): 1232–1236.

Fjeldbo, G., Sweetnam, D.N., Brenkle, J.P., Christiansen, E., Farless, D., Mehta, J., Seidel, B.L., Michael, W., Jr., Wallio, A., and Grossi, M. (1977). Viking radio occultation measurements of the Martian atmosphere and topography: primary mission coverage. *Journal of Geophysical Research* 82: 4317–4324.

Florensky, C.P., Ronca, L.B., and Basilevsky, A.T. (1977). Geomorphic degradations on the surface of Venus: an analysis of Venera 9 and Venera 10 data. *Science* 196 (4292): 869–871.

Folkner, W.M., Iess, L., Anderson, J.D., Asmar, S.W., Buccino, D.R., Durante, D., Feldman, M., Gomez Casajus, L., Gregnanin, M., Milani, A. and Parisi, M., 2017. Jupiter gravity field estimated from the first two Juno orbits. *Geophysical Research Letters*, 44(10): 4694–4700.

Folkner, W.M., Dehant, V., Le Maistre, S., Yseboodt, M., Rivoldini, A., Van Hoolst, T., Asmar, S.W. and Golombek, M.P., 2018. The rotation and interior structure experiment on the InSight mission to Mars. *Space Science Reviews*, 214(5): 1–16.

Folkner, W.M. (1995). DSN Station Locations and Uncertainties. JPL TDA Progress Report, 42–128. Pasadena: Jet Propulsion Laboratory, California Institute of Technology, 1–34. July–September.

Folkner, W.M. (1996). DSN station locations and uncertainties. *TDA Progress Report* 42.

Folkner, W.M., Asmar, S.W., Border, J.S., Franklin, G.W., Finley, S.G., Gorelik, J., Johnston, D., Kerzhanovich, V.V., Lowe, S.T., Preston, R.A., Bird, M.K., Dutta-Roy, R., Allison, M., Atkinson, D.H., Edenhofer, P., Plettemeier, D., and Tyler, G.L. (2006). Winds on Titan from ground-based tracking of the Huygens probe. *Journal of Geophysical Research* 111: E07S02. doi:10.1029/2005JE002649.

Folkner, W.M. and Border, J.S. (1992). Orbiter-orbiter and orbiter-lander tracking using same-beam interferometry. TDA progress report, 74–86.

Folkner, W.M., Iess, L., Anderson, J.D., Asmar, S.W., Buccino, D.R., Durante, D., Feldman, M., Gomez Casajus, L., Gregnanin, M., Milani, A., Parisi, M., Park, R.S., Serra, D., Tommei, G., Tortora, P., Zannoni, M., Bolton, S.J., Connerney, J.E.P., and Levin, S.M. (2017). Jupiter gravity field estimated from the first two Juno orbits. *Geophysical Research Letters.*

Folkner, W.M., Kahn, R.D., Preston, R.A., Yoder, C.F., Standish, E.M., Williams, J.G., Edwards, C.D., Hellings, R.W., Eubanks, T.M., and Bills, B.G. (1997a). Mars Dynamics from Earth-Based Tracking of the Mars Pathfinder Lander. *Journal of Geophysical Research: Planets* 102 (E2): 4057–4064.

Folkner, W.M., Preston, R.A., Border, J.S., Navarro, J., Wilson, W.E. and Oestreich, M. (1997c). Earth-based radio tracking of the Galileo probe for Jupiter wind estimation. *Science* 275 (5300): 644–646.

Folkner, W.M., Steppe, J.A., and Oliveau, S.H. (1993). Earth Orientation Parameter File Description and Usage. Pasadena, CA: Interoffice Memorandum 335.1-11-93 (internal document), Jet Propulsion Laboratory, California Institute of Technology.

Folkner, W.M., Williams, J.G., Boggs, D.H., Park, R.S. and Kuchynka, P. (2014). The planetary and lunar ephemerides DE430 and DE431. *Interplanetary Network Progress Report* 196 (1).

Folkner, W.M., Woo, R., and Nandi, S. (1998). Ammonia Abundance in Jupiter's Atmosphere Derived from the Attenuation of the Galileo Probe's Radio Signal. *Journal of Geophysical Research: Planets* 103 (E10): 22847–22855.

Folkner, W.M., Yoder, C.F., Yuan, D.-N., Standish, E.M., and Preston, R.A. (1997b). Interior Structure and Seasonal Mass Redistribution of Mars from Radio Tracking of Mars Pathfinder. *Science* 278 (5344): 1749–1752.

Fortes, A.D. (2012). Titan's internal structure and the evolutionary consequences. *Planetary and Space Science* 60 (1): 10–17.

Fu, R.R., Ermakov, A.I., Marchi, S., Castillo-Rogez, J.C., Raymond, C.A., Hager, B.H., Zuber, M.T., King, S.D., Bland, M.T., De Sanctis, M.C. and Preusker, F. (2017). The interior structure of Ceres as revealed by surface topography. *Earth and Planetary Science Letters* 476: 153–164.

Gair, J.R., Vallisneri, M., Larson, S.L., and Baker, J.G. (2013). Testing general relativity with low-frequency, space-based gravitational-wave detectors. *Living Reviews in Relativity* 16 (7): http://www.livingreviews.org/lrr-2013-7.

Garmier, R., Barriot, J.-P. (2001). Ellipsoidal harmonic expansions of the gravitational potential: Theory and application. *Celestial Mechanics and Dynamical Astronomy* 79: 235–275.

Garmier, R., Barriot, J.P., Konopliv, A.S. and Yeomans, D.K. (2002). Modeling of the Eros gravity field as an ellipsoidal harmonic expansion from the NEAR Doppler tracking data. *Geophysical Research Letters* 29 (8): 72–1.

Garmier, R., Barriot, J.P., Konopliv, A.S. and Yeomans, D.K. (2002). Modeling of the Eros gravity field as an ellipsoidal harmonic expansion from the NEAR Doppler tracking data. *Geophysical Research Letters* 29 (8): 72–1.

Gaskell, R.W., Barnouin-Jha, O.S., Scheeres, D.J., Konopliv, A.S., Mukai, T., Abe, S., Saito, J., Ishiguro, M., Kubota, T., Hashimoto, T. and Kawaguchi, J. (2008). Characterizing and navigating small bodies with imaging data. *Meteoritics & Planetary Science* 43 (6): 1049–1061.

Genova, A., Goossens, S., Lemoine, F.G., Mazarico, E., Neumann, G.A., Smith, D.E. and Zuber, M.T. (2016). Seasonal and static gravity field of Mars from MGS, Mars Odyssey and MRO radio science. *Icarus* 272: 228–245.

Genova, A. and Iess, L. (2013). The geodesy of Mercury with the missions MESSENGER and BepiColombo. Dissertation. University of Rome.

Genova, A., Mazarico, E., Goossens, S., Lemoine, F.G., Neumann, G.A., Smith, D.E., and Zuber, M.T. (2018). Solar system expansion and strong equivalence principle as seen by the NASA MESSENGER mission. *Nature communications* 9 (1): 1–9.

Goldstein, R.M. (1969). Superior conjunction of Pioneer-6. *Science* 166: 598–601. doi:10.1126/science.166.3905.598.

Goossens, S., Matsumoto, K., Liu, Q., Kikuchi, F., Sato, K., Hanada, H., Ishihara, Y., Noda, H., Kawano, N., Namiki, N., Iwata, T., Lemoine, F.G., Rowlands, D.D., Harada, Y., and Chen, M. (2011). Lunar gravity field determination using SELENE same-beam differential VLBI tracking data. *Journal of Geodesy* 85 (4): 205–228.

Grall, R.R., Coles, W.A., Klinglesmith, M.T., Breen, A.R., Williams, P.J.S., Markkanen, J., and Esser, R. (1996). Rapid acceleration of the polar solar wind. *Nature* 379: 429–432.

Grasset, O., Dougherty, M.K., Coustenis, A., Bunce, E.J., Erd, C., Titov, D., Blanc, M., Coatesg, A., Drossartc, P., Fletcherh, L.N., Hussmanni, H., Jaumanni, R., Kruppj, N., Lebretonk, J.-P., Prieto-Ballesterosl, O., Tortoram, P., Tosin, F., and Van Hoolst, T. (2013). JUpiter ICy moons Explorer (JUICE): an ESA mission to orbit Ganymede and to characterise the Jupiter system. *Planetary and Space Science* 78: 1–21.

Greeley, R. (1973). Mariner 9 photographs of small volcanic structures on Mars. *Geology* 1 (4): 175–180.

Green, D.W. (1980). Validation of roundtrip charged-particle calibrations derived from S-and X-band Doppler via DRVID measurements. *The Deep Space Network Progress Report* 42–55, November–December 1979, Jet Propulsion Laboratory, California Institute of Technology, Pasadena, CA, 15 February 1980.

Gregnanin, M., Bertotti, B., Chersich, M., Fermi, M., Iess, L., Simone, L., Tortora, P., and Williams, J.G. (2012). Same beam interferometry as a tool for the investigation of the lunar interior. *Planetary and Space Science* 74 (1): 194–201.

Gudim, M.A., Gawronski, W., Hurd, W.J., Brown, P.R. and Strain, D.M. (1999). Design and performance of the monopulse pointing system of the DSN 34-meter beam-waveguide antennas. *Telecommunications and Mission Operations Progress Report* 42: 138.

Haibo, Y., Lili, Q., Shaowu, D., Wei, L. and Hong, Z. (2009, April). The hydrogen maser and cesium clocks in time keeping at NTSC. In: *2009 IEEE International Frequency Control Symposium Joint with the 22nd European Frequency and Time forum* 639–642. IEEE.

Hajj, G.A., Ao, C.O., Iijima, B.A., Kuang, D., Kursinski, E.R., Mannucci, A.J., Meehan, T.K., Romans, L.J., De La Torre Juarez, M., and Yunck, T.P. (2004). CHAMP and SAC-C atmospheric occultation results and intercomparisons. *Radio Science* 42 (2): 18. doi:10.1029/2003JD003909.

Hajj, G.A., Kursinski, E.R., Romans, L.J., Bertiger, W.I., and Leroy, S.S. (2002). A technical description of atmospheric sounding by GPS occultation. *Journal of Atmospheric and Solar-Terrestrial Physics* 64 (4): 451–469.

Han, S.C., Jekeli, C., and Shum, C.K. (2004). Time-variable aliasing effects of ocean tides, atmosphere, and continental water mass on monthly mean GRACE gravity field. *Journal of Geophysical Research: Solid Earth* 109 (B4).

Harmon, J.K. and Coles, W.A. (1983). Spectral broadening of planetary radar signals by the solar wind. *Astrophysical Journal* 270: 748–757.

Hartop, R. (1987). Microwave component time delays for the 70-meter antennas. *TDA Progress Report 42* 89: 73–75.

Haskins, C.B., Angert, M.P., Sheehi, E.J., Millard, W.P., Adams, N. and Hennawy, J.R. (2016, March). The Frontier software-defined radio for the solar probe plus mission. In *2016 IEEE Aerospace Conference* 1–11. IEEE.

Haskins, C.B. and DeBoy, C.C. (2007). Deep-space transceivers—an innovative approach to spacecraft communications. *Proceedings of the IEEE* 95 (10): 2009–2018.

Häusler, B., Pätzold, M., Tyler, G.L., Simpson, R.A., Bird, M.K., Dehant, V., Barriot, J.P., Eidel, W., Mattei, R., Remus, S., and Selle, J. (2006). Radio science investigations by VeRa onboard the Venus Express spacecraft. *Planetary and Space Science* 54 (13–14), 1315–1335.

Hayati, S., Volpe, R., Backes, P., Balaram, J., Welch, R., Ivlev, R., Tharp, G., Peters, S., Ohm, T., Petras, R. and Laubach, S. (1997, April). The Rocky 7 rover: a Mars sciencecraft prototype. In: *Proceedings of international conference on robotics and automation* 3: 2458–2464. IEEE.

Heiskanen, W.A. and Moritz, H. (1967). Physical geodesy. *Bulletin Géodésique (1946–1975)* 86 (1): 491–492.

Helled, R., Anderson, J.D., Schubert, G. and Stevenson, D.J. (2011). Jupiter's moment of inertia: A possible determination by Juno. *Icarus* 216 (2): 440–448.

Helled, R., Nettelmann, N., and Guillot, T. (2020). Uranus and Neptune: Origin, evolution and internal structure. *Space Science Reviews* 216 (3): 1–26.

Hellings, R.W., Callahan, P.S., Anderson, J.D., and Moffet, A.T. (1981). Spacecraft Doppler gravity-wave detection; II. Results. *Physical Review D* 23: 844–851.

Hemingway, D., Iess, L., Tajeddine, R. and Tobie, G. (2018). The interior of Enceladus. *Enceladus and the icy moons of Saturn* 57–77.

Hemingway, D.J., Zannoni, M., Tortora, P., Nimmo, F., and Asmar, S.W. (2016). Dione's internal structure inferred from Cassini gravity and topography. #1903: 1314. *47th Lunar and Planetary Science Conference*, The Woodlands, TX: (21–25 March 2016).

Hemmati, H., (ed). (2006). Deep Space Optical Communications. Hoboken, NJ: Wiley. Used Wiley-specific formatting.

Hewish, A., Scott, P.F., and Wills, D. (1964). Interplanetary Scintillation of Small Diameter Radio Sources. *Nature* 203: 1214–1217.

Hinson, D.P. (1984). Magnetic field orientations in Saturn's upper ionosphere inferred from Voyager radio occultations. *Journal of Geophysical Research: Space Physics* 89: 65–73.

Hinson, D.P. (1986). Strong scintillations during atmospheric occultations: Theoretical intensity spectra. *Radio science* 21 (2): 257–270.

Hinson, D.P., Asmar, S.W., Kahan, D.S., Akopian, V., Spiga, A., Schofield, J.T., Kleinbohl, A., Abdou, W.A., Lewis, S.R., Paik, M., and Maalouf, S.G. (2014). Initial results from radio occultation measurements with the Mars Reconnaissance Orbiter: A nocturnal mixed layer in the tropics and comparisons with polar profiles from the Mars Climate Sounder. *Icarus* 243: 91–103.

Hinson, D.P., Flasar, F.M., Kliore, A.J., Schinder, P.J., Twicken, J.D., and Herrera, R.G. (1997). Jupiter's ionosphere: results from the first Galileo radio occultation experiment. *Geophysical Research Letters* 24: 2107–2110.

Hinson, D.P., Kliore, A.J., Flasar, F.M., Twicken, J.D., Schinder, P.J., and Herrera, R.G. (1998). Galileo radio occultation measurements of Io's ionosphere and plasma wake. *Journal of Geophysical Research – Space Physics* 103: 29343–29357.

Hinson, D.P., Pätzold, M., Wilson, R.J., Häusler, B., Tellmann, S. and Tyler, G.L. (2008). Radio occultation measurements and MGCM simulations of Kelvin waves on Mars. *Icarus* 193 (1): 125–138.

Hinson, D.P., Tyler, G.L., Hollingsworth, J.L., and Wilson, R.J. (2001). Radio occultation measurements of forced atmospheric waves on Mars. *Journal of Geophysical Research: Planets* 106: 1463–1480.

Hodges, R.E., Chahat, N.E., Hoppe, D.J. and Vacchione, J.D. (2016, June). The Mars Cube One deployable high gain antenna. In: *2016 IEEE International Symposium on Antennas and Propagation (APSURSI)* 1533–1534. IEEE.

Hoffman-Wellenhof, B., Lichtenegger, H., and Collins, J. (1997). GPS Theory and Practice. 4th edn. Wein: Springer

Hofstadter, M., Simon, A., Atreya, S., Banfield, D., Fortney, J., Hayes, A., Hedman, M., Hospodarsky, G., Mandt, K., Masters, A., and Showalter, M. (2017). A vision for ice giant exploration.

Hofstadter, M., Simon, A., Atreya, S., Banfield, D., Fortney, J.J., Hayes, A., Hedman, M., Hospodarsky, G., Mandt, K., Masters, A., and Showalter, M. (2019). Uranus and Neptune missions: A study in advance of the next Planetary Science Decadal Survey. *Planetary and Space Science* 177: 104680.

Howard, H.T., Eshleman, V.R., Hinson, D.P., Kliore, A.J., Lindal, G.F., Woo, R., Bird, M.K., Volland, H., Edenhofer, P., Pätzold, M., and Porsche, H. (1992). Galileo radio science investigations. *Space Science Reviews* 60: 565–590.

Howard, H.T., Tyler, G.L., Esposito, P.B., Anderson, J.D., Reasenberg, R.D., Shapiro, I.I., Fjeldbo, G., Kliore, A.J., Levy, G.S., Brunn, D.L., Dickinson, R., Edelson, R.E., Martin, W.L., Postal, R.B., Seidel, B., Sesplaukis, T.T., Shirley, D.L., Stelzried, C.T., Sweetnam, D.N., Wood, G.E., and Zygielbaum, A.L. (1974a). Mercury: results on mass, radius, ionosphere, and atmosphere from Mariner 10 dual-frequency radio signals. *Science* 185 (4146): 179–180.

Howard, H.T., Tyler, G.L., Fjeldbo, G., Kliore, A.J., Levy, G.S., Brunn, D.L., Dickenson, R., Edelson, R.E., Martin, W.L., Postal, R.B., Seidel, B.L., Sesplaukis, T.T., Shirley, D.L., Stelzried, C.T., Sweetnam, D.N., Ziegielbaum, A.I., Esposito, P.B., Anderson, J.D., Shapiro, I.I., and Reasenberg, R.D. (1974b). Venus: mass, gravity field, atmosphere, and ionosphere as measured by the Mariner 10 dual-frequency radio system. *Science* 183: 1297–1301.

Hubbard, W.B., Nellis, W.J., Mitchell, A.C., Holmes, N.C., Limaye, S.S., and McCandless, P.C. (1991). Interior structure of Neptune: comparison with Uranus. *Science* 253 (5020): 648–651.

Hulse, R.A. and Taylor, J.H. (1975). Discovery of a pulsar in a binary system. *Astrophysical Journal* 195: L51–L53.

Ichikawa, T. and Kato, T. (1999). The study of Delta VLBI and doppler navigation for interplanetary spacecraft. *Advances in Space Research* 23 (4): 739–746.

Iess, L. (2013). 3GM: Gravity and geophysics of Jupiter and the Galilean Moons. *European Planetary Science Congress* September 2013.

Iess, L. and Armstrong, J.W. (1996). Spacecraft Doppler experiments (1997). In: Proceedings of the International Conference, Gravitational Waves: Sources and Detectors, (eds. I. Ciufolini and F. Fidecaro) 323–329. Cascina, Italy (19 March 1996). Singapore: World S'cientific; New Jersey: River Edge. http://trs-new.jpl.nasa.gov/dspace/handle/2014/22760?mode=full&submit_simple=Show+full+item+record.

Iess, L., Armstrong, J.W., Bertotti, B., Wahlquist, H.D., and Estabrook, F.B. (1997). Search for Gravitational Wave Bursts by Simultaneous Doppler Tracking of Three Interplanetary Spacecraft. *15th International Conference on General Relativity and Gravitation*, Pune, India, 16 December 1997.

Iess, L., Asmar, S. W., Cappuccio, P., Cascioli, G., De Marchi, F., di Stefano, I., ... & Zannoni, M. (2021). Gravity, geodesy and fundamental physics with BepiColombo's MORE investigation. *Space Science Reviews* 217 (1): 1–39.

Iess, L. and Asmar, S.W. (2007). Probing Space-Time in the Solar System: from Cassini to BepiColombo. *International Journal of Modern Physics D* 16 (12a): 2117–2126. 10.1142/S0218271807011449.

Iess, L., Asmar, S.W., Mercolino, M., McManamon, P., and Palmerini, G. (2004). Preliminary results from the tracking of SMART-1 Ka-band. ESA 3rd Tracking,

Telemetry & Command Systems for Space Applications Workshop, Darmstadt: Germany (September 2004).

Iess, L., Asmar, S.W., and Tortora, P. (2009). MORE: an advanced tracking experiment for the exploration of Mercury with the mission BepiColombo. *Acta Astronautica* 65 (5): 666–675.

Iess, L. and Armstrong, J. W. Spacecraft Doppler Experiments. In: Ciufolini, I. and Fidecaro, F., eds., *Gravitational Waves: Sources and Detectors*, Proceedings of the international conference, Cascina, Italy: 19–23 March, 1996, 323–329. (World Scientific, Singapore; River Edge, NJ, 1997).

Iess, L. and Boscagli, G. (2001). Advanced radio science instrumentation for the mission BepiColombo to Mercury. *Planetary and Space Science* 49 (14–15): 1597–1608.

Iess, L., Folkner, W.M., Durante, D., Parisi, M., Kaspi, Y., Galanti, E., Guillot, T., Hubbard, W.B., Stevenson, D.J., Anderson, J.D., and Buccino, D.R. (2018). Measurement of Jupiter's asymmetric gravity field. *Nature* 555 (7695): 220–222.

Iess, L., Jacobson, R.A., Ducci, M., Stevenson, D.J., Lunine, J.I., Armstrong, J.W., Asmar, S.W., Racioppa, P., Rappaport, N.J., and Tortora, P. (2012). The tides of Titan. *Science* 337: 457–459. doi:10.1126/science.1219631.

Iess, L., Militzer, B., Kaspi, Y., Nicholson, P., Durante, D., Racioppa, P., Anabtawi, A., Galanti, E., Hubbard, W., Mariani, M.J. and Tortora, P. (2019). Measurement and implications of Saturn's gravity field and ring mass. *Science* 364 (6445).

Iess, L., Rappaport, N.J., Jacobson, R.A., Racioppa, P., Stevenson, D.J., Tortora, P., Armstrong, J.W., and Asmar, S.W. (2010). Gravity Field, Shape and Moment of Inertia of Titan. *Science* 327: 1367–1369. doi:10.1126/science.1182583.

Iess, L., Rappaport, N.J., Tortora, P., Lunine, J.I., Armstrong, J.W., Asmar, S.W., Somenzi, L., and Zingoni, F. (2007a). *Gravity Field and Interior of Rhea from Cassini Data Analysis*. Icarus 190 (2): 585–593. doi:10.1016/j.icarus.2007.03.027.

Iess, L., Stevenson, D.J., Parisi, M., Hemingway, D., Jacobson, R.A., Lunine, J.I., Nimmo, F., Armstrong, J.W., Asmar, S.W., Ducci, M., and Tortora, P. (2014). The Gravity Field and Interior Structure of Enceladus. *Science* 344 (6179): 78–80. doi:10.1126/science.1250551.

Iess, L., Tortora, P., Madde, R., Mercolino, M., and Asmar, S.W. (2007b). The tracking system of the BepiColombo radio science experiment (MORE). *4th ESA International Workshop on Tracking, Telemetry and Command Systems for Space Applications*, Darmstadt, Germany: (11–14 September 2007).

Imamura, T., Toda, T., Tomiki, A., Hirahara, D., Hayashiyama, T., Mochizuki, N., Yamamoto, Z.I., Abe, T., Iwata, T., Noda, H., and Futaana, Y. (2011). Radio occultation experiment of the Venus atmosphere and ionosphere with the Venus orbiter Akatsuki. *Earth, Planets and Space* 63 (6): 493–501.

Imbriale, W.A., Bhanji, A.M., Blank, S., Lobb, V.B., Levy, R. and Rocci, S.A. (1987). Ka-band (32-GHz) performance of 70-meter antennas in the Deep Space Network. The Telecommunications and Data Acquisition Progress Report 42-88 126.

Ishimaru, A. (1978). Wave Propagation and Scattering in Random Media. New York: Academic Press.

Ivanov-Kholodny, G.S., Kolosov, M.A., Savich, N.A., Alexandrov, Y.N., Vasilyev, M.B., Vyshlov, A.S., Dubrovin, V.M., Zaitsev, A.L., Michailov, A.V., Petrov, G.M., and Samovol, V.A. (1979). Daytime ionosphere of Venus as studied with Veneras 9 and 10 dual-frequency radio occultation experiments. *Icarus* 39 (2): 209–213.

Ivins, E.R., Watkins, M.M., Yuan, D.N., Dietrich, R., Casassa, G., and Rülke, A. (2011). On-land ice loss and glacial isostatic adjustment at the Drake Passage: 2003–2009. *Journal of Geophysical Research: Solid Earth* 116 (B2).

Jacobson, R.A. (2001). The gravity field of the Jovian system and the orbits of the regular Jovian satellites. *Bulletin of the American Astronomical Society* 33: 1039.

Jacobson, R.A., Antreasian, P.G., Bordi, J.J., Criddle, K.E., Ionasescu, R., Jones, J.B., Mackenzie, R.A., Meek, M.C., Parcher, D., Pelletier, F.J., Owen, W.M., Jr., Roth, D.C., Roundhill, I.M., and Stauch, J.R. (2006). The gravity field of the Saturnian system from satellite observations and spacecraft tracking data. *The Astronomical Journal* 132 (6): 2520.

Jacobson, R.A., Spitale, J., Porco, C.C., Beurle, K., Cooper, N.J., Evans, M.W. and Murray, C.D. (2007). Revised orbits of Saturn's small inner satellites. *The Astronomical Journal* 135 (1): 261.

Jaffe, A.H. and Backer, D.C. (2003). Gravitational waves probe the coalescence rate of massive black hole binaries. *Astrophysical Journal* 583: 616–631.

Janardhan, P., Bird, M.K., Edenhofer, P., Wohlmuth, R., Plettemeier, D., Asmar, S.W., Pätzold, M., and Karl, J. (1999). Coronal velocity measurements with Ulysses: multi-link correlation studies during two superior conjunctions. *Solar Physics* 184 (1): 157–172.

Janle, P. and Erkul, E. (1991). Gravity studies of the Tharsis area on Mars. *Earth, Moon, and Planets* 53 (3): 217–232.

Jaumann, R., Williams, D.A., Buczkowski, D.L., Yingst, R.A., Preusker, F., Hiesinger, H., Schmedemann, N., Kneissl, T., Vincent, J.B., Blewett, D.T. and Buratti, B.J. (2012). Vesta's shape and morphology. *Science* 336 (6082): 687–690.

Jenkins, G.M. and Watts, D.G. (1969). Spectral Analysis and Its Applications. San Francisco, CA: Holden-Day.

Jenkins, J.M. (1993). Variations in the 13 cm Opacity Below the Main Cloud Layer in the Atmosphere of Venus Inferred from Pioneer–Venus Radio Occultation Studies, 1978–1987. Georgia Institute of Technology Dissertation.

Jenkins, J.M., Steffes, P.G., Hinson, D.P., Twicken, J.D., and Tyler, G.L. (1994). Radio occultation studies of the Venus atmosphere with the Magellan spacecraft: 2. Results from the October 1991 experiments. *Icarus* 110: 79–94.

Jokipii, J.R. (1973). Turbulence and scintillations in the interplanetary plasma. *Annual Review of Astronomy and Astrophysics* 11 (1): 1–28. doi:10.1146/annurev.aa.11.090173.000245.

Jongeling, A., Tanner, A., Border, J., Long, E. and Lin, H. (2018). December. In-line Water Vapor Radiometer for Precision Deep Space Doppler Tracking. *AGU Fall Meeting Abstracts* 2018: P41F–3789.

Kahan, D. S., Folkner, W. M., Buccino, D. R., Dehant, V., Le Maistre, S., Rivoldini, A., ... & Marty, J. C. (2021). Mars precession rate determined from radiometric tracking of the InSight Lander. *Planetary and Space Science* 199: 105208.

Kakinuma, T. (1977). Observations of interplanetary scintillation: solar wind velocity measurements. In: Study of Travelling Interplanetary Phenomena, (eds. M.A. Shea and D.F. Smart), 101–118. *Proceeding of the L.D. De Feiter Memorial Symposium*, Tel Aviv, Israel (7–10 June 1977). Dordrecht, Netherlands: Reidel.

Karatekin, Ö. and Asmar, S.W. (2011). Entry trajectory reconstruction using Phoenix radio link. 8th *International Planetary Probe Workshop*, Norfolk, VA (June 2011).

Karatekin, Ö., Gerbal, N., Van Hove, B., Ferri, F., Asmar, S.W., Firre, D., and Denis, M., and AMELIA team. (2017). Preliminary analysis of entry and descent radio communications of ExoMars 2016 Schiaparelli. *14th International Planetary Probe Workshop*, The Hague, Netherlands: (June 2017).

Kaspi, Y., Galanti, E., Showman, A.P., Stevenson, D.J., Guillot, T., Iess, L., and Bolton, S.J. (2020). Comparison of the deep atmospheric dynamics of Jupiter and Saturn in light of the Juno and Cassini gravity measurements. *Space Science Reviews* 216 (5): 1–27.

Kato, M., Sasaki, S., Tanaka, K., Iijima, Y. and Takizawa, Y. (2008). The Japanese lunar mission SELENE: Science goals and present status. *Advances in Space Research* 42 (2): 294–300.

Kantha, L., Barlow, E., Axelrad, P., Hooper, D., and Cornman, L. (2013). January. Turbulence retrievals from radio occultations. In 2013 US National Committee of URSI National Radio Science Meeting (USNC-URSI NRSM) (pp. 1–1). IEEE.

Kaufman, J.J. (1976). Measurements of the Solar Wind Velocity and Compact Radio Source Structure from 74 MHz Interplanetary Scintillations. Ph. D. Thesis, University of California San Diego.

Kaufman, W.J. (1970). Redshift fluctuations arising for gravitational waves. *Nature* 327: 157–159.

Kaula, W.M. (1963). Determination of the Earth's gravitational field. *Reviews of Geophysics* 1 (4): 507–551.

Kaula, W.M. (1966). Theory of Satellite Geodesy. Waltham, MA: Blaisdell.

Karimi, S. and Lewis, K.W. (2018). March. Why Do Mascon Basins Occur on the Moon? *Lunar and Planetary Science Conference* 2083: 2543.

Keihm, S.J. (1995). Water Vapor Radiometer Measurements of the Tropospheric Delay Fluctuations at Goldstone over a Full Year. The Telecommunications and Data Acquisition Progress Report 42-122, April–June 1995, Jet Propulsion Laboratory, California Institute of Technology, Pasadena, CA, 15 August 1995. https://ipnpr.jpl.nasa.gov/progress_report/42–122/122J.pdf

Keihm, S.J., Tanner, A., and Rosenberger, H. (2004). Measurements and Calibration of Tropospheric Delay at Goldstone from the Cassini Media Calibration System. In: The Interplanetary Network Progress Report 42–158. Pasadena, CA: Jet Propulsion Laboratory, California Institute of Technology. 15 August 2004. https://ipnpr.jpl.nasa.gov/progress_report/42-158/158A.pdf.

Kerzhanovich, V.V., Marov, M.Y., and Rozhdestvensky, M.K. (1972). Data on dynamics of the subcloud Venus atmosphere from Venera spaceprobe measurements. *Icarus* 17 (3): 659–674.

Kerzhanovich, V.V., Antsibor, N.M., Kustodiev, V.D., Makarov, Y.F., Matsygorin, I.A., Molotov, E.P., Sorokin, V.P., Sukhanov, K.G., Tikhonov, V.F., Karyagin, V.P., and Mottsulev, B.I. (1983). Vertical profiles of wind velocity in the Venus atmosphere from Doppler measurements of Venera 13 and Venera 14. *Kosmicheskie Issledovaniia* 21: 211–217.

Kim, J. and Lee, S.W. (2009). Flight performance analysis of GRACE K-band ranging instrument with simulation data. *Acta Astronautica* 65 (11–12): 1571–1581.

Kim, J. and Tapley, B.D. (2002). Error analysis of a low-low satellite-to-satellite tracking mission. *Journal of Guidance, Control, and Dynamics* 25 (6): 1100–1106.

Kinman, P.W. and Asmar, S.W. Two-Way Coherent Doppler Error Due to Solar Corona. (1988). *The Telecommunications and Data Acquisition Progress Report* 42-95, 28–31. July–September 1988. Jet Propulsion Laboratory, California Institute of Technology, Pasadena, CA, 15 November 1988. https://tda.jpl.nasa.gov/progress_report/42-95/95C.PDF.

Klipstein, W.M., Arnold, B.W., Enzer, D.G., Ruiz, A.A., Tien, J.Y., Wang, R.T. and Dunn, C.E. (2013). The lunar gravity ranging system for the Gravity Recovery and Interior Laboratory (GRAIL) mission. *Space Science Reviews* 178 (1): 57–76.

Kliore, A., Levy, G.S., Cain, D.L., Fjeldbo, G., and Rasool, S.I. (1967). Atmosphere and ionosphere of Venus from the Mariner V S-band radio occultation measurement. *Science* 158 (3809): 1683–1688.

Kliore, A.J., Anabtawi, A., Herrera, R.G., Asmar, S.W., Nagy, A.F., Hinson, D.P., and Flasar, F.M. (2002). Ionosphere of Callisto from Galileo radio occultation observations. *Journal of Geophysical Research – Space Physics* 107 (A11): SIA–19. doi:10.1029/2002JA009365.

Kliore, A.J., Anderson, J.D., Armstrong, J.W., Asmar, S.W., Hamilton, C.L., Rappaport, N.J., Wahlquist, H.D., Ambrosini, R., Bertotti, B., Flasar, F.M., French, R.G., Iess, L., Marouf, E.A., and Nagy, A.F. (2004). Cassini radio science. *Space Science Reviews* 115: 1–70.

Kliore, A.J. and Cain, D.L. (1968). Mariner V and the radius of Venus. *Journal of Atmospheric Science* 25: 549–554.

Kliore, A.J., Cain, D.L., and Hamilton, T.W. (1964). Determination of some physical properties of the atmosphere of mars from changes in the doppler signal of a spacecraft on an earth-occultation trajectory. Technical Report 32-674, Pasadena, CA: Jet Propulsion Laboratory, California Institute of Technology.

Kliore, A.J., Cain, D.L., Fjeldbo, G., Seidel, B.L., Sykes, M.J., and Rasool, S.I. (1972). The atmosphere of Mars from Mariner 9 radio occultation measurements. *Icarus* 17 (2): 484–516.

Kliore, A.J., Cain, D.L., Levy, G.S., Eshleman, V.R., Fjeldbo, G., and Drake, F.D. (1965). Occultation experiment results of the first direct measurement of Mars atmosphere and ionosphere. *Science* 149: 1243–1248.

Kliore, A.J., Fjeldbo, G., Seidel, B.L., and Rasool, S.I. (1969). Mariners 6 and 7: radio occultation measurements of the atmosphere of Mars. *Science* 166: 1393–1397.

Kliore, A.J., Hinson, D.P., Flasar, F.M., Nagy, A.F., and Cravens, T.E. (1997). The ionosphere of Europa from Galileo radio occultations. *Science* 277: 355–357.

Kliore, A. J., Hinson, D. P., Flasar, F. M., Nagy, A. F., & Cravens, T. E. (1997). The ionosphere of Europa from Galileo radio occultations. *Science* 277 (5324): 355–358.

Kliore, A.J., Lindal, G.F., Sweetnam, D.N., Hotz, H.B., and McDonough, T.R. (1980a). Vertical structure of the ionosphere and upper neutral atmosphere of Saturn from Pioneer radio occultation. *Science* 207: 446–449.

Kliore, A.J., Luhmann, J.G., and Zhang, M.H. (1991). The effect of the solar cycle on the maintenance of the nightside ionosphere of Venus. *Journal of Geophysical Research: Space Physics* 96: 11065–11071.

Kliore, A.J., Nagy, A.F., Asmar, S.W., Anabtawi, A., El Barbinis, D., Fleischman, D.S., Kahan, and Klose, J. (2014). The ionosphere of Saturn as observed by the Cassini radio science system. *Geophysical Research Letters* 41 (16): 5778–5782. doi:10.1002/2014GL060512.

Kliore, A.J., Nagy, A.F., Marouf, E.A., French, R.G., Flasar, F.M., Rappaport, N.J., Anabtawi, A., Asmar, S.W., Kahan, D.S., Barbinis, E., Goltz, G.L., Fleischman, D.U., and Rochblatt, D.J. (2008). First results from the Cassini radio occultations of the Titan ionosphere. *Journal of Geophysical Research: Space Physics* 113: doi:10.1029/2007JA012965.

Kliore, A.J. and Patel, I.R. (1980). Vertical structure of the atmosphere of Venus from Pioneer Venus orbiter radio occultations. *Journal of Geophysical Research: Space Physics* 85 (A13): 7957–7962.

Kliore, A.J., Patel, I.R., Lindal, G.F., Sweetnam, D.N., Hotz, H.B., Waite, J.H., Jr., and McDonough, T.R. (1980b). Structure of the ionosphere and atmosphere of Saturn from Pioneer 11 Saturn radio occultation. *Journal of Geophysical Research* 85: 5857–5870.

Kliore, A.J., Patel, I.R., Nagy, A.F., Cravens, T.E., and Gombosi, T.I. (1979). Initial observations of the nightside ionosphere of Venus from Pioneer Venus orbiter radio occultations. *Science* 205 (4401): 99–102.

Kliore, A.J. and Woiceshyn, P.M. (1976). Structure of the atmosphere of Jupiter from Pioneer 10 and 11 radio occultation measurements. IAU Colloquium 30: Jupiter Studies of the Interior, Atmosphere, Magnetosphere and Satellites, 216–237.

Klipstein, W.M., Seidel, D.J., White, J.A., and Young, B.C. (2001). Optical frequency standard development in support of NASA's gravity-mapping missions. *Frequency Control Symposium, Proceedings of the 2001 IEEE International*, 108–110.

Koehler, R.L. (1968). Radio propagation measurements of pulsed plasma streams from the Sun using Pioneer spacecraft. *Journal of Geophysical Research* 73: 4883–4894. doi:10.1029/JA073i015p04883.

Kolosov, M.A., Yakovlev, O.I., Efimov, A.I., Rogal'sky, V.I., Razmanov, V.M., and Shtrykov, V.K. (1982). Decimeter radio wave propagation in the turbulent plasma near the sun, using Venera 10 spacecraft. *Radio Science* 17 (03): 664–674.

Konopliv, A.S., Asmar, S.W., Bills, B.G., Mastrodemos, N., Park, R.S., Raymond, C.A., Russell, C.T., Smith, D.E., and Zuber, M.T. (2011a). The Dawn gravity investigation at Vesta and Ceres. *Space Science Reviews* doi:10.1007/s11214-011-9794-8.

Konopliv, A.S., Asmar, S.W., Carranza, E., Sjogren, W.L., and Yuan, D.-N. (2001). Recent gravity models as a result of the Lunar Prospector mission. *Icarus* 150 (1): 1–18.

Konopliv, A.S., Asmar, S.W., Folkner, W.M., Karatekin, Ö., Nunes, D.C., Smrekar, S.E., Yoder, C.F., and Zuber, M.T. (2011). Mars high resolution gravity fields from MRO, Mars seasonal gravity, and other dynamical parameters. *Icarus* 211 (1): 401–428.

Konopliv, A.S., Asmar, S.W., Folkner, W.M., Karatekin, Ö., Nunes, D.C., Smrekar, S.E., Yoder, C.F., and Zuber, M.T. (2011b). Mars high resolution gravity fields from MRO, Mars seasonal gravity, and other dynamical parameters. *Icarus* 211 (1): 401–428. doi:10.1016/j.icarus.2010.10.004.

Konopliv, A.S., Asmar, S.W., Park, R.S., Bills, B.G., Centinello, F., Chamberlin, A.B., Ermakov, A., Gaskell, R.W., Rambaux, N., Raymond, C.A., and Russell, C.T. (2014a). The Vesta gravity field, spin pole and rotation period, landmark positions, and ephemeris from the Dawn tracking and optical data. *Icarus* 240: 103–117.

Konopliv, A.S., Banerdt, W.B., and Sjogren, W.L. (1999). Venus gravity: 180th degree and order model. *Icarus* 139 (1): 3–18.

Konopliv, A.S., Park, R.S. and Ermakov, A.I. (2020a). The Mercury gravity field, orientation, love number, and ephemeris from the MESSENGER radiometric tracking data. *Icarus* 335: 113386.

Konopliv, A.S., Park, R.S., Rivoldini, A., Baland, R.M., Le Maistre, S., Van Hoolst, T., Yseboodt, M. and Dehant, V. (2020b). Detection of the Chandler wobble of Mars from Orbiting Spacecraft. *Geophysical Research Letters* 47 (21): e2020GL090568.

Konopliv, A.S., Park, R.S., and Folkner, W.M. (2016). An improved JPL Mars gravity field and orientation from Mars Orbiter and Lander tracking data. *Icarus* 274: 253–260.

Konopliv, A.S., Park, R.S., Vaughan, A.T., Bills, B.G., Asmar, S.W., Ermakov, A.I., Rambaux, N., Raymond, C.A., Castillo-Rogez, J.C., Russell, C.T., and Smith, D.E. (2018). The Ceres gravity field, spin pole, rotation period and orbit from the Dawn radiometric tracking and optical data. *Icarus* 299: 411–429.

Konopliv, A.S., Park, R.S., Yuan, D.-N., Asmar, S.W., Watkins, M.M., Williams, J.G., Fahnestock, E., Kruizinga, G., Paik, M., Strekalov, D., Harvey, N., Smith, D.E., and Zuber, M.T. (2013). The JPL lunar gravity field to spherical harmonic degree 660 from the GRAIL primary mission. *Journal of Geophysical Research: Planets* 118 (7): 1415–1434. doi:10.1002/jgre.20097.

Konopliv, A.S., Park, R.S., Yuan, D.-N., Asmar, S.W., Watkins, M.M., Williams, J.G., Fahnestock, E., Kruizinga, G., Paik, M., Strekalov, D., Harvey, N., Smith, D.E., and Zuber, M.T. (2014b). JPL high resolution lunar gravity fields from the GRAIL primary and extended mission. *Geophysical Research Letters* doi:10.1002/2013GL059066.

Konopliv, A.S. and Sjogren, W.L. (1995). The JPL Mars Gravity Field, Mars50c, Based upon Viking and Mariner 9 Doppler tracking data. JPL Publication 95-5, Pasadena CA: Jet Propulsion Laboratory, California Institute of Technology. Also NASA CR-198881.

Konopliv, A.S. and Yoder, C.F. (1996). Venusian k2 tidal love number from Magellan and PVO tracking data. *Geophysical Research Letters* 23 (14): 1857–1860.

Konopliv, A.S., Yoder, C.F., Standish, E.M., Yuan, D.-N., and Sjogren, W.L. (2006). A global solution for the Mars static and seasonal gravity, Mars orientation, Phobos and Deimos masses, and Mars ephemeris. *Icarus* 182 (1): 23–50.

Kornfeld, R.K., Bruvold, D.D., Craig, M.L., Asmar, S.W., and Ilott, P. (2011). Reconstruction of entry, descent and landing communications for the Phoenix Mars Lander. *Journal of Spacecraft and Rockets* 48 (5): 822–835. doi:10.2514/1.47909.

Kusters, J.A. (1981, October). The SC cut crystal-An overview. In: *1981 Ultrasonics Symposium* 402–409. IEEE.

Kramm, U., Nettelmann, N., Redmer, R. and Stevenson, D.J. (2011). On the degeneracy of the tidal Love number k2 in multi-layer planetary models: application to Saturn and GJ 436b. *Astronomy & Astrophysics* 528: A18.

Kraus, J.D., Tiuri, M., Räisänen, A.V., and Carr, T.D., (1966). Radio Astronomy Vol. 66. New York: McGraw-Hill.

Krisher, T.P., Anderson, J.D., Morabito, D.D., Asmar, S.W., Borutzki, S.E., Delitsky, M.L., Densmore, A.C., Eshe, P.M., Lewis, G.D., Maurer, M.J., Roth, D.C., Son, Y.H., Spilker, T.R., Sweetnam, D.N., Taylor, A.H., Tyler, G.L., Gresh, D.L., and Rosen, P.A. (1991). Radio range measurements of coronal electron densities at 13-centimeter and 3.6-centimeter wavelengths during the 1988 solar conjunction of Voyager 2. *Astrophysical Journal* 375: L57–L60.

Krogh, F.T. (1973). Changing step size in the Integration of differential equations using Modified Divided Differences. JPL Technical Memorandum No. 312, Section 914, Pasadena, CA: Jet Propulsion Laboratory, California Institute of Technology.

Kursinski, E.R., Hajj, G.A., Hardy, K.R., Schofield, J.T., and Linfield, R. (1997). Observing Earth's atmosphere with radio occultation measurements. *Journal of Geophysical Research: Atmospheres* 102 (D19): 23429–23465.

Kuchynka, P., Folkner, W.M., Konopliv, A.S., Parker, T.J., Park, R.S., Le Maistre, S. and Dehant, V. (2014). New constraints on Mars rotation determined from radiometric tracking of the Opportunity Mars Exploration Rover. *Icarus* 229: 340–347.

Lainey, V., Casajus, L.G., Fuller, J., Zannoni, M., Tortora, P., Cooper, N., Murray, C., Modenini, D., Park, R.S., Robert, V. and Zhang, Q. (2020). Resonance locking in giant planets indicated by the rapid orbital expansion of Titan. *Nature Astronomy* 4 (11): 1053–1058.

Lainey, V., Jacobson, R.A., Tajeddine, R., Cooper, N.J., Murray, C., Robert, V., Tobie, G., Guillot, T., Mathis, S., Remus, F. and Desmars, J. (2017). New constraints on Saturn's interior from Cassini astrometric data. *Icarus* 281: 286–296.

Lambeck, K. (1988). Geophysical Geodesy: The Slow Deformations of the Earth. Clarendon, NY: Oxford.

Lawson, C.L. and Hanson, R.J. (1995). Solving Least Squares Problems. SIAM Classics in Applied Mathematics, 15. Society for Industrial and Applied Mathematics, Philadelphia, PA.

Lee, D. (2002). Measurement Error for Two-Way Coherent Doppler for MRO. Inter-Office Memorandum, Jet Propulsion Laboratory, Pasadena, CA: California Institute of Technology.

Lee, A. and Hanover, G. (2005, August). Cassini spacecraft attitude control system flight performance. In: *AIAA Guidance, Navigation, and Control Conference and Exhibit* 6269.

Le Maistre, S., Rosenblatt, P., Rivoldini, A., Dehant, V., Marty, J.C. and Karatekin, O. (2012). Lander radio science experiment with a direct link between Mars and the Earth. *Planetary and Space Science*, 68(1), pp.105–122.

Le Maistre, S., Rosenblatt, P., Dehant, V., Marty, J.C. and Yseboodt, M. (2018). Mars rotation determination from a moving rover using Doppler tracking data: What could be done? *Planetary and Space Science* 159: 17–27.

Lemoine, F.G., Goossens, S., Sabaka, T.J., Nicholas, J.B., Mazarico, E., Rowlands, D.D., Loomis, B.D., Chin, D.S., Caprette, D.S., Neumann, G.A., Smith, D.E., and Zuber, M.T. (2013). High-Degree Gravity Models from GRAIL Primary Mission Data. *Journal of Geophysical Research: Planets* 118: 1676–1698. doi:10.1002/jgre.20118.

Lemoine, F.G., Smith, D.E., Rowlands, D.D., Zuber, M.T., Neumann, G., Chinn, D.S., and Pavlis, D.E. (2001). An improved solution of the gravity field of Mars (GMM-2 B) from Mars Global Surveyor. *Journal of Geophysical Research* 106 (E10): 23359–23376.

Levy, G.S., Sato, T., Seidel, B.L., Stelzried, C.T., Ohlson, J.E., and Rusch, W.V.T. (1969). Pioneer-6 measurement of transient faraday rotation phenomena observed during solar occultation. *Science* 166: 596–598.

Li, B., Rodell, M., Zaitchik, B.F., Reichle, R.H., Koster, R.D., and van Dam, T.M. (2012). Assimilation of GRACE terrestrial water storage into a land surface model: Evaluation and potential value for drought monitoring in western and central Europe. *Journal of Hydrology* 446: 103–115.

Lieske, J.H., Lederle, T., Fricke, W., and Morando, B. (1977). Expressions for the precession quantities based upon the IAU (1976) system of astronomical constants. *Astronomy and Astrophysics* 58: 1–16.

Lillis, R.J., Mitchell, D., Montabone, L., Heavens, N., Harrison, T., Stuurman, C., Guzewich, S., England, S., Withers, P., Chaffin, M., and Curry, S. (2021). MOSAIC: A Satellite Constellation to Enable Groundbreaking Mars Climate System Science and Prepare for Human Exploration. *The Planetary Science Journal* 2 (5): 211.

Lindal, G.F. (1992). The atmosphere of Neptune: An analysis of radio occultation data acquired with Voyage 2. *The Astronomical Journal* 103: 967–982.

Lindal, G.F., Hotz, H.B., Sweetnam, D.N., Shippony, Z., Brenkle, J.P., Hartsell, G.V., Spear, R.T., and Michael, W.H. (1979). Viking radio occultation measurements of the atmosphere and topography of Mars: data acquired during 1 Martian year of tracking. *Journal of Geophysical Research: Solid Earth* 84: 8443–8456.

Lindal, G.F., Wood, G.E., Hotz, H.B., Sweetnam, D.N., Eshleman, V.R., and Tyler, G.L. (1983). The atmosphere of Titan: An analysis of the Voyager 1 radio occultation measurements. *Icarus* 53: 348–363.

Linscott, I.R., Bird, M.K., Hinson, D.P., Pätzold, M., and Tyler, G.L. (2011). The Radioscience Experiment on New Horizons. *2011 XXXth URSI General Assembly and Scientific Symposium*, IEEE, pp. 1–4.

Lipa, B. and Tyler, G.L. (1979). Statistical and computational uncertainties in atmospheric profiles from radio occultation: Mariner 10 at Venus. *Icarus* 39: 192–208.

Little, L.T. and Ekers, R.D. (1971). A method for analyzing drifting random patterns in astronomy and geophysics. *Astronomy and Astrophysics* 10: 306–309.

Long, D., Longuevergne, L., and Scanlon, B.R. (2014). Uncertainty in evapotranspiration from land surface modeling, remote sensing, and GRACE satellites. *Water Resources Research* 50 (2): 1131–1151.

Long, D., Scanlon, B.R., Longuevergne, L., Sun, A.Y., Fernando, D.N., and Save, H. (2013). GRACE satellite monitoring of large depletion in water storage in response to the 2011 drought in Texas. *Geophysical Research Letters* 40 (13): 3395–3401.

Lorell, J. (1969). The shape of the moon's gravity field as determined from the Lunar Orbiter tracking data. *Moon* 1: 132.

Love, A.E.H. (1911). Some Problems of Geodynamics. Cambridge, UK: Cambridge University Press.

Lovelace, R.V.E. (1970). Theory and Analysis of Interplanetary Scintillations. Ph. D. Thesis, Ithaca, NY: Cornell University.

Lu, C. and Czanderna, A.W. (eds.) (2012). Applications of piezoelectric quartz crystal microbalances. Elsevier.

Mackenzie, R.A., Iess, L., Tortora, P., and Rappaport, N.J. (2008). A non-hydrostatic Rhea. *Geophysical Research Letters* 35 (5): L05204. doi:10.1029/2007GL032898.

Mannucci, A.J., Wilson, B.D., Yuan, D.-N., Ho, C.H., Lindqwister, U.J., and Runge, T.F. (1998). A global mapping technique for GPS-derived ionospheric total electron content measurements. *Radio Science* 33 (3): 565–582. May–June 1998.

Mannucci, A., Asmar, S., Lazio, J., Ao, C., Williamson, W., Preston, R., and Renno, N. (2018). Globally distributed and high-resolution Mars atmospheric profiles from small spacecraft constellations. In 42nd COSPAR Scientific Assembly (Vol. 42).

Marians, M. (1975). Computed scintillation spectra for strong turbulence. *Radio Science* 10: 115–119.

Marouf, A.E.A.A. (1975). The Rings of Saturn: Analysis of a Bistatic-radar Experiment. Stanford University.

Marouf, E.A., Flasar, F.M., French, R.G., Kliore, A.J., Nagy, A.F., Rappaport, N.J., McGhee, C.A., Schinder, P.J., Simpson, R.A., Anabtawi, A., Asmar, S.W., Barbinis, E., Fleischman, D.U., Goltz, G.L., Kahan, D.S., Kern, A., and Rochblatt, D. (2006). Evidence for likely liquid hydrocarbons on Titan's surface from Cassini radio science bistatic scattering observations. *AGU Fall Meeting 2006*, Abstract P11A- 07.

Marouf, E.A., French, R.G., Rappaport, N.J., Kliore, A.J., Flasar, F.M., Nagy, A.F., McGhee, C., Schinder, P., Anabtawi, A., Asmar, S.W., Barbinis, E., Fleischman, D., Goltz, G., Johnston, D., Rochblatt, D., Thomson, T., and Wong, K. (2005). Structure of Saturn's rings from Cassini diametric radio occultations. *37th Annual Meeting of the Division of Planetary Sciences*, England.

Marouf, E.A., French, R.G., Rappaport, N.J., Thomson, F., McGhee, C., Asmar, S.W., and Johnston, D. (2004). Cassini SOI radio occultation of Saturn's rings. *Division of Planetary Sciences 36th Annual Meeting*, Louisville, KY.

Marouf, E.A., Rappaport, N.J., French, R.G., Simpson, R.A., Kliore, A.J., McGhee, C.A., Schinder, P.J., and Anabtawi, A. (2008). Dielectric constant of Titan's south polar region from Cassini radio science bistatic scattering observations. *AGU Fall Meeting 2008*, Abstract P21A-1319.

Marouf, E.A., Tortora, P., Asmar, S.W., Folkner, W.M., Hinson, D.P., Iess, L., Linscott, I.R., Lorenz, R., and Mueller-Wodarg, I. (2010). Potential spacecraft-to-spacecraft radio observations with EJSM: wave of the future? *American Geophysical Union Fall Meeting*.

Marouf, E.A., Tyler, G.L., and Eshleman, V.R. (1982). Theory of radio occultation by Saturn's rings. *Icarus* 49: 161–193.

Marouf, E.A., Tyler, G.L., and Rosen, P.A. (1986). Profiling Saturn's rings by radio occultations. *Icarus* 68: 120–166.

Marouf, E.A., Tyler, G.L., Zebker, H.A., Simpson, R.A., and Eshleman, V.R. (1983). Particle size distribution in Saturn's rings from Voyager 1 radio occultation. *Icarus* 54: 189–211.

Marouf, E.A. and Wong, K.K. (2012). Documentation of a subset of Cassini radio occultation profiles of Saturn's rings to be archived with the NASA PDS by the end of September 2012. http://pds-rings.seti.org/cassini/rss/index.html (Updated January 2019).

Marty, B., Guillot, T., Coustenis, A., Achilleos, N., Alibert, Y., Asmar, S.W., Atkinson, D.H., Atreya, S., Babasides, G., Baines, K., Balint, T., Banfield, D., Barber, S., Bézard, B., Bjoraker, G.L., Blanc, M., Bolton, S., Chanover, N., Charnoz, S., Chassefière, E., Colwell, J.E., Deangelis, E., Dougherty, M., Drossart, P., Flasar, F.M., Fouchet, T., Frampton, R., Franchi, I., Gautier, D., Gurvits, L., Hueso, R., Kazeminejad, B., Krimigis, T., Jambon, A., Jones, G., Langevin, Y., Leese, M., Lellouch, E., Lunine, J., Milillo, A., Mahaffy, P., Mauk, B., Morse, A., Moreira, M., Moussas, X., Murray, C., Mueller-Wodarg, I., Owen, T.C., Pogrebenko, S., Prangé, R., Read, P., Sanchez-Lavega, A., Sarda, P., Stam, D., Tinetti, G., Zarka, P., and Zarnecki, J. (2009). KRONOS: exploring the depths of Saturn with probes and remote sensing through an international mission. *Experimental Astronomy* 23 (3): 947–976. 10.1007/s10686-008-9094-9.

Matsumoto, K., Goossens, S., Ishihara, Y., Liu, Q., Kikuchi, F., Iwata, T., Namiki, N., Matsumoto, K., Noda, H., Hanada, H., Kawano, N., Lemoine, F.G., and Rowlands, D.D. (2010). An improved lunar gravity field model from SELENE and historical tracking data: revealing the farside gravity features. *Journal of Geophysical Research: Planets* 115 (E6): doi:10.1029/2009JE003499.

Mazarico, E., Barker, M.K., Neumann, G.A., Smith, D.E. and Zuber, M.T. (2016). March. Mercury's Rotational State from the Mercury Laser Altimeter. *Lunar and Planetary Science Conference* 1903: 2062.

Mazarico, E., Neumann, G.A., Rowlands, D.D. and Smith, D.E. (2010). Geodetic constraints from multi-beam laser altimeter crossovers. *Journal of Geodesy* 84 (6): 343–354.

McCarthy, D.D. (1992). *IERS Technical Note# 13*. Observatoire de Paris. Central Bureau of IERS. Paris, France.

McGovern, P.J., Solomon, S.C., Smith, D.E., Zuber, M.T., Simons, M., Wieczorek, M.A., Phillips, R.J., Neumann, G.A., Aharonson, O. and Head, J.W. (2002). Localized gravity/

topography admittance and correlation spectra on Mars: Implications for regional and global evolution. *Journal of Geophysical Research: Planets* 107 (E12):19–1.

McMahon, J.W., Scheeres, D.J., Hesar, S.G., Farnocchia, D., Chesley, S., and Lauretta, D. (2018). The OSIRIS-REx radio science experiment at Bennu. *Space Science Reviews* 214 (1): 1–41.

McGovern, P.J., Solomon, S.C., Smith, D.E., Zuber, M.T., Simons, M., Wieczorek, M.A., Phillips, R.J., Neumann, G.A., Aharonson, O. and Head, J.W. (2004). Correction to "Localized gravity/topography admittance and correlation spectra on Mars: Implications for regional and global evolution". *Journal of Geophysical Research: Planets* 109 (E7).

Melbourne, W.G. (2005). Radio Occultations Using Earth Satellites: A Wave Theory Treatment. Hoboken, NJ: Wiley. Followed Wiley Style.

Melosh, H.J. (2011). Planetary surface processes, 13 vols. Cambridge University Press.

Melosh, H.J., Freed, A.M., Johnson, B.C., Blair, D.M., Andrews-Hanna, J.C., Neumann, G.A., Phillips, R.J., Smith, D.E., Solomon, S.C., Wieczorek, M.A. and Zuber, M.T. (2013). The origin of lunar mascon basins. *Science* 340 (6140): 1552–1555.

Mignard, F. and Klioner, S.A. (2009). Gaia: Relativistic modelling and testing. *Proceedings of the International Astronomical Union* 5 (S261): 306–314.

Milani, A., Vokrouhlický, D., Villani, D., Bonanno, C., and Rossi, A. (2002). Testing general relativity with the BepiColombo radio science experiment. *Physical Review D* 66 (8): 082001.

Milsom, J. (2018). The hunt for earth gravity: A history of gravity measurement from Galileo to the 21st Century. Springer.

Misner, C.W., Thorne, K.S., and Wheeler, J.A. (1973). Gravitation. San Francisco, CA: W.H. Freeman.

Mizuno, E.I., Kawashima, N., Takano, T. and Rosen, P.A. (1992). Voyager radio science: Observations and analysis of Neptune's atmosphere. *IEICE TRANSACTIONS on Communications* 75 (7): 665–672.

Modenini, D. and Tortora, P. (2014). Pioneer 10 and 11 orbit determination analysis shows no discrepancy with Newton-Einstein laws of gravity. *Physical Review D* 90 (2): 022004.

Moore, W.B. and Schubert, G. (2003). The tidal response of Ganymede and Callisto with and without liquid water oceans. *Icarus* 166 (1): 223–226.

Moore, W.B., Schubert, G., Anderson, J.D. and Spencer, J.R. (2007). The interior of Io. In: Io after Galileo, 89–108. Berlin, Heidelberg: Springer.

Morabito, D.D. and Asmar, S.W. (1995). Radio Science Performance Analysis Software. Telecommunications and Data Acquisitions Progress Report 42-120, October–December 1994, Jet Propulsion Laboratory, California Institute of Technology, Pasadena, CA, 15 February 1995.

Moyer, T.D. (2003). Formulation for Observed and Computed Values of Deep Space Network Data Types for Navigation. Hoboken, NJ: Wiley. Followed Wiley Style.

Mudgway, D.J. (2001). Uplink-downlink: A history of the nasa deep space network, 1957–1997 (Vol. 4227). National Aeronautics and Space Administration.

Muhleman, D.O. and Anderson, J.D. (1981). Solar wind electron densities from Viking dual frequency radio measurements. *Astrophysical Journal* 247: 1093–1101.

Muhleman, D.O., Esposito, P.B., and Anderson, J.D. (1977). Electron-density profile of outer corona and interplanetary medium from Mariner-6 and Mariner-7 time-delay measurements. *Astrophysical Journal* 211: 943–957.

Muller, P.M. and Sjogren, W.I. (1968). Mascons: lunar mass concentrations. *Science* 161: 680–684.

Nagy, A.F. and Cravens, T.E. (1988). Hot oxygen atoms in the upper atmospheres of Venus and Mars. *Geophysical Research Letters* 15: 433–435.

Nagy, A.F., Kliore, A.J., Marouf, E.A., French, R., Flasar, F.M., Rappaport, N.J., Anabtawi, A., Asmar, S.W., Johnston, D., Barbinis, E., Goltz, G., and Fleischman, D. (2006). First results from the ionospheric radio occultations of Saturn by the Cassini spacecraft. *Journal of Geophysical Research* 111: A06310. doi:10.1029/2005JE002640.

Namiki, N., Iwata, T., Matsumoto, K., Hanada, H., Noda, H., Goossens, S., Ogawa, M., Kawano, N., Asari, K., Tsuruta, S., Ishihara, Y., Liu, Q., Kikuchi, F., Ishikawa, T., Sasaki, S., Aoshima, C., Kurosawa, K., Sugita, S., and Takano, T. (2009). Farside gravity field of the Moon from four-way Doppler measurements of SELENE (Kaguya). *Science* 323 (5916): 900–905.

Navarro, R., Armstrong, J.W., Rogstad, S., Asmar, S.W., Taylor, G., Miyatake, C., Kahan, D., Klose, J., and Tinto, M. (2010). The digital design and testing of a prototype radio atmospheric sounding and scattering instrument. *IPN Progress Report.*

Nerem, R.S., Jekeli, C., and Kaula, W.M. (1995). Gravity field determination and characteristics: retrospective and prospective. *Journal of Geophysical Research: Solid Earth* 100 (B8): 15053–15074.

Neumann, G.A., Zuber, M.T., Smith, D.E., Wieczorek, M.A., Head, J.W., Baker, D.M.H., Lemoine, F.G., Sabaka, T.J., Mazarico, E., Goossens, S., Solomon, S.C., Melosh, H.J., Phillips, R.J., Asmar, S.W., Konopliv, A.S., Williams, J.G., Sori, M.M., Soderblom, J., Miljkovic, K., Andrews-Hanna, J.C., Nimmo, F., and Kiefer, W.S. (2015). Lunar impact basins revealed by the Gravity Recovery and Interior Laboratory (GRAIL) mission. *Science Advances* 1 (9): e1500852.

Neumann, G.A., Zuber, M.T., Wieczorek, M.A., McGovern, P.J., Lemoine, F.G., and Smith, D.E. (2004). Crustal structure of Mars from gravity and topography. *Journal of Geophysical Research: Planets* 109 (E8).

Nieto, M.M. and Turyshev, S.G. (2004). Finding the origin of the Pioneer anomaly. *Classical and Quantum Gravity* 21 (17): 4005.

Nimmo, F. (2002). Admittance estimates of mean crustal thickness and density at the Martian hemispheric dichotomy. *Journal of Geophysical Research: Planets* 107 (E11): 27–1.

Nimmo, F., Bills, B.G., Thomas, P.C., and Asmar, S.W. (2010). Geophysical implications of the long-wavelength topography of Rhea. *Journal of Geophysical Research: Planets* 115 (E10).

Nimmo, F., Bills, B.G. and Thomas, P.C. (2011). Geophysical implications of the long-wavelength topography of the Saturnian satellites. *Journal of Geophysical Research: Planets* 116 (E11).

Norton, J.R. and Cloeren, J.M. (1994). Precision quartz oscillators and their use aboard satellites. *Johns Hopkins APL Technical Digest* 15 (1): 30–37.

Notaro, V., Durante, D., Iess, L., & Bolton, S. J. (2021). Determination of jupiter's mass from juno radio tracking data. *Journal of Guidance, Control, and Dynamics* 44 (5): 1062–1067.

Notaro, V., Durante, D. and Iess, L. (2019). On the determination of Jupiter's satellite-dependent Love numbers from Juno gravity data. *Planetary and Space Science* 175: 34–40.

Notaro, V., Iess, L., Armstrong, J.W., and Asmar, S.W. (2020). Reducing Doppler noise with multi-station tracking: The Cassini test case. *Acta Astronautica* 173: 45–52.

Null, G.W., Lau, E.L., Biller, E.D., and Anderson, J.D. (1981). Saturn gravity results obtained from Pioneer 11 tracking data and Earth-based Saturn satellite data. *The Astronomical Journal* 86: 456–468.

O'Dea, J.A., Jenet, F.A., Cheng, T.-H., Buu, C.M., Beroiz, M., Asmar, S.W., and Armstrong, J.W. (2014). Bright microwave pulses from PSR B0531+21 observed with a prototype transient survey receiver. *The Astronomical Journal* 147: 100. doi:10.1088/0004-6256/147/5/100.

O'Leary, B.T., Campbell, M.J. and Sagan, C. (1969). Lunar and planetary mass concentrations. *Science* 165 (3894): 651–657.

Ohlson, J.E., Levy, G.S., and Stelzried, C.T. (1974). Tracking polarimeter for measuring solar and ionospheric faraday-rotation of signals from deep space probes. In *IEEE Transactions on Instrumentation and Measurement*, .pp. 167–177. IM23.

Otoshi, T.Y. and Franco, M.M. (1992). DSS-13 Beam Waveguide Antenna Frequency Stability. *The Telecommunications and Data Acquisition Progress Report* 42–110, pp. 151–159. Jet Propulsion Laboratory, California Institute of Technology, Pasadena, CA, April–June 1992. http://ipnpr.jpl.nasa.gov/progress_report/42-110/110title.htm (accessed 15 April 1992).

Ostro, S.J. (1993). Planetary radar astronomy. *Reviews of Modern Physics* 65 (4): 1235.

Ostro, S.J., Scott, R., Hudson, Nolan, M.C., Margot, J.L., Scheeres, D.J., Campbell, D.B., Magri, C., Giorgini, J.D. and Yeomans, D.K. (2000). Radar observations of asteroid 216 Kleopatra. *Science* 288 (5467): 836–839.

Paik, M. and Asmar, S.W. (2010). Detecting High Dynamics Signals From Open-Loop Radio Science Investigations. *Proceedings of the IEEE* 99 (5): 881–888. 5643089.

Paik, M., Border, J.S., Esterhuizen, S., and Shin, D.K., (2018, December). Advanced Ranging Instrumentation. In AGU Fall Meeting 2018. AGU.

Palguta, J., Anderson, J.D., Schubert, G., and Moore, W.B. (2006). Mass anomalies on Ganymede. *Icarus* 180 (2): 428–441.

Pappalardo, R.T., Senske, D., Korth, H., Klima, R., Vance, S., and Craft, K. (2017, September). The Europa Clipper Mission: Exploring the Habitability of a Unique Icy World. In European Planetary Science Congress (pp. EPSC2017–304).

Parisi, M., Galanti, E., Finocchiaro, S., Iess, L., and Kaspi, Y. (2016). Probing the depth of Jupiter's great red spot with the Juno gravity experiment. *Icarus* 267: 232–242.

Parisi, M., Kaspi, Y., Galanti, E., Durante, D., Bolton, S. J., Levin, S. M., ... & Wong, M. H. (2021). The depth of Jupiter's Great Red Spot constrained by Juno gravity overflights. *Science* 374 (6570): 964–968.

Park, R.S., Asmar, S.W., Buffington, B.B., Bills, B.G., Campagnola, S., Chodas, P.W., Folkner, W.M., Konopliv, A.S., and Petropoulos, A.E. (2011b). Detecting tides and gravity at Europa from multiple close flybys. *Geophysical Research Letters* doi:10.1029/2011GL049842.

Park, R.S., Asmar, S.W., Fahnestock, E.G., Konopliv, A.S., Lu, W., and Watkins, M.M. (2012). Gravity recovery and interior laboratory simulations of static and temporal gravity field. *Journal of Spacecraft and Rockets* 49: 390–400.

Park, R.S., Konopliv, A.S., Asmar, S.W., Bills, B.G., Gaskell, R.W., Raymond, C.A., Smith, D.E., Toplis, M.J., and Zuber, M.T. (2014). Gravity field expansion in ellipsoidal harmonic and polyhedral internal representations applied to Vesta. *Icarus* 240: 118–132.

Park, R.S., Konopliv, A.S., Bills, B.G., Rambaux, N., Castillo-Rogez, J.C., Raymond, C.A., Vaughan, A.T., Ermakov, A.I., Zuber, M.T., Fu, R.R., and Toplis, M.J. (2016). A partially differentiated interior for (1) Ceres deduced from its gravity field and shape. *Nature* 537: 515–517.

Pätzold, M., Edenhofer, P., Bird, M.K., and Volland, H. (1993). The Giotto encounter with Comet P/Grigg-Skjellerup-First results from the Giotto Radio-Science Experiment. *Astronomy and Astrophysics* 268: L13–L16.

Pätzold, M., Andert, T., Hahn, M., Asmar, S.W., Barriot, J.P., Bird, M.K., Häusler, B., Peter, K., Tellmann, S., Grün, E., and Weissman, P.R. (2016a). A homogeneous nucleus for Comet 67P/Churyumov–Gerasimenko from its gravity field. *Nature* 530 (7588): 63–65.

Pätzold, M., Häusler, B., Tyler, G.L., Andert, T., Asmar, S.W., Bird, M.K., Dehant, V., Hinson, D.P., Rosenblatt, P., Simpson, R.A., and Tellmann, S. (2016b). Mars express 10 years at Mars: Observations by the Mars express radio science experiment (MaRS). *Planetary and Space Science* 127: 44–90.

Pätzold, M., Andert, T.P., Asmar, S.W., Anderson, J.D., Barriot, J.-P., Bird, M.K., Häusler, B., Hahn, M., Tellmann, S., Sierks, H., Lamy, P., and Weiss, B. (2011). Asteroid (21) Lutetia – low mass, high density. *Science* 334 (6055): 491–492. doi:10.1126/science.12093892011.

Pätzold, M., Andert, T.P., Häusler, B., Tellmann, S., Anderson, J.D., Asmar, S.W., Barriot, J.-P., and Bird, M.K. (2010). Pre-flyby estimates on the precision of the mass determination of Asteroid (21) Lutetia from Rosetta radio tracking. *Astronomy and Astrophysics* doi:10.1051/0004-6361/201014325.

Pätzold, M., Andert, T.P., Tyler, G.L., Asmar, S.W., and Häusler, B. (2014). Phobos mass determination from the very close flyby of Mars Express in 2010. *Icarus* doi:dx.doi.org/10.1016/j.icarus.2013.10.021.

Pätzold, M., Bird, M.K., Edenhofer, P., Asmar, S.W., and McElrath, T.P. (1995). Dual-frequency radio sounding of the solar corona during the 1995 conjunction of the Ulysses spacecraft. *Geophysical Research Letters* 22: 3313–3316.

Pätzold, M., Bird, M.K., Volland, H., Edenhofer, P., Asmar, S.W., and Brenkle, J.P. (1992). Coronal sounding with Ulysses: preliminary results from the first solar conjunction. In: Solar Wind Seven, (ed. R. Schwenn), 237–240. Oxford, UK: Pergamon Press.

Pätzold, M., Bird, M.K., Volland, H., Levy, G.S., Seidel, B.L., and Stelzried, C.T. (1987). The mean coronal magnetic field determined from Helios faraday rotation measurements. *Solar Physics* 109: 91–105.

Pätzold, M., Hahn, M., Tellmann, S., Häusler, B., Bird, M.K., Tyler, G.L., Asmar, S.W., and Tsurutani, B.T. (2012). Coronal density structures and CMEs: superior solar conjunctions of Mars Express, Venus Express, and Rosetta: 2004, 2006, and 2008. *Solar Physics* doi:10.1007/s11207-012-9991-y.

Pätzold, M., Häusler, B., Bird, M.K., Tellmann, S., Mattei, R., Asmar, S.W., Dehant, V., Eidel, W., Imamura, T., Simpson, R.A., and Tyler, G.L. (2007a). The structure of Venus' middle atmosphere and ionosphere. *Nature* doi:10.1038/nature06239.

Pätzold, M., Häusler, B., Wenmacher, A., Aksnes, K., Anderson, J.D., Asmar, S.W., Barriot, J.-P., Bird, M.K., Boehnhardt, H., Eidel, W., Grun, E., Ip, W., Marouf, E.A., Morley, T., Neubauer, F.M., Rickman, H., Thomas, N., Tsurutani, B., Wallis, M., Wickramasinghe, N., Mysen, E., Olsen, O., Remus, S., Tellmann, S., Andert, T., Carone, L., Fles, M., Stanzel, C., Audenrieth-Kersten, I., Gahr, A., Muller, A., Stupar, D., and Walter, C. (2007b). Rosetta radio science investigations. *Space Science Reviews* doi:10.1007/s11214-006-9117-7.

Pätzold, M., Häusler, B., Tyler, G.L., Andert, T., Asmar, S.W., Bird, M.K., Dehant, V., Hinson, D.P., Rosenblatt, P., Simpson, R.A. and Tellmann, S. (2016). Mars express 10 years at Mars: observations by the Mars express radio science experiment (MaRS). *Planetary and Space Science* 127: 44–90.

Pätzold, M., Karl, J., and Bird, M.K. (1996). Coronal radio sounding with Ulysses: dual-frequency phase scintillation spectra in coronal holes and streamers. *Astronomy and Astrophysics* 316: 449–456.

Patzold, M., Tellmann, S., Andert, T., Carone, L., Fels, M., Schaa, R., Stanzel, S., Audenrieth-Kersten, I., Gahr, A., Muller, A.-L., Stracke, B., Stupar, D., Walter, C., Haussler, B., Remus, S., Selle, J., Griebrel, H., Eidel, W., Asmar, S.W., Goltz, G., Kahan, D.S., Barriot, J.-P., Dehant, V., Beuthe, M., Rosenblatt, P., Karatekin, O., Lainey, V., Tyler, G.L., Hinson, D.P., Simpson, R.A., and Twicken, J. (2009). MaRS: the Mars Express radio science experiment. *European Space Agency Special Publication No. 1291* 217–245 (June): 2009

Percival, D.B. and Walden, A.T. (1993). Spectral Analysis for Physical Applications. Cambridge, UK: Cambridge University Press.

Perry, M.E., Kahan, D.S., Barnouin, O.S., Ernst, C.M., Solomon, S.C., Zuber, M.T., Smith, D.E., Phillips, R.J., Srinivasan, D.K., Oberst, J., and Asmar, S.W. (2011). Measurement of the radius of Mercury by radio occultation during the MESSENGER flybys. *Planetary and Space Science* 59: 1925–1931. doi:10.1016/j.pss.2011.07.022.

Perry, M.E., Neumann, G.A., Phillips, R.J., Barnouin, O.S., Ernst, C.M., Kahan, D.S., Solomon, S.C., Zuber, M.T., Smith, D.E., Hauck, S.A. and Peale, S.J. (2015). The low-degree shape of Mercury. *Geophysical Research Letters* 42 (17): 6951–6958.

Phinney, R.A. and Anderson, D.L. (1968). On the occultation method for studying planetary atmospheres. *Journal of Geophysical Research* 73: 1819–1827.

Pilinski, E.B. and Lee, A.Y. (2009). Pointing-stability performance of the Cassini spacecraft. *Journal of Spacecraft and Rockets* 46 (5):.1007–1015.

Pines, S. (1973). Uniform representation of the gravitational potential and its derivatives. *AIAa Journal* 11 (11): 1508–1511.

Plaut, J.J., Barabash, S., Bruzzone, L., Dougherty, M., Erd, C., Fletcher, L., Gladstone, R., Grasset, O., Gurvits, L., Hartogh, P., Hussmann, H., Iess, L., Jaumann, R., Langevin, Y., Palumbo, P., Piccioni, G., Titov, D., and Wahlund, J.-E. (2014). Jupiter icy moons explorer (JUICE): science objectives, mission and instruments. *Lunar and Planetary Science Conference* 45: 2717.

Poulsen, S.K. (2009). Tidal deformation of the Solid Earth-A finite difference discretization. Doctoral dissertation, thesis. University of Copenhagen.

Prestage, J.D. and Weaver, G.L. (2007). Atomic clocks and oscillators for deep-space navigation and radio science. *Proceedings of the IEEE* 95 (11): 2235–2247. doi:10.1109/JPROC.2007.905130 November 2007.

Preston, R.A. (1974). Dual-Spacecraft Radio Metric Tracking. May and June 1974, 51. JPL DEEP SPACE NETWORK PROGRESS REPORT 42–22.

Preston, R.A., Hildebrand, C.E., Purcell, G.H., Ellis, J., Stelzried, C.T., Finley, S.G., Sagdeev, R.Z., Linkin, V.M., Kerzhanovich, V.V., Altunin, V.I., and Kogan, L.R. (1986). Determination of Venus winds by ground-based radio tracking of the VEGA balloons. *Science* 231 (4744): 1414–1416.

Pugh, M., Kuperman, I., Aguirre, F., Mojaradi, H., Spurgers, C., Kobayashi, M., Satorius, E. and Jedrey, T. (2017, March). The Universal Space Transponder: A next generation software defined radio. In *2017 IEEE Aerospace Conference* 1–14. IEEE.

Ragozzine, D. and Wolf, A.S. (2009). Probing the interiors of very hot Jupiters using transit light curves. *The Astrophysical Journal* 698 (2): 1778.

Rambaux, N., Konopliv, A.S., Asmar, S.W., and Park, R.S. (2014). The rotational motion of Vesta. *IAU-Symposium: Complex Planetary Systems*, Namur, Belgium (July 2014).

Rappaport, N., Iess, L., Wahr, J., Lunine, J., Armstrong, J.W., Asmar, S.W., Tortora, P., Ardito, A., Di Benedetto, M., Racioppa, P., Rapino, G., and Somenzi, L. (2008). Can Cassini detect a subsurface ocean in Titan from gravity. *Icarus* doi:10.1016/j.icarus.2007.11.024.

Rappaport, N.J., Asmar, S.W., Castillo, J., and Iess, L. (2004). Phoebe's mass and interior from Cassini flyby. The Saturn Universe: A Cassini Workshop, Capri, Italy: (October 2004).

Rappaport, N.J., Iess, L., Tortora, P., Anabtawi, A., Asmar, S.W., Somenzi, L., and Zingoni, F. (2007). Mass and interior of Enceladus from Cassini data analysis. *Icarus* 190: 175–178. doi:10.1016/j.icarus.2007.03.025.

Rappaport, N.J., Konopliv, A.S., Kucinskas, A.B., and Ford, P.G. (1999). An improved 360 degree and order model of Venus topography. *Icarus* 139 (1): 19–31.

Rappaport, N.J. and Plaut, J.J. (1994). A 360-degree and order model of Venus topography. *Icarus* 112 (1): 27–33.

Rebold, T.A., Tinto, M., Asmar, S.W., and Kursinski, E.R. (1997). Neptune Revisited: synthesizing Coherent Doppler from Voyager's Non-coherent Downlink. Telecommunications and Data Acquisition Progress Report 42–131, Pasadena, CA: Jet Propulsion Laboratory, California Institute of Technology. 15 November 1997.

Resch, G.M., Clark, J.E., Keihm, S.J., Lanyi, G.E., Naudet, C.J., Riley, A.L., Rosenberger, H.W. and Tanner, A.B. (2001). The media calibration system for Cassini radio science: Part II. TMO Progress Report, 42–145.

Resch, G.M., Keihm, S.J., Lanyi, G.E., Linfield, R.P., Naudet, C.J., Riley, A.L., Rosenberger, H.W., and Tanner, A.B. (2002). The Media Calibration System for Cassini Radio Science: part III. The Interplanetary Progress Report 42–148, Jet Propulsion Laboratory, California Institute of Technology, Pasadena, CA, 15 February 2002. https://ipnpr.jpl.nasa.gov/progress_report/42-148/148H.pdf

Richie-Halford, A.C., Iess, L., Tortora, P., Armstrong, J.W., Asmar, S.W., Woo, R., Habbal, S.R., and Morgan, H. (2009). Space-Time Localization of Heliospheric Plasma Turbulence Using Multiple Spacecraft Links. *Space Weather* 7: S12003. doi:10.1029/2009SW000499.

Rickett, B.J. (1990). Radio propagation through the turbulent interstellar plasma. *Annual Review of Astronomy and Astrophysics* 28: 561–605.

Riley, A.L., Antsos, D., Armstrong, J.W., Kinman, P., Wahlquist, H.D., Bertotti, B., Comoretto, G., Pernice, B., Carnicella, G. and Giordani, R. (1990). Cassini Ka-band precision Doppler and enhanced telecommunications system study. Report by NASA-JPL and Agenzia Spaziale Italiana.

Rohlfs, K. (1986). Tools of Radio Astronomy. Berlin: Springer.

Rosborough, G.W. and Lemoine, F.G. (1991). Sensitivity studies of Mars orbiters for Mars gravity recovery. *Journal of the Astronautical Sciences* 39 (3): 327–357.

Rosen, P.A. (1989). Waves in Saturn's rings probed by radio occultation. Ph.D. Dissertation, Appendix B, Stanford, CA: Stanford University.

Rosenblatt, P., Bruinsma, S.L., Müller-Wodarg, I.C.F., Häusler, B., Svedhem, H., and Marty, J.C. (2012). First ever in situ observations of Venus' polar upper atmosphere density using the tracking data of the Venus Express Atmospheric Drag Experiment (VExADE). *Icarus* 217 (2), 831–838.

Roth, D., Antreasian, P., Bordi, J., Criddle, K., Ionasescu, R., Jacobsen, R., Jones, J., Meek, M.C., Roundhill, I. and Stauch, J. (2005). Cassini orbit resconstruction from Jupiter to Saturn.

Rumsey, V.H. (1975). Scintillations due to a concentrated layer with a powerlaw turbulence spectrum. *Radio Science* 10: 107–114.

Rusch, W.V.T. and Stelzried, C.T. (1972). Net-field polarization in a magnetically biased plasma. *Radio Science* 7: 1131–1141. doi:10.1029/RS007i012p01131.

Russell, C.T., Raymond, C.A., Coradini, A., McSween, H.Y., Nathues, A., De Sanctis, M.C., Jaumann, R., Konopliv, A.S., Asmar, S.W., Park, R.S., Zuber, M.T., Smith, D.E., Toplis, M.J., Mottola, S., Roatsch, T., Scully, J.E.C., Tricarico, P., Christensen, U.R., Feldman, W.C., Lawrence, D.J., McCoy, T., Prettyman, T.H., Reedy, R.C., Sykes, M.E., and Titus, T.N. (2012). Dawn at Vesta: testing the protoplanetary paradigm. *Science* doi:10.1126/science.1219381.

Sagdeev, R.Z., Linkin, V.M., Blamont, J.E., and Preston, R.A. (1986). The VEGA Venus balloon experiment. *Science* 231 (4744): 1407–1408.

Sagdeev, R.Z., Linkin, V.M., Kerzhanovich, V.V., Lipatov, A.N., Shurupov, A.A., Blamont, J.E., Crisp, D., Ingersoll, A.P., Elson, L.S., Preston, R.A. and Hildebrand, C.E. (1986). Overview of VEGA Venus balloon in situ meteorological measurements. *Science* 231 (4744): 1411–1414.

Satorius, E., Jedrey, T., Bell, D., Devereaux, A., Ely, T., Grigorian, E., Kuperman, I. and Lee, A. (2006). The electra radio. Autonomous Software-Defined Radio Receivers for Deep Space Applications 19.

Satorius, E.H., Shah, B.N., Bruvold, K.N. and Bell, D.J. (2011, March). Adaptation of the Electra radio to support multiple receive channels. In: *2011 Aerospace Conference* 1–10 IEEE.

Savich, N.A. (1981). Comparative analysis of Venusian ionosphere dual-frequency radio soundings with the satellites Venera-9, 10 and Pioneer-Venus. *Advances in Space Research* 1 (9): 17–25.

Savich, N.A., Andreev, V.E., Vyshlov, A.S., Gavrik, A.L., and Konofalov, V.P. (1986a and 1986b). Dayside ionosphere of Venus according to Venera 15 and 16 radio-occultation data in 1983. *Radiotekhnika i Elektronika* 31: 2113–2120.

Savich, N.A., Andreev, V.E., Vyshlov, A.S., Gavrik, A.L., and Konofalov, V.P. (1986). Nighttime ionosphere of Venus according to two-frequency radio occultation using the Venera 15, 16 spacecraft in 1983. *Radiotekhnika i Elektronika* 31: 433–439.

Savich, N.A., Samovol, V.A., Vasilyev, M.B., Vyshlov, A.S., Samoznaev, L.N., Sidorenko, A.I. and Shtern, D.Y. (1976). The nighttime ionosphere of Mars from Mars-4 and Mars-5 radio occultation dual-frequency measurements. *Solar-Wind Interaction with the Planets Mercury, Venus, and Mars* 41–46.

Schinder, P.J., Flasar, F.M., Marouf, E.A., French, R.G., McGhee, C.A., Kliore, A.J., Rappaport, N.J., Barbinis, E., Fleischman, D., and Anabtawi, A. (2012). The structure of Titan's atmosphere from Cassini radio occultations: occultations from the prime and equinox missions. *Icarus* 221: 1020–1031.

Schinder, P.J., Flasar, F.M., Marouf, E.A., French, R.G., McGhee, C.A., Kliore, A.J., Rappaport, N.J., Nagy, A.F., Anabtawi, A., Asmar, S.W., Barbinis, E., Fleischman, D., and Goltz, G. (2006). The atmosphere of Titan from Cassini radio occultations. AAS Division of Planetary Sciences Meeting, Pasadena, CA: (October 2006).

Schinder, P. J., Flasar, F. M., Marouf, E. A., French, R. G., McGhee, C. A., Kliore, A. J., ... & Anabtawi, A. (2012). The structure of Titan's atmosphere from Cassini radio occultations: Occultations from the Prime and Equinox missions. *Icarus* 221 (2): 1020–1031.

Schubert, G., Zhang, K., Kivelson, M.G., and Anderson, J.D. (1996). The magnetic-field and internal structure of Ganymede. *Nature* 384 (6609): 544–545.

Scott, S.L., Rickett, B.J., and Armstrong, J.W. (1983). The velocity and density spectrum of the solar wind from simultaneous three-frequency IPS observations. *Astronomy and Astrophysics* 123 (1983): 191–206.

Seidelmann, P.K. (1982). 1980 IAU theory of nutation: the final report of the IAU Working Group on nutation. *Celestial Mechanics* 27: 79–106.

Seidelmann, P.K., Abalakin, V.K., Bursa, M., Davies, M.E., De Bergh, C., Lieske, J.H., Oberst, J., Simon, J.L., Standish, E.M., Stooke, P., and Thomas, P.C. (2002). Report of the IAU/IAG Working Group on cartographic coordinates and rotational elements of the planets and satellites: 2000. *Celestial Mechanics and Dynamical Astronomy* 82 (1): 83–111.

Shambayati, S. (2007, March). The struggle for Ka-band: NASA's gradual move towards using 32-GHz Ka-band for deep space missions. In: *2007 IEEE Aerospace Conference* 1–21. IEEE.

Shapiro, I.I. (1964). Fourth test of general relativity. *Physical Review Letters* 13: 789–791.

Shapiro, I.I. (1967). New method for the detection of light deflection by solar gravity. *Science* 157 (3790): 806–808.

Sheard, B.S., Heinzel, G., Danzmann, K., Shaddock, D.A., Klipstein, W.M. and Folkner, W.M. (2012). Intersatellite laser ranging instrument for the GRACE follow-on mission. Journal of Geodesy 86(12): 1083–1095.

Shida, T. (1912). On the Body Tides of the Earth, A Proposal for the International Geodetic Association. *Proceedings of the Tokyo Mathematico-Physical Society. 2nd Series* 6 (16): 242–258.

Showalter, M.R. (1991). Visual detection of 1981S13, Saturn's eighteenth satellite, and its role in the Encke Gap. *Nature* 351: 709–713.

Simons, M., Solomon, S.C., and Hager, B.H. (1997). Localization of gravity and topography: constraints on the tectonics and mantle dynamics of Venus. *Geophysical Journal International* 131 (1): 24–44.

Simon, M.K., Vilnrotter, V., Mileant, A. and Hinedi, S. (1996). Optimum Strategies for Monitoring the Operational Status of a Spacecraft. The Telecommunications and Data Acquisition Progress Report 42-127, July–September 1996, 1–51.

Simpson, R.A. (2007). Planetary exploration. In: Advances in Bistatic Radar, (eds. N.J. Willis and H.D. Griffiths), Chapter 5. Raleigh, NC: SciTech Publishing Inc.

Simpson, R.A., Lichtenberg, C., and Asmar, S.W. (1994). Clementine bistatic radar experiment. *26th Annual Meeting of the Division of Planetary Sciences of the American Astronomical Society*, Washington DC: (October 1994).

Simpson, R.A., Tyler, G.L., Patzold, M., Hausler, B., and Asmar, S.W. (2009a). "Bistatic radar probing of planetary surfaces," 2009 IEEE Radar Conference, 2009: 1–4, doi: 10.1109/RADAR.2009.4976965

Simpson, R.A., Tyler, G.L., Pätzold, M., Häusler, B., and Asmar, S.W. (2009b). Dual-frequency bistatic radar probing of Mars: potential and pitfalls for depth sounding at centimeter wavelengths. *American Geophysical Union Conference Fall.*

Simpson, S.A., Tyler, G.L., Patzold, M., Häusler, B., Asmar, S.W., and Sultan-Salem, A.K. (2011). Polarization in bistatic radar probing of planetary surfaces: application to Mars Express data. *Proceedings of the IEEE* 99 (5): 858–874.

Sjogren, W.L. (1979). Mars gravity: High-resolution results from Viking Orbiter 2. *Science* 203 (4384): 1006–1010.

Sjogren, W.L., Bills, B.G., Birkeland, P.W., Esposito, P.B., Konopliv, A.S., Mottinger, N.A., Ritke, S.J., and Phillips, R.J. (1983). Venus gravity anomalies and their correlations with topography. *Journal of Geophysical Research: Solid Earth* 88 (B2): 1119–1128.

Sjogren, W.L., Bills, B.G., and Mottinger, N.A. (1984). Venus: Ishtar gravity anomaly. *Geophysical Research Letters* 11 (5): 489–491.

Sjogren, W.L., Phillips, R.J., Birkeland, P.W. and Wimberly, R.N. (1980). Gravity anomalies on Venus. *Journal of Geophysical Research: Space Physics* 85 (A13): 8295–8302.

Sjogren, W.L. and Wimberley, R.N. (1981). Mars: Hellas Planitia gravity analysis. *Icarus* 45 (2): 331–338.

Sjogren, W.L., Gottlieb, P., Muller, P.M., and Wollenhaupt, W., 1972. Lunar gravity via Apollo 14 Doppler radio tracking. *Science* 175 (4018): 165–168.

Slade, M.A., Benner, L.A. and Silva, A. (2010). Goldstone solar system radar observatory: Earth-based planetary mission support and unique science results. *Proceedings of the IEEE* 99 (5): 757–769.

Smith, D.E., Lerch, F.J., Nerem, R.S., Zuber, M.T., Patel, G.B., Fricke, S.K., and Lemoine, F.G. (1993). An improved gravity model for Mars: Goddard Mars model 1. *Journal of Geophysical Research: Planets* 98 (E11): 20871–20889.

Smith, D.E., Zuber, M.T., Phillips, R.J., Solomon, S.C., Hauck, S.A., Lemoine, F.G., Mazarico, E., Neumann, G.A., Peale, S.J., Margot, J.-L., Johnson, C.L., Torrence, M.H., Perry, M.E., Rowlands, D.D., Goossens, S., Head, J.W., and Taylor, A.H. (2012). Gravity field and internal structure of Mercury from MESSENGER. *Science* 336 (6078): 214–217.

Smith, D.E., Zuber, M.T., Mazarico, E., Genova, A., Neumann, G.A., Sun, X., Torrence, M.H., and Mao, D.D. (2018). Trilogy, a planetary geodesy mission concept for measuring the expansion of the solar system. *Planetary and Space Science* 153: 127–133.

Smith, D.E., Zuber, M.T., Solomon, S.C., Phillips, R.J., Head, J.W., Garvin, J.B., Banerdt, W.B., Muhleman, D.O., Pettengill, G.H., Neumann, G.A., Lemoine, F.G., Abshire, J.B., Aharonson, O., Brown, C.D., Hauck, S.A., Ivanov, A.B., McGovern, P.J., Zwally, H.J., and Duxbury, T.C. (1999). The global topography of Mars and implications for surface evolution. *Science* 284 (5419): 1495–1503.

Smrekar, S.E. and Phillips, R.J. (1991). Venusian highlands: geoid to topography ratios and their implications. *Earth and Planetary Science Letters* 107 (3): 582–597.

Solomon, S.C., McNutt, Jr, R.L., Gold, R.E., Acuña, M.H., Baker, D.N., Boynton, W.V., Chapman, C.R., Cheng, A.F., Gloeckler, G., Head, Iii, J.W., and Krimigis, S.M. (2001). The MESSENGER mission to Mercury: Scientific objectives and implementation. *Planetary and Space Science* 49 (14–15): 1445–1465.

Sori, M.M., James, P.B., Johnson, B.C., Soderblom, J.M., Solomon, S.C., Wieczorek, M.A. and Zuber, M.T. (2018). Isostatic compensation of the lunar highlands. *Journal of Geophysical Research: Planets* 123 (2): 646–665.

Steffes, P.G. (1985). Laboratory measurements of the microwave opacity and vapor pressure of sulfuric acid vapor under simulated conditions for the middle atmosphere of Venus. *Icarus* 64 (3): 576–585.

Steffes, P.G., Jenkins, J.M., Austin, R., Asmar, S.W., Lyons, D.T., Seale, E.H., and Tyler, G.L. (1994). Radio occultation studies of the Venus atmosphere with the Magellan spacecraft: 1. Experimental description and performance. *Icarus* 110: 71–78.

Stelzried, C.T., Levy, G.S., Sato, T., Rusch, V.V.T., Ohlson, J.E., Shatten, K.H., and Wilcox, J.M. (1970). Quasi-stationary coronal magnetic field and electron density determined from a faraday rotation experiment. *Solar Physics* 14: 440–456.

Stern, S.A., Bagenal, F., Ennico, K., Gladstone, G.R., Grundy, W.M., McKinnon, W.B., Moore, J.M., Olkin, C.B., Spencer, J.R., Weaver, H.A., and Young, L.A. (2015). The Pluto system: initial results from its exploration by New Horizons. *Science* 350 (6258): 1815.

Stern, S.A., Weaver, H.A., Spencer, J.R., and Elliott, H.A. (2018). The New Horizons Kuiper Belt Extended Mission. *Space Science Reviews* 214 (4): 1–23.

Stevenson, D.J., Iess, L., Parisi, M., Ducci, M., Lunine, J.I., and Asmar, S.W. (2013). The interpretation of Enceladus gravity. *American Geophysical Union Conference Fall.*

Swenson, S. and Wahr, J. (2009). Monitoring the water balance of Lake Victoria, East Africa, from space. *Journal of Hydrology* 370 (1–4): 163–176.

Takamoto, M., Hong, F.-L., Higashi, R., and Katori, H. (2005). An optical lattice clock. *Nature* 435: 321–324.

Takeuchi, H. and Saito, M. (1972). Seismic surface waves. *Methods in Computational Physics* 11: 217–295.

Takeuchi, H., Yoshikawa, K., Takei, Y., Oki, Y., Kikuchi, S., Ikeda, H., Soldini, S., Ogawa, N., Mimasu, Y., Ono, G. and Terui, F. (2020). The deep-space multi-object orbit determination system and its application to Hayabusa2's asteroid proximity operations. *Astrodynamics* 4 (4): 377–392.

Tanner, A.B. (1998). Development of a high-stability water vapor radiometer. *Radio Science* 33 (2): 449–462.

Tapley, B.D., Schutz, B.E., and Born, G.H. (2004). Statistical Orbit Determination. Boston, MA: Elsevier.

Tatarski, V.I. (1961). Wave Propagation in a Turbulent Medium. New York: Dover.

Taylor, J. (ed.) (2016). Deep Space Communications. John Wiley & Sons.

Tellmann, S., Pätzold, M., Häusler, B., Hinson, D.P. and Tyler, G.L. (2013). The structure of Mars lower atmosphere from Mars Express Radio Science (MaRS) occultation measurements. *Journal of Geophysical Research: Planets* 118 (2): 306–320.

Thomas, P.C., Armstrong, J.W., Asmar, S.W., Burns, J.A., Denk, T., Giese, B., Helfenstein, P., Iess, L., Johnson, T., McEwen, A., Nicolaisen, L., Porco, C., Rappaport, N.J., Richardson, J., Somenzi, L., Tortora, P., Turtle, E., and Veverka, J. (2007). Hyperion's sponge-like appearance. *Nature* doi:10.1038/nature05779.

Thomas, P.C., Davies, M.E., Colvin, T.R., Oberst, J., Schuster, P., Neukum, G., Carr, M.H., McEwen, A., Schubert, G., Belton, M.J.S. and Team, G.I. (1998). The shape of Io from Galileo limb measurements. *Icarus* 135 (1): 175–180.

Thomson, F., Asmar, S.W., and Oudrhiri, K. (2005). Limitations on the use of the power-law S(f) to compute Allan variance. *IEEE Transactions on Ultrasonics Ferroelectronics & Frequency Control* 52 (9): 1468–1472. doi:10.1109/TUFFC.2005.1516018.

Thompson, P.F., Abrahamson, M., Ardalan, S. and Bordi, J. (2014). Reconstruction of Earth flyby by the Juno spacecraft.

Thorne, K.S. (1987). Gravitational Radiation. In: Three Hundred Years of Gravitation, (eds. S.W. Hawking and W. Israel), 330–458. Cambridge, UK; New York: Cambridge University Press.

Thornton, C.L. and Border, J.S. (2003). Radiometric Tracking Techniques for Deep-Space Navigation. Hoboken, NJ: Wiley. Formatted Wiley Style.

Tinto, M. (2002). The Cassini Ka-band gravitational wave experiment. *Classical and Quantum Gravity* 19: 1767–1773.

Toth, V.T. and Turyshev, S.G. (2012). Finding the source of the pioneer anomaly. *IEEE Spectrum* 49 (12): 38–62.

Tortora, P., Iess, L., Bordi, J.J., Ekelund, J.E., and Roth, D.C. (2004). Precise Cassini navigation during solar conjunctions through multifrequency plasma calibrations. *Journal of Guidance, Control and Dynamics* 27 (2): 251–257.

Tortora, P., Iess, L., and Ekelund, J.E. (2002). Accurate navigation of deep dpace probes using multifrequency links: The Cassini breakthrough during solar conjunction experiments. Presented at *World Space Congress, AIAA*, Houston, TX (10–19 October 2002).

Tortora, P., Zannoni, M., Hemingway, D., Nimmo, F., Jacobson, R.A., Iess, L., and Parisi, M. (2016). Rhea gravity field and interior modeling from Cassini data analysis. *Icarus* 264: 264–273.

Tortora, P., Zannoni, M., Gutierrez, G., Martino, P. and Carnelli, I. (2019. September. Didymos Gravity Science through Juventas Satellite-to-Satellite Doppler Tracking. *EPSC-DPS Joint Meeting 2019* 2019: EPSC–DPS2019.

Turcotte, D. and Schubert, G. (2014). Geodynamics. Cambridge University Press.

Turyshev, S.G. and Toth, V.T. (2010). The Pioneer anomaly. *Living Reviews in Relativity* 13 ((lrr-2010-4)): https://link.springer.com/article/10.12942/lrr-2010-4 (accessed 1 September 2010).

Tsuda, Y., Yoshikawa, M., Saiki, T., Nakazawa, S. and Watanabe, S.I. (2019). Hayabusa2–Sample return and kinetic impact mission to near-earth asteroid Ryugu. *Acta Astronautica* 156: 387–393.

Turyshev, S.G., Toth, V.T., Kinsella, G., Lee, S.C., Lok, S.M. and Ellis, J. (2012). Support for the thermal origin of the Pioneer anomaly. *Physical review letters* 108 (24): 241101.

Tyler, G.L., and Howard, H. (1973). Dual-frequency bistatic-radar investigations of the Moon with Apollos 14 and 15. *Journal of Geophysical Research* 78 (23): 4852–4874.

Tyler, G., Balmino, G., Hinson, D., Sjogren, W., Smith, D., Simpson, R., Asmar, S.W., Priest, P., and Twiken, J. (2001). Radio science observations with Mars Global Surveyor: orbit insertion through one year of mapping orbit. *Journal of Geophysical Research* 106 (E10): 23,348.

Tyler, G.L. (1987). Radio propagation experiments in the outer solar system with Voyager. *Proceedings of the IEEE* 75 (10): 1404–1431.

Tyler, G.L., Balmino, G., Hinson, D.P., Sjogren, W.L., Smith, D.E., Woo, R., Asmar, S.W., Connally, M.J., Hamilton, C.L., and Simpson, R.A. (1992). Radio science investigations with Mars Observer. *Journal of Geophysical Research* 97: 7759–7779.

Tyler, G.L., Brenkle, J.P., Komarek, T.A., and Zygielbaum, A.I. (1977). The Viking solar corona experiment. *Journal of Geophysical Research* 82: 4335–4340.

Tyler, G.L., Linscott, I.R., Bird, M.K., Hinson, D.P., Strobel, D.F., Pństzold, M., Summers, M.E. and Sivaramakrishnan, K. (2009). The New Horizons radio science experiment (REX). New Horizons 217–259.

Tyler, G.L., Linscott, I.R., Cheng, A., and Asmar, S.W. (1993). Pluto fast flyby uplink radio occultation instrument. abstract. *Pluto and Charon Conference*, Flagstaff, AZ (10–14 July 1993).

Tyler, G.L., Sweetnam, D.N., Anderson, J.D., Borutzki, S.E., Campbell, J.K., Eshleman, V.R., Gresh, D.L., Gurrola, E.M., Hinson, D.P., Kawashima, N., Kursinski, E.R., Levy, G.S., Lindal, G.F., Lyons, J.R., Marouf, E.A., Rosen, P.A., Simpson, R.A., and Wood, G.E. (1989). Voyager radio science observations of Neptune and Triton. *Science* 246 (4936): 1466–1473.

Vacchione, J.D., Kruid, R.C., Prata, A., Amaro, L.R. and Mittskus, A.P. (2012, March). Telecommunications antennas for the Juno Mission to Jupiter. In: *2012 IEEE Aerospace Conference* 1–16. IEEE.

Van Allen, J.A. (2003). Gravitational assist in celestial mechanics—a tutorial. *American Journal of Physics* 71 (5): 448–451.

Vening Meinesz, F. A. (1951). A remarkable feature of the Earth's topography. *Proceedings Koninklijke Nederlandse Academie van Wetenschap- pen Series B Physical Sciences* 54: 1–2.

Verma, A.K. and Margot, J.L. (2018). Expected precision of Europa Clipper gravity measurements. *Icarus* 314: 35–49.

Vig, J.R. (2001). Quartz crystal resonators and oscillators. US Army Communications-Electronics Command, January.

Villard Jr, O.G., Peterson, A.M., Manning, L.A., and Eshleman, V.R. (1953). Extended-range radio transmission by oblique reflection from meteoric ionization. *Journal of Geophysical Research* 58 (1): 83–94.

Volland, H., Bird, M.K., Levy, G.S., Stelzried, C.T., and Seidel, B.L. (1977). Helios-1 faraday-rotation experiment – results and interpretations of solar occultations in 1975. *Journal of Geophysics – Zeitschrieft Für Geophysik* 42: 659–672.

Vyshlov, A.S. (1976). Preliminary results of circumlunar plasma research by the Luna 22 spacecraft. *Space Research* xvi: 945–949.

Wahl, S.M., Hubbard, W.B., Militzer, B., Guillot, T., Miguel, Y., Movshovitz, N., Kaspi, Y., Helled, R., Reese, D., Galanti, E. and Levin, S. (2017). Comparing Jupiter interior structure models to Juno gravity measurements and the role of a dilute core. *Geophysical Research Letters* 44 (10): 4649–4659.

Wahlquist, H.D. (1987). The Doppler response to gravitational waves from a binary star source. *General Relativity and Gravitation* 19: 1001–1113.

Wahlquist, H.D., Anderson, J.D., Estabrook, F.B., and Thorne, K.S. (1977). Recent JPL work on gravity wave detection and solar system relativity experiments. *Atti Dei Covegni Lincei* 34: 335–350.

Wahr, J.M. (1981). The forced nutations of an elliptical, rotating, elastic, and oceanless Earth. *Geophysical Journal of the Royal Astronomical Society* 64: 705–727.

Walls, F.L. and Vig, J.R. (1995). Fundamental limits on the frequency stabilities of crystal oscillators. *IEEE transactions on ultrasonics, ferroelectrics, and frequency control* 42 (4): 576–589.

Wallis, R.E., Weaver, G.L., Reinhart, M.J. and Cheng, S. (2005, March). An advanced synthesized ultra-stable oscillator for spacecraft applications. In: *2005 IEEE Aerospace Conference* 952–960. IEEE.

Watanabe, S., Hirabayashi, M., Hirata, N., Hirata, N., Noguchi, R., Shimaki, Y., Ikeda, H., Tatsumi, E., Yoshikawa, M., Kikuchi, S., and Yabuta, H. (2019). Hayabusa2 arrives at the carbonaceous asteroid 162173 Ryugu—a spinning top–shaped rubble pile. *Science* 364(6437): 268–272.

Watkins, M.M., Davis, E.S., Melbourne, W.G., Yunck, T.P., Sharma, J., and Tapley, B.D. (1995). GRACE: A new mission concept for high resolution gravity field mapping.

Watkins, M.M., Wiese, D.N., Yuan, D.-N., Boening, C., and Landerer, F.W. (2015). Improved methods for observing Earth's time variable mass distribution with GRACE using spherical cap mascons. *Journal of Geophysical Research: Solid Earth* 120 (4): 2648–2671.

Watts, A.B. (2001). Isostasy and Flexure of the Lithosphere. Cambridge University Press.

Weaver, G.L. and Kantsiper, B.L. (2007, March). Developments toward a disciplined timekeeping system for lunar and planetary navigation. In: *2007 IEEE Aerospace Conference* 1–9. IEEE.

Wertz, J.R. and Sarzi-Amade, N. (2011). Time and Date Systems. *Atlantic* 300 (4): 3.

Wieczorek, M.A. (2007). Gravity and topography of the terrestrial planets. *Treatise on Geophysics* 10: 165–206.

Wieczorek, M.A. and Phillips, R.J. (1998). Potential anomalies on a sphere: Applications to the thickness of the lunar crust. *Journal of Geophysical Research: Planets* 103 (E1): 1715–1724.

Wieczorek, M.A., Neumann, G.A., Nimmo, F., Kiefer, W.S., Taylor, G.J., Melosh, H.J., Phillips, R.J., Solomon, S.C., Andrews-Hanna, J.C., Asmar, S.W., Konopliv, A.S., Lemoine, F.G., Smith, D.E., Watkins, M.M., Williams, J.G., and Zuber, M.T. (2012). The crust of the Moon as seen by GRAIL. *Science* 339 (6120): 671–675. doi:10.1126/science.1231530.

Will, C.M. (1971). Theoretical frameworks for testing relativistic gravity. III. Conservation laws, Lorentz invariance, and values of the PPN parameters. *The Astrophysical Journal* 169: 125.

Will, C.M. (1993). Theory and Experiment in Gravitational Physics. Cambridge, UK: Cambridge University Press.

Will, C.M. (2006). The confrontation between general relativity and experiment: a centenary perspective. *Progress of Theoretical Physics Supplemental* 163: 146–162. doi:10.1143/PTPS.163.146.

Will, C.M. (2014). The confrontation between general relativity and experiment. *Living Reviews in Relativity* 17 (llr-2014-4).

Williams, J.G., Konopliv, A.S., Boggs, D.H., Park, R.S., Yuan, D.-.N., Lemoine, F.G., Goossens, S., Mazarico, E., Nimmo, F., Weber, R.C., Asmar, S.W., Melosh, H.J., Neumann, G.A., Phillips, R.J., Smith, D.E., Solomon, S.C., Watkins, M.M., Wieczorek, M.A., Andrews-Hanna, J.C., Head, J.W., Kiefer, W.S., Matsuyama, I., McGovern, P.J., Taylor, G.J., and Zuber, M.T. (2014). Lunar interior properties from the GRAIL mission. *Journal of Geophysical Research: Planets* 119 (7): 1546–1578.

Williams, J.P., Nimmo, F., Moore, W.B. and Paige, D.A. (2008). The formation of Tharsis on Mars: What the line-of-sight gravity is telling us. *Journal of Geophysical Research: Planets* 113 (E10).

Williams, J.G., Turyshev, S.G., and Boggs, D.H. (2004). Progress in lunar laser ranging tests of relativistic gravity. *Physical Review Letters* 93 (26): 261101.

Williams, J.G., Turyshev, S.G., and Boggs, D.H. (2006). Lunar laser ranging tests of the equivalence principle with the Earth and Moon. *Advances in Space Research* 37: 67–71.

Withers, P., Felici, M., Mendillo, M., Moore, L., Narvaez, C., Vogt, M.F. and Jakosky, B.M. (2018). First ionospheric results from the MAVEN radio occultation science experiment (ROSE). *Journal of Geophysical Research: Space Physics* 123 (5): 4171–4180.

Withers, P. and Vogt, M.F. (2017). Occultations of astrophysical radio sources as probes of planetary environments: A case study of Jupiter and possible applications to exoplanets. *The Astrophysical Journal* 836 (1): 114.

Withers, P., Felici, M., Mendillo, M., Moore, L., Narvaez, C., Vogt, M.F. and Jakosky, B.M. (2018). First ionospheric results from the MAVEN radio occultation science experiment (ROSE). *Journal of Geophysical Research: Space Physics* 123 (5): 4171–4180.

Wohlmuth, R., Plettemeier, D., Edenhofer, P., Bird, M.K., and Asmar, S.W. (1997a). Analysis of galileo doppler measurements to determine coronal electron density fluctuations and plasma propagation velocities during solar occultations 1994 and 1995.

In: The Three Galileos: The Man, The Spacecraft, The Telescope (ed. C Barbieri, T. V. Johnson, A. M. Sohus, J. H. Rahe), 421–428. Dordrecht, Netherlands: Springer.

Wohlmuth, R., Plettemeier, D., Edenhoffer, P., Bird, M.K., Pätzold, M., and Asmar, S.W. (1997b). Measurement of the propagation speed of plasma inhomogeneities in the solar corona using and uplink/downlink cross-correlation method. *Radio Science* 32: 617–628.

Won, L. and Lee, A. (2001). *AACS Support for the Gravity Wave Experiment*. JPL Internal Memorandum SCO-01-044, 2001 (internal document), Jet Propulsion Laboratory, California Institute of Technology, Pasadena, CA, 29 October 2001.

Woo, R. (1977). Measurement of the solar wind using spacecraft radio signals. In: Study of Travelling Interplanetary Phenomena, (ed. M.A. Shea, D.F. Smart and S.T. Wu), 81–101. Dordrecht, Netherlands: D. Reidel.

Woo, R. (1993). Spacecraft radio scintillation and solar system exploration. In: Wave Propagation in Random Media (Scintillation), (eds. V.I. Tatarskii, A. Ishimaru and V.U. Zavorotny), 50–83. Bellingham, Washington: SPIE/IOP Publishing.

Woo, R. (1996). 1985 Voyager 2 radio ranging measurements of coronal density: asymmetry in the radio profiles explained. *Astrophysical Journal* 458: L87–L90.

Woo, R. (1988). A synoptic study of Doppler scintillation transients in the solar wind. *Journal of Geophysical Research: Space Physics* 93 (A5): 3919–3926.

Woo, R. (2010). Revealing the true solar corona: imaging may have inadvertently led astrophysicists astray in understanding the Sun. *American Scientist* 98: 212–219.

Woo, R. and Armstrong, J.W. (1979). Spacecraft radio scattering observations of the power spectrum of electron density fluctuations in the solar wind. *Journal of Geophysical Research: Space Physics* 84: 7288–7296.

Woo, R. and Armstrong, J.W. (1981). Measurements of a solar flare-generated shock wave at 13.1 RO. *Nature* 292: 608–610.

Woo, R. and Armstrong, J.W. (1992). Observations of the electron density spectrum near the Sun during spectral broadening transients. In: Solar Wind Seven, (eds. E. Marsch and R. Schwenn), 583–586. *Proceeding of the 3rd COSPAR Colloquium*, Goslar, Germany (16–21 September 1991). Oxford, UK: Pergamon Press.

Woo, R. and Armstrong, J.W. (1992). Observations of large-scale structure in the inner heliosphere with Doppler scintillation measurements. In: Solar Wind Seven 319–322). Pergamon.

Woo, R., Armstrong, J.W., Bird, M.K., and Pätzold, M. (1995). Variation of fractional electron density fluctuations near 0.1 AU from the Sun observed by Ulysses dualfrequency ranging measurements. *Geophysical Research Letters* 22: 329–332.

Woo, R., Armstrong, J.W., and Ishimaru, A. (1980). Radio occultation measurements of turbulence in the Venus atmosphere by Pioneer Venus. *Journal of Geophysical Research: Space Physics* 85: 8031–8038.

Woo, R., Armstrong, J.W., Sheeley, N.R., Howard, R.A., Koomen, M.J., and Michels, D.J. (1985). Doppler scintillation observation of interplanetary shocks within 0.3 AU. *Journal of Geophysical Research* 90: 154–162.

Woo, R. and Habbal, S.R. (1999). Imprint of the Sun on the solar wind. *Astrophysical Journal* 510: L69–L72.

Woo, R., Yang, F.C., Yip, K.W., and Kendall, W.B. (1976). Measurements of large scale density fluctuations in the solar wind using dual frequency phase scintillations. *Astrophysical Journal* 210: 568–574.

Wood, G.E., Asmar, S.W., Rebold, T.A., and Lee, R.A. (1997). Mars Pathfinder entry, descent, and landing communications. Telecommunications and Data Acquisition Progress Report 42–131, Jet Propulsion Laboratory, California Institute of Technology, Pasadena, CA, 15 November 1997.

Wyithe, J.S.B. and Lobe, A. (2003). Low-frequency gravitational waves from massive black hole binaries: predictions for LISA and pulsar timing Arrays. *Astrophysical Journal* 590: 691–706.

Yakovlev, O.I. (1985). Propagation of radio waves in space. *Moscow Izdatel Nauka* January Issue. Moscow, USSR (in Russian; English translation available on books.google.com).

Yakovlev, O.I. (2002). Space Radio Science. Earth Space Institute Book Series. Oxfordshire, UK: Taylor and Francis, CRC Press. (25 April 2002).

Yakovlev, O.I. and Efimov, A.I. (1966). Fluctuation Characteristics and Radiowave Spectrum in Interplanetary Space. *Radiotekh. Elektron.* 11: *2064–6 (Nov. 1966)*.

Yeomans, D.K., Antreasian, P.G., Cheng, A., Dunham, D.W., Farquhar, R.W., Gaskell, R.W., Giorgini, J.D., Helfrich, C.E., Konopliv, A.S., McAdams, J.V., and Miller, J.K. (1999). Estimating the mass of Asteroid 433 Eros during the NEAR spacecraft flyby. *Science* 285 (5427): 560–561.

Yeomans, D.K., Barriot, J.P., Dunham, D.W., Farquhar, R.W., Giorgini, J.D., Helfrich, C.E., Konopliv, A.S., McAdams, J.V., Miller, J.K., Owen, W.M., and Scheeres, D.J. (1997). Estimating the mass of Asteroid 253 Mathilde from tracking data during the NEAR flyby. *Science* 278 (5346): 2106–2109.

Yeomans, D.K., Antreasian, P.G., Barriot, J.P., Chesley, S.R., Dunham, D.W., Farquhar, R.W., Giorgini, J.D., Helfrich, C.E., Konopliv, A.S., McAdams, J.V. and Miller, J.K. (2000). Radio science results during the NEAR-Shoemaker spacecraft rendezvous with Eros. *Science* 289 (5487): 2085–2088.

Yoder, C.F. (1995a). Astrometric and geodetic properties of Earth and the Solar System. *Global Earth Physics: A Handbook of Physical Constants* 1 (88): 1.

Yoder, C.F. (1995b). Venus' free obliquity. *Icarus* 117 (2): 250–286.

Yoder, C.F., Konopliv, A.S., Yuan, D.-N., Standish, E.M., and Folkner, W.M. (2003). Fluid core size of Mars from detection of the solar tide. *Science* 300 (5617): 299–303.

Yuan, D.N. (1991). The determination and error assessment of the Earth's gravity field model. Ph.D. thesis, University of Texas at Austin.

Yuan, D.-.N., Sjogren, W.L., Konopliv, A.S., and Kucinskas, A.B. (2001). Gravity field of Mars: a 75th degree and order model. *Journal of Geophysical Research: Planets* 106 (E10): 23377–23401.

Yuen, J.H. (1983). Telecommunications systems design. In: Deep Space Telecommunications Systems Engineering, 1–22. Boston, MA: Springer.

Zebker, H.A., Marouf, E.A., and Tyler, G.L. (1985). Saturn's rings: particle size distribution for thin layer models. *Icarus* 64: 531–548.

Zuber, M.T., Smith, D., Solomon, S.C., Muhleman, D.O., Head, J.W., Garvin, J.B., Abshire, J.B. and Bufton, J.L. (1992). The Mars Observer laser altimeter investigation. *Journal of Geophysical Research: Planets* 97 (E5): 7781–7797.

Zuber, M.T., Smith, D.E., Phillips, R.J., Solomon, S.C., Neumann, G.A., Hauck, S.A., Peale, S.J., Barnouin, O.S., Head, J.W., Johnson, C.L. and Lemoine, F.G. (2012). Topography of the northern hemisphere of Mercury from MESSENGER laser altimetry. *Science* 336 (6078): 217–220.

Zuber, M.T., Lemoine, F.G., Smith, D.E., Konopliv, A.S., Smrekar, S.E., and Asmar, S.W. (2007a). Mars Reconnaissance Orbiter Radio Science Gravity Investigation. *Journal of Geophysical Research: Planets* 112 (E5): doi:10.1029/2006JE002833.

Zuber, M.T., Phillips, R.J., Andrews-Hanna, J.C., Asmar, S.W., Konopliv, A.S., Lemoine, F.G., Plaut, J.J., Smith, D.E., and Smrekar, S.E. (2007b). Density of Mars' south polar layered deposits. *Science* 317 (5845): 1718–1719.

Zuber, M.T. and Smith, D.E. (2018, September). Reviewing the contribution of GRAIL to lunar science and planetary missions. European Planetary Science Congress EPSC2018-575.

Zuber, M.T., Smith, D.E., Lehman, D.H., Hoffman, T.L., Asmar, S.W., and Watkins, M.M. (2013a). Gravity Recovery and Interior Laboratory (GRAIL): mapping the lunar interior from crust to core. *Space Science Reviews* 178: 3–24. doi:10.1007/s11214-012-9952-7.

Zuber, M.T., Smith, D.E., Neumann, G.A., Goossens, S., Andrews-Hanna, J.C., Head, J.W., Kiefer, W.S., Asmar, S.W., Konopliv, A.S., Lemoine, F.G., Matsuyama, I., Melosh, H.J., McGovern, P.J., Nimmo, F., Phillips, R.J., Solomon, S.C., Taylor, G.J., Watkins, M.M., Wieczorek, M.A., Williams, J.G., Jansen, J.C., Johnson, B.C., Keane, J.T., Mazarico, E., Miljković, K., Park, R.S., Soderblom, J.M., and Yuan, D.-N. (2016). Gravity field of the Orientale Basin from the Gravity Recovery and Interior Laboratory Mission. *Science* 354 (6311): 438–441.

Zuber, M.T., Smith, D., Solomon, S.C., Muhleman, D.O., Head, J.W., Garvin, J.B., Abshire, J.B. and Bufton, J.L. (1992). The Mars Observer laser altimeter investigation. *Journal of Geophysical Research: Planets* 97 (E5): 7781–7797.

Zuber, M.T., Smith, D.E., Mazarico, E.M., Genova, A., Neumann, G.A. and Sun, X. (2017, December). Trilogy, an Interplanetary Constellation to Measure the Rate of Mass Loss by the Sun. AGU Fall Meeting Abstracts 2017: P11F-08.

Zuber, M.T., Solomon, S.C., Phillips, R.J., Smith, D.E., Tyler, G.L., Aharonson, O., Balmino, G., Banerdt, W.B., Head, J.W., Johnson, C.L. and Lemoine, F.G. (2000). Internal structure and early thermal evolution of Mars from Mars Global Surveyor topography and gravity. *Science* 287 (5459): 1788–1793.

Zuber, M.T., Smith, D.E., Watkins, M.M., Asmar, S.W., Konopliv, A.S., Lemoine, F.G., Melosh, J.H., Neumann, G.A., Phillips, R.J., Solomon, S.C., and Wieczorek, M.A. (2013b). Gravity field of the Moon from the Gravity Recovery and Interior Laboratory (GRAIL) Mission. *Science* 339: 668–671. doi:10.1126/science.1231507.

Acronyms and Abbreviations

1D	One-dimensional
2D	Two-dimensional
3D	Three-dimensional
3GM	Gravity and Geophysics of Jupiter and the Galilean Moons
AACS	Attitude and Articulation Control Subsystem
ACE	Advanced Composition Explorer
ACE	Call Sign for Project Real-time Mission Controller
ACQ	Acquisition
ACS	Antenna Control System
ADC	Analog-to-digital Conversion
AGC	Automatic Gain Control
AMC	Advanced Media Calibration Subsystem
AMELIA	ExoMars Atmospheric Mars Entry and Landing Investigations and Analysis
AMMOS	Advanced Multi-mission Operations System
AMS	Antenna Microwave System
AOS	Acquisition of Signal
APA	Antenna Pointing Assembly
APL	Applied Physics Laboratory
ARI	Advanced Ranging Instrument
ASCII	American Standard Code for Information Interchange
ASI	Agenzia Spaziale Italiana (Italian Space Agency)
ASIC	Application-specific Integrated Circuit
ATDF	Archival Tracking Data File
ATLO	Assembly, Test, and Launch Operations
AU	Astronomical Unit
AutoNav	Autonomous Navigation
AUX	Auxiliary
Aux Osc	Auxiliary Oscillator
AWVR	Advanced Water Vapor Radiometer
AZ	Azimuth
BER	Bit-error Rate
BLF	Best-lock Frequency

Radio Science Techniques for Deep Space Exploration, First Edition. Sami W. Asmar.

BOA	Beginning of Activity
BOT	Beginning of Track
BPF	Band Pass Filter
BPM	Bits Per Minute
BPS	Bits Per Second
BPSK	Binary Phase-shift Keying
BSR	Bistatic Radar
BVA	Boîtier à Vieillissement Amélioré (enclosure with improved aging)
BVE	Block-V Exciter
BVR	Block-V Receiver
BW	Bandwidth
BWG	Beam Wave Guide (antenna)
Caltech	California Institute of Technology
CAN	Canberra (Deep Space Communications Complex)
CBE	Current Best Estimate
CCSDS	Consultative Committee for Space Data Systems
CCT	Central Communications Terminal
C&DH	Command and Data Handling
CDR	Central Data Recorder
CDR	Critical Design Review
CDS	Command and Data Subsystem
CDSCC	Canberra Deep Space Communications Complex
CDU	Command Detector Unit
CFDP	CCSDS (Consultative Committee for Space Data Systems) File Delivery Protocol
CG	Center of Gravity
CKF	Spacecraft Orientation C-Kernel File
CL	Closed Loop
CLR	Closed-loop Receiver
CLTU	Command Link Transmission Unit
cm	Centimeter(s)
CM	Center of Mass
CMA	Command Modulator Assembly
CMC	Complex Monitor and Control
Cmd	Command
CME	Coronal Mass Ejection
CNR	Carrier-to-noise Ratio
COI	Center of Integration
Comm	Communication, or on a Diplexer, Common Port
Conscan	Conical Scanning (Antenna Pointing Mode)
COSMIC	Constellation Observing System for Meteorology, Ionosphere, and Climate
CPA	Command Processor Assembly
CSAC	Chip-scale Atomic Clock

CSIRO	Commonwealth Scientific and Industrial Research Organization (Australia)
CSO	Cryogenic Sapphire Oscillator
CTS	Coaxial Transfer Switch (Also Referred to as CXS)
CTT	Compatibility Test Trailer
CTX	Context Camera
CW	Continuous Wave
D/A	Digital to Analog
DAC	Digital-to-analog Converter
dB	Decibel
dBc	dB Below Carrier
dB–Hz	Decibel–Hertz
dBi	Decibel with Respect to Isotropic Gain
dBic	Decibel with Respect to Isotropic Antenna
dBm	Decibels Referenced to Milli-Watts
DART	Double Asteroid Redirection Test
DC	Direct Current; Steady-state Bias
DCD	Data Capture and Display
DCO	Digital Control Oscillator
DCPC	Downlink Channel Processing Cabinet
DCS	Digital Communications Subsystem
DCT	Design Control Table
DDOR	Delta Differential One-way Ranging
DDS	Direct Digital Synthesis
DEA	Descent Engine Assembly
DEA	Digital Electronics Assembly
DEC	Declination
deg	Degree
delta-DOR	Delta Differential One-way Ranging
DESCANSO	Deep Space Communications and Navigation Systems Center of Excellence
DKF	Deep Space Network Keyword Files
D/L	Downlink
DMC	DSCC Monitor and Control Subsystem
DMD	DSCC Media Calibration Subsystem
DN	Data Number
DOR	Differential One-way Ranging
DOWR	Dual One-way Range
DOY	Day of Year
DPDU	Digital Power Distribution Unit
DPM	Digital Processing Module (in Small Deep Space Transponder)
DR	Discrepancy Report
DSAC	Deep Space Atomic Clock
DSCC	Deep Space Communications Complex
DSDST	Descent Stage Small Deep Space Transponder
DSMS	Deep Space Mission Systems

DSN	Deep Space Network
DSOC	Deep Space Optical Communications
DSS	Deep Space Station
DSS-14	70-m Deep Space Station at Goldstone, California
DSS-15	34-m Deep Space Station at Goldstone, California (Decommissioned)
DSS-24	34-m Deep Space Station at Goldstone, California
DSS-25	34-m Deep Space Station at Goldstone, California
DSS-26	34-m Deep Space Station at Goldstone, California
DSS-30	45-m Misasa Deep Space Antenna (Japan)
DSS-34	34-m Deep Space Station at Canberra, Australia
DSS-35	34-m Deep Space Station at Canberra, Australia
DSS-36	34-m Deep Space Station at Canberra, Australia
DSS-43	70-m Deep Space Station at Canberra, Australia
DSS-45	34-m Deep Space Station at Canberra, Australia (decommissioned)
DSS-48	64-m Usuda Deep Space Station (Japan)
DSS-49	64-m Parkes Radio Telescope (Australia)
DSS-54	34-m Deep Space Station at Madrid, Spain
DSS-55	34-m Deep Space Station at Madrid, Spain
DSS-56	34-m Deep Space Station at Madrid, Spain
DSS-63	70-m Deep Space Station at Madrid, Spain
DSS-65	34-m Deep Space Station at Madrid, Spain
DSS-69	65-m Sardinia Deep Space Antenna (Italy)
DST	Deep Space Transponder
DTE	Direct-to-Earth
DTF	DSN Test Facility
DTK	DSCC Tracking Subsystem
DTR	Digital Tape Recorder
DTT	Downlink Telemetry and Tracking Subsystem (in Deep Space Network)
DWE	Doppler Wind Experiment
E_b/N_0	Bit-energy-to-noise Spectral-density Ratio
EDA	EDL (Entry, Descent, and Landing) Data Analysis
EDL	Entry, Descent, and Landing
EH&A	Engineering Health and Accountability (Analysis)
EIRP	Effective Isotropic Radiated Power
EL	Elevation
ELT	Electra Lite Transponder
EM	Electromagnetic
EM	Electromagnetic Compatibility
EMI	Electromagnetic Interference
Eng/HK	Engineering/Housekeeping
EOA	End of Activity
EOL	End of Life
EOM	End of Mission
EOP	Earth Orientation Parameters

EOP	Earth Orientation Platform
EOS	Equations of State
EOT	End of Track
EPC	Electronic Power Converter
EPOCh	Extrasolar Planet Observations and Characterization
EPOXI	(Combination of Two Mission Names) Extrasolar Planet Observations and Characterization (EPOCh), and the Flyby of Comet Hartley 2, Called the Deep Impact Extended Investigation (DIXI)
EPS	Electrical Power System
ERT	Earth Received Time
Es	Energy Per Telemetry Symbol
E_s/N_0	Symbol Energy-to-noise Spectral-density Ratio
ESA	European Space Agency
EST	Eastern Standard Time
ESTEC	European Space Administration Research and Technology Centre
ESTRACK	European Space Tracking (network)
ESOC	European Space Operations Center
ET	Ephemeris Time
ETX	Exciter Transmitter
EU	Engineering Unit
EUT	Electra UHF (Ultrahigh Frequency) Transceiver
eV	Electron Volt
EVR	Event Report
Ex	Exciter (Provides the Radio Frequency Drive to the Transmitter)
EXC	Exciter
F1	Fundamental Frequency of the Small Deep Space Transponder (SDST)
FA	Flight Acceptance
FAROT	Faraday Rotation
FCD	Feedback Concatenated Decoder
FD	Flight Dynamics
FDS	Flight Data Subsystem
FEI	Frequency Electronics, Inc.
FEM	Flight-engineering Model
FET	Field Effect Transistor
FFA	Front-end Filter Assembly
FFT	Fast Fourier Transform
FPA	Fault-protection Algorithm
FPGA	Field Programmable Gate Array
FR	Faraday Rotation
FRO	Frequency Reference Offset
FRPLL	Faraday Rotation Phase-locked Loop
FS	Flight System
FSC	Full-spectrum Combiner
FSK	Frequency-shift Keying
FSP	Full-spectrum Processor Subsystem

FSP-R	Full-spectrum Processor-radio Science
FSR	Full-spectrum Recorder
FSS	Frame Synchronizer System (subsystem in Voyager)
FSS	Frequency Selective Surface (on Voyager S-/X-band Antenna (SXA)
FSU	Filtering and Switch Unit
FSW	Flight Software
f	Foot
FTE	Full-time Equivalent
FTP	File Transfer Protocol
FTS	Frequency and Timing Subsystem
FY	Fiscal Year
g	Gram
gal	Galileo (Unit of Acceleration)
GaAs	Gallium Arsenide
Gb	Gigabit
GB	Gigabyte
GCF	Ground Communications Facility
GCN	Ground Communications Network
GDS	Goldstone (Deep Space Communications Complex)
GDS	Ground Data System
GDSCC	Goldstone Deep Space Communications Complex (DSCC)
GEM	Galileo Europa Mission
GFE	Government Furnished Equipment
GHz	Giga-Hertz
GM	Gravitational constant multiplied by planetary mass
GMM	Galileo Millennium Mission
GMSK	Gaussian-filtered Minimum-shift Keying
GMT	Greenwich Mean Time
GNC	Guidance, Navigation, and Control
GOCE	Gravity Field and Steady-state Ocean Circulation Explorer
GPS	Global Positioning System
GR	General Relativity
GRACE	Gravity Recovery and Climate Experiment
GRACE-FO	Gravity Recovery and Climate Experiment Follow On
GRAIL	Gravity Recovery and Interior Laboratory
GRV	Gravity
G/S	Ground Station
GS&E	General Science and Engineering
GSE	Ground Support Equipment
GSFC	Goddard Space Flight Center
GSOC	German Space Operations Center
G/T	Ratio of Antenna Gain to Noise Temperature
GTP	Generalized Telecom Predictor
GTR	General Theory of Relativity
GUI	Graphical User Interface
GW	Gravitational Wave

GWE	Gravitational Wave Experiment
GWU	George Washington University
H	Hydrogen
HA	Hour Angle
Hazcam	Hazard Camera
HCD	Hardware Command Decoder
HDO	Half-duplex Overlay
HEF	High-efficiency (as in 34-m HEF Antennas)
HEMT	High Electron Mobility Transistor (amplifier)
Hexem	Unit of Electron Column Density (1 hexem = 1 TEC Unit = 1 TECU = 10^{16} electrons/m^2)
HGA	High-gain Antenna
HGAG	High-gain Antenna Gimbal
hh:	mm hours:minutes
HiRISE	High-resolution Imaging Science Experiment
HK	Housekeeper
HLAN	High-speed Local Area Network
HP	High Power
HPCW	High-priority Comm Window
HPM	Hours Past Midnight
HRS	Heat Rejection System (or subsystem)
HSB	High (Angular Tracking) Speed Beam Waveguide (BWG)
HSD	High-speed Data
HVPS	High-voltage Power Supply
Hz	Hertz
IAU	International Astronomical Union
ICA	IF (interface) channel assembly
ICD	Interface Control Document
ICRF	International Celestial Reference Frame
ID	Identification
IDC	Intermediate Frequency to Digital Converter
IERS	International Earth Rotation and Reference Systems Service
IF	Intermediate Frequency
IFMS	Intermediate Frequency and Modem System
IMP	Interplanetary Monitoring Platform
IMU	Inertial-measurement Unit
IND	Interplanetary Network Directorate (now IPN-ISD)
InSight	Seismic Investigations, Geodesy and Heat Transport
ION	Ionospheric Calibration File
ION	Media Calibration Ionosphere (file)
I&Q	In-phase and Quadrature-phase
IR	Infrared
ISA	Incident, Surprise, Anomaly (report)
ISO	Isolator (at Traveling Wave Tube Amplifier Output)
ISS	Imaging Science Subsystem
ITE	Impact through Egress
ITRF	Terrestrial Reference Frame

IUS	Inertial Upper Stage
JAXA	Japan Aerospace Exploration Agency
JOI	Jupiter Orbit Insertion
JPL	Jet Propulsion Laboratory
JSX	Jupiter Saturn Explorer (Previous Name for Voyager 2 Spacecraft)
JUICE	Jupiter Icy Moons Explorer
K	Kelvin
Ka-band	Frequencies in the Range 26.5–40 GHz with 32–35 GHz Used for Deep Space Communications Band
KaPA	Ka-band Power Amplifier
KaT	Ka-band Translator
KaTS	Ka-band Translator Subsystem (including power amplifier)
kb	Kilobit
Note:	telecommunications Uses Bits (b); Data Storage Uses Bytes (B), Which Are 8 b/B.
kbps	Thousand(s) of Bits Per Second
KEX	Ka-band Exciter
kg	Kilogram
KHA	Ka-band Horn Antenna
kHz	Kilo-Hertz
km	Kilometer
km/h	Kilometers Per Hour
km^2/s^2	Kilometers Squared Per Second Squared
kN	Kilo-Newton
KUPL	Ka-band Uplink Subsystem
kW	Kilo-Watt
LAN	Local Area Network
LaRa	Lander Radio Science for Obtaining the Rotation and Orientation of Mars
LASCO	Large Angle and Spectrometric Coronagraph
L-band	Frequency Range between 390–1550 MHz
LC	Inductance Capacitance Filter
LCP	Left (Hand)-circular Polarization; Left Circularly Polarized
LEM	Lower Equipment Module
LGA	Low-gain Antenna
LGAX	Low-gain Antenna Directed along the +*x*-axis
LGAZ	Low-gain Antenna directed along the +*z*-axis
LGAZ–LGA	Low-gain Antenna directed along the –*z*-axis
LH	Left Hand
LHCP	Left-hand Circular Polarized (or Polarization)
LICIACube	Light Italian CubeSat for Imaging of Asteroid
LIGO	Laser Interferometer Gravitational-wave Observatory
LISA	Laser Interferometer Space Antenna
LLR	Lunar Laser Ranging
LMA	Lockheed Martin Aerospace
LMC	Link Monitor and Control
LMST	Local Mean Solar Time

LNA	Low-noise Amplifier
LO	Local Oscillator
LOLA	Lunar Orbiter Laser Altimeter
LOS	Loss of Signal
LOS	Line of Sight
LP	Low Power
LPA	Lander Petal Actuator (MER)
LPE	Low-power Electronics
LPF	Low-pass Filter
LRO	Lunar Reconnaissance Orbiter
LST	Local Solar Time
LTST	Local True Solar Time
m	Meter(s)
MAD	Madrid DSCC (Deep Space Communications Complex)
MAG	Magnetometer
MarCO	Mars Cube One
MaRS	Mars Express Radio Science
Mas	Milliarcseconds
MAVEN	Mars Atmosphere and Volatile Evolution (Mission)
MB	Megabyte
Mbit	(Note: Megabit Telecommunications Uses Bits [b]; Data Storage Uses Bytes [B], Which Are 8 b/B)
Mbps	Megabit Per Second
MBR	Mars Balloon Relay (protocol)
MCD/FS	Maximum-likelihood Convolutional Decoder Frame Synchronizer (Subsystem)
MCS	Media Calibration System
MCS	Mars Climate Sounder (MRO)
MDA	Metric Data Assembly
mdeg	Millidegree
MDNS	Mission Design and Navigation Subsystem
MDS	Modulation-demodulation Subsystem
MDSCC	Madrid Deep Space Control Center (DSCC)
MECO	Main Engine Cutoff
MEP	Mars Exploration Program
MER	Mars Exploration Rover
MER-A	Mars Exploration Rover "Spirit"
MER-B	Mars Exploration Rover "Opportunity"
MES	Main Engine Start
MESSENGER	Mercury Surface, Space Environment, Geochemistry, and Ranging (Mission)
MEX	Mars Express
MFS	Multifunctional Structure (Experiment)
MFSK, M-FSK	Multiple Frequency Shift Keying (Tones)
MGA	Medium-Gain Antenna
mGal	0.001 Galileo, which is 1 cm/s^2
Mgate	Million Gates

MGDS	Multimission Ground Data System
MGS	Mars Global Surveyor
MGSS	Multi-Mission Ground System
MHz	Mega-Hertz
MIC	Microwave-integrated Circuit
MIL-71	Name of Deep Space Network Station at Cape Canaveral (not an Acronym)
MIL-STD	Military Standard (Usually Followed by a Document Number)
MIRAGE	Multiple Interferometric Ranging Analysis Using GPS Ensemble
MIT	Massachusetts Institute of Technology
MLE	Mars Landing Engine
MM	Multimission
MMO	(JPL) Mission Management Office
MMRTG	Multimission Radioisotope Thermoelectric Generator
Mm/s	Millimeter Per Second
MMX	Martian Moons Exploration
MOC	Mars Orbiter Camera
MOD	Modulation
MOI	Mars Orbit Insertion
MoI	Moment of Inertia
MOLA	Mars Observer Laser Altimeter
MOM	Mars Orbiter Mission
MON	Monitor and Control Subsystem
MON	Monitor System
MONTE	Mission-analysis, Operations, and Navigation Toolkit Environment
MORE	Mercury Orbiter Radio Science Experiment
MOS	Mission Operations System
MP	Modem Processor
MPa	Mega-Pascal
MPCS	Mission Data Processing and Control Subsystem
MPF	Mars Pathfinder
MPO	Mercury Planetary Orbiter
MPST	Mission Planning and Sequencing Team
MPTS	Multipurpose Tracking System
mrad	Milliradian
MRO	Mars Reconnaissance Orbiter
MRO	Memory Readout
MS	Master of Science
m/s	Meter(s) Per Second
ms	Millisecond
MSA	Mission-support Area
MSAP	Multimission System Architecture Platform
MSAT	Mobile Satellite
MSL	Mars Science Laboratory
MSPA	Multiple Spacecraft Per Aperture (or Antenna)

Msps	Megasymbols Per Second
MSR	Mars Sample Return
MSSS	Malin Space Science Systems
MST	Monitor Sample Time
MTC	Mars Time Coordinated
mGal	milliGal (see Gal/gal)
mW	milliWatt
N	Newton
Na	Sodium
NAIF	Navigation and Ancillary Information Facility
NASA	National Aeronautics and Space Administration
NAV	Navigation
NCO	Numerically Controlled Oscillator
NEAR	Near Earth Asteroid Rendezvous
NISN	NASA Integrated Services Network
NIST	National Institute of Standards and Technology
NIU	Network Interface Unit
NMC	Network Monitor and Control
NMoI	Normalized Moment of Inertia
NMP	New Millennium Program
NNO	New Norcia ESA Ground Station
NNR-NEWVAL1	No-net-rotation (Model of Plate Velocities)
noap	No a Priori
NOCC	Network Operations and Control System
NOCC RT	Network Operations Control Center Real Time (System)
NOP	Network Operations Plan
nossf	No Small Forces File
NPO	NPO Energomash (Russian Manufacturer)
NPP	Network Planning and Preparation System
nrad	Nanoradians
NRAO	National Radio Astronomy Observatory
NRZ	Non-return to Zero
ns	Nanosecond
NS	NOCC Support Subsystem
NSP	Network Simplification Project
NVM	Nonvolatile Memory
NVM/CAM	Nonvolatile Memory/Camera
OCI	Operator Control Input
OCM	Organic Check Material
OCVCXO	Oven-controlled Voltage-controlled Crystal Oscillator
OCXO	Oven-controlled Crystal Oscillator
OD	Orbit Determination
ODB	Operational Database
ODF	Orbit Data File
ODP	Orbit Determination Program
ODY	Odyssey
OEM	Orbit Ephemeris Message

OL	Open Loop
OLR	Open-loop Receiver
ONC	Optical Navigation Camera Experiment (MRO)
ORSC	Orbiter Relay State Change
Osc	Oscillator
OSIRIS-REx	Origins, Spectral Interpretation, Resource Identification, Security, Regolith Explorer
OWLT	One-way Light Time
P-file	Prediction File
P/FR	Problem/Failure Report
Pa	Pascal
PASM	Power Actuation and Switching Module
P_c	Carrier Power
P_c/N_0	Carrier Power-to-noise Spectral-density Ratio
PCK	Planetary Constants
PCM	Pulse Code Modulation
PD	Passive Device (in Solid-state Power Amplifier, a Microwave Coupler or Combiner)
PDF	Portable Document Format
P_d/N_0	Data Power-to-noise Spectral-density Ratio
PDS	Planetary Data System
PDT	Pointing Design Tool
PDU	Protocol Data Unit
PEDL	Pre-entry, Descent, and Landing
PEPE	Plasma Experiment for Planetary Exploration
PI	Principal Investigator
PLAR	Post-launch Assessment Review
PLF	Payload Fairing
PLL	Phase-locked Loop
PLS	Plasma Science
PM	Post Meridian
PMP	Payload Mounting Module
PMS	Propulsion Module Subsystem
PN	Pseudo-noise
Pol	Polarizer
POLR	Portable Open-loop Receiver
POR	Power-on Reset
PPI	Planetary Plasma Interactions
ppm	Parts Per Million
PPM	Precision Power Monitor
PPN	Parameterized Post-Newtonian
PPS	Photopolarimeter System
PPS	Power and Pyrotechnics Subsystem
P_r	Downlink Ranging Power
PRA	Planetary Radio Astronomy
P_r/N_0	Ranging Power-to-noise Spectral-density Ratio
PROBA	Project for On-board Autonomy

Prox-1	Proximity-1 Protocol
PRSR	Portable Radio Science Receiver
PR/TSA	Project Requirements/TMOD (Telecommunications and Mission Operations Directorate; Now Called IPN-ISD)
PSA	Planetary Science Archive
psf	Pounds Per Square Foot
PSM	Power Supply Module (MRO)
PSS	Parachute Support Structure
P_t/N_0	Total Power-to-noise Spectral-density Ratio
P_{tone}	Power in Differential One-way Ranging (DOR) Tone
PUHF	Parachute UHF (Ultrahigh Frequency) Antenna
PWS	Plasma Wave Spectrometer (Galileo)
PWS	Plasma Wave System
QPSK	Quadrature Phase-shift Keying
RA	Right Ascension
RAAR	Relay Antenna Assembly Receive
RAAT	Relay Antenna Assembly Transmit
rad	Radian (57.3°)
RAD	Radiation Assessment Detector
RAD	Rocket-assisted Deceleration (System in MER)
RAMP	Rover Avionics Mounting Plate
RASSI	Radio Atmospheric Sounding and Scattering Instrument
RAT	Rock Abrasion Tool
RC	Radar Controller
RC	Request Command
RCC	Receiver Control Computer
RCE	Rover Computer Element
RCP	Receiver Channel Processor
RCP	Right (Hand)-circular Polarization; Right-circularly Polarized
RCS	Reaction Control System
RCVR	Receiver
R&D	Research and Development
RED	Rover Equipment Deck (MER)
Reff	Effective Information Rate
REL	Relativity
REM	Rover Electronics Module
REMS	Rover Environmental Monitoring Station (Instrument to Measure Meteorological Conditions and Ultraviolet Near the Rover)
REU	Remote Electronics Unit (Also Called MREU for MSL Remote Electronics Unit)
REX	Radio Science Experiment (New Horizons Mission)
RF	Radio Frequency
RFDU	Radio Frequency Distribution Unit
RFIS	Radio Frequency Instrument Subsystem
RFM	Radio Frequency Module (Ultrahigh Frequency; MRO)
RFPDU	Radio Frequency Power Distribution Unit
RFS	Radio Frequency Subsystem

RFSTLC	Radio Frequency Subsystem Tracking-loop Capacitor
RH	Right Hand
RHCP	Right-hand Circular Polarized (or Polarization)
RID	Radio Frequency to Intermediate-frequency Downconverter
RISE	Rotation and Interior Structure Experiment (InSight)
R_j	Jupiter Radii
RM	Rotation Measure
RMDC	Radio Metric Data Conditioning
RMS	Root Mean Square
rng	Ranging
RNS	Reliable Network Server; Service
RO	Radio Occultation
ROS	Rosetta
RP	Radar Processor
RP	Rocket Propellant or Refined Petroleum (kerosene)
RPAM	Rover Power and Analog Module
RPDU	RF (Radio Frequency) Power Distribution Unit (in TDS)
RPFA	Rover Pyro Fire Assembly
rpm	Revolutions Per Minute
RRA	Relay Receiving (or Radio) Antenna
RRH	Relay Radio Hardware
RRH	Relay Receiving Hardware
RRP	Receiver and Ranging Processor
RS	Radio Science
RS	Reed–Solomon (code)
RSB	Radio Science Beacon
RSDL	RS (Reed–Solomon) Downlink
RSDS	Raw Science Data Server
RSDST	Rover Small Deep Space Transponder
RSM	Remote Sensing Mast
RSOS	Root of the Sum of Squares
RSR	Radio Science Receiver
RSS	Radio Science Subsystem
RSSG	Radio Science Systems Group
RSTI	Radio Science Transponder Instrument
R/T	Receive/Transmit
RT	Remote Terminal
RTG	Radioisotope Thermonuclear Generator
RTLT	Round-trip Light Time
RTN	Return
ru	Range Unit
R-UCTS	Rover UHF (Ultrahigh Frequency) Coax Transfer Switch
RUHF	Rover UHF (Ultrahigh Frequency) Antenna
RUSO	Receiver Ultrastable Oscillator
RWA	Reaction-wheel Assembly
R-WTS	Rover Waveguide Transfer Switch
Rx	Receive

s	Second
SA	Sample Acquisition
SAIC	Science Applications International Corporation
SAM	Sample Analysis at Mars (Instrument for Chemical and Isotopic Analysis of Acquired Samples)
SAMPEX	Solar Anomalous and Magnetospheric Particle Explorer
SAR	Synthetic Aperture Radar
SASF	Spacecraft Activity Sequence File
S-band	Frequencies in the Range 2–4 GHz (Spacecraft Uplink Frequency of ~2115 MHz and Downlink Frequency of ~2295 MHz in Deep Space Communications)
SBT	S-band Transmitter
SC	Solar Corona
S/C	Spacecraft
SC	Stress-compensated
SCaN	Space Communications and Navigation (Office)
SCE	Solar Conjunction Experiment
SCET	Spacecraft Event Time
SCID	Spacecraft Identifier
SCMF	Spacecraft Command Message File
SCO	Solar Corona Occultation
SCO	Solar Corona Observation
SD	Software-defined Radio
SDST	Small Deep Space Transponder
SEC	Sun–Earth–Craft (Angle)
SEC	Systematic Error Correction
SELENE	Selenological and Engineering Explorer (Kaguya Mission)
SEP	Solar Electric Propulsion
SEP	Sun–Earth–Probe (Angle)
SeqGen	Sequence Generation (Spacecraft Activity Planning Software)
SES	Spacecraft Engineering Subsystem
SET	Spacecraft Engineering Telemetry (File)
SETI	Search for Extraterrestrial Intelligence
S-EXC	S-band Exciter
SFDU	Standard Formatted Data Unit
SFOC	Space Flight Operations Center
SFOS	Spaceflight Operations Schedule
SHARAD	Shallow (Subsurface) Radar (MRO)
SHP	Shape Modeling
sigma	Spelled Out Form of Greek Character, σ (for 1 Standard Deviation)
SIM	Simulation
SIM	Space Interferometry Mission (Proposed Space Radio Telescope; the Program Was Cancelled in 2010)
SIP	Sequence Implementation Process
SIS	Software Interface Specification
SIT	Select in Test

SKA	Square-kilometer Array
S&L	Standards and Limits
SLC-17A	Space Launch Complex 17A
SLE	Space Link Extension
SMART	Small Missions for Advanced Research in Technology
SMAP	Soil Moisture Active Passive Mapping (Satellite)
SMM	Spacecraft Mode Manager
S/N	Signal-to-noise Ratio
SNR	Signal-to-noise Ratio
SNT	System Noise Temperature
SOAS	Science Operations Analysis Subsystem
SOC	State of Charge
SODF	SFDU Orbit Data File
SOE	Sequence of Events
SOHO	Solar and Heliospheric Observatory
sol	Martian Day
SOLR	Spacecraft Open-loop Receivers
SOP	Science Operation Plan
SOST	Satellites Orbiter Science Team
SOWG	Science Operations Working Group
SP	Science Planning
SP	Service Package
SPaH	Sample Preparation and Handling
SPARC	Scalable Processor Architecture
SPC	Signal Processing Center
SPC	Stereo-photo-clinometry
SPD	S-band Polarization Diversity
SPE	Static Phase Error
SPE	Sun–Probe–Earth (Angle)
SPICE	Spacecraft, Planet, Instrument, C-matrix, Events (Kernel Files)
SPIE	Society of Photo-optical Instrumentation Engineers
SPK	Spacecraft and Planetary Kernel
SPM	Seconds Past Midnight
SPP	Science Planning Process
SPS	Service Preparation Subsystem
sps	Symbols Per Second
SR	Sweep Range
SRA	Sequential Ranging Assembly
SRB	Solid Rocket Booster
SRBJ	Solid Rocket Booster Jettison
SRC	Subreflector Controller
S-RCVR	S-band Receiver
SRIF	Square Root Information
SRM	Solid Rocket Motor
SRU	Stellar-reference Unit
SSA	Solid State Amplifier
SSA	Sun-sensor Assembly (Deep Space 1)

SSD	Symbol Stream Distribution
SSI	Solid-state Imaging
SSK	Soft Symbol Controller
SSNR	Symbol Signal-to-noise Ratio
SSO	Sufficiently Stable Oscillator
SSPA	Solid-state Power Amplifier
SSR	Solid-state Recorder
SSUP	Science Sequence Update Process
S_{T}/N_0	Ratio to Product of Power in Data Sidebands (S) and Bit Duration (T) to Noise Spectral Density; Equivalent to E_b/N_0
STDF	SFDU Tracking Data File
STDN	NASA Spaceflight Tracking and Data Network
STE	System Test Equipment
STEREO	Solar Terrestrial Relations Observatory
STS	Space Transportation System
STS-34	Space Transpiration System 34 (Shuttle Atlantis)
S-TWTA	S-band Traveling-wave Tube Amplifier
SUFR	Straighten Up and Fly Right
SUPAERO	École nationale supérieure de l'aéronautique et de l'espace
SVT	Sequence Virtual Team
SXA	S- and X-band antenna
sync	Synchronization
TBD	To Be Determined
TBOT	Telecom at Beginning of Tracks
TC	Transmit Command
TCA	Telemetry Channel Assembly (Voyager)
TCB	Barycentric Coordinate Time
TCM	Trajectory Correction Maneuver
TCS	Temperature Control Subsystem
TCVCXO	Temperature-compensated Voltage-controlled Crystal Oscillator
TCXO	Temperature-compensated Crystal Oscillator
TDA	Telecommunications and Data Acquisition (previous name for Deep Space Network)
TDB	Barycentric Dynamical Time
TDDS	Tracking Data Delivery Subsystem
TDL	Telecom Development Lab
TDM	Time-division Multiplexing
TDM	Tracking Data Message
TDRSS	Tracking and Data Relay Satellite System
TDS	Terminal Descent Sensor
TDSA	TDS Antenna Array
TDSD	TDS Digital Stack
TDSR	TDS Radio Frequency Stack
TEC	Total Electron Content
TEOT	Telecom at End of Tracks
TFP	Telecom Forecaster Predictor
TFREQ	Transmitted Frequency (Nominal for Noncoherent Mode)

TGO	Trace Gas Orbiter
TID	Time Insertion and Distribution Assembly
TIP	Target Interface Point
TIRS	Transverse Impulse Rocket System (in MER)
TLGA	Tilted Low-gain Antenna
TLM	Spacecraft Engineering Telemetry
TLMGEN	Telemetry-rate Generator
TLP	Telemetry Processor
TMOD	Telecommunications and Mission Operations Directorate (Now Called IPN-ISD)
TMU	Telemetry Modulation Unit
TNF	Tracking and Navigation File (SODF and STDF)
TOAST	Telecom Orbit Analysis and Simulation Tool
TOST	Titan Orbiter Science Team
T/R	Transmit/Receive; Transmitter/Receiver
TRcycR	Cycle Time
TRK	Tracking System
TRM	Transmit/Receive Module
TRO	Media Calibration Troposphere (file)
TRO	Tropospheric Calibration File
TSAC	Tracking Systems Analytic Calibration
TSB	Telecom Support Board
TT&C	Telemetry, Tracking, and Command
TTE	Turn to Entry
TTS	Test and Telemetry System
TUSO	Transmitter Ultrastable Oscillator
T/W	Ratio of Thrust and Weight on Pad
TW	Truth Window
TWM	Traveling Wave Maser
TWNC	Two-way Noncoherent (Mode, TWNC-on or TWNC-off)
TWTA	Traveling Wave Tube Amplifier
Tx	Transmit (or Transmitter)
TXR	Transmitter (Subsystem)
UA	University of Arizona
UCTS	UHF (Ultrahigh Frequency) Coax Transfer Switch
UCXS	UHF (Ultrahigh Frequency) Coaxial Transfer Switch
UDIM	Up/Down IF (Intermediate Frequency) Module
UDMM	Up/Down MIC (Microwave Integrated Circuit) Module
UEM	Upper Equipment Module
UHF	Ultrahigh Frequency (300 MHz–3 GHz)
UL, U/L	Uplink
UPA	Uplink Processor Assembly
UPL	Uplink Subsystem
URA	Uplink Ranging Assembly
USA	United States of America
USO	Ultrastable Oscillator
UST	Universal Space Transponder

UTC	Universal Coordinated Time
UTCS	UHF Transceiver Coaxial Switch
UV	Ultraviolet
UVS	Ultraviolet Spectrometer
VCO	Voltage-controlled Oscillator
VCTCXO	Voltage-controlled Temperature-compensated Crystal Oscillator
VCXO	Voltage-controlled Crystal Oscillator
VEEGA	Venus–Earth–Earth Gravity Assist
VEGA	Venus and Halley (Mission)
VEX	Venus Express
VF	Video-band Frequency
VG	Group Velocity
VGR	Voyager
VLA	Very Large Array (Near Socorro, New Mexico)
VLBI	Very Long Baseline Interferometry
VP	Phase Velocity
VSR	Very Long Baseline Interferometry (VLBI) Science Receiver
VSWR	Voltage Standing Wave Ratio
W	Watt
W/m^2	Watts Per Square Meter
WVR	Water Vapor Radiometer
X-band	Radio Frequencies from 7 to 12.5 GHz (Spacecraft Downlink Frequency of ~8415 MHz in Deep Space Communications)
XBT	X-band Transmitter
X-EXC	X-band Exciter
XO	Crystal Oscillator
XPA	X-band Power Amplifier
XSDC	X- to S-band Downconverter
XTR	X-band Transmit Receive
X-TWTA	X-band Traveling-wave Tube Amplifier
ZDD	Zero-delay Device

Index

Radio Science Techniques for Deep Space Exploration, First Edition. Sami W. Asmar.

Printed and bound by CPI Group (UK) Ltd, Croydon, CR0 4YY

17/09/2026

14964324-0002